COMPUTATIONAL SUBSURFACE HYDROLOGY

Fluid Flows

COMPUTATIONAL SUBSURFACE HYDROLOGY

Fluid Flows

by

Gour-Tsyh (George) Yeh
The Pennsylvania State University
University Park, Pennsylvania U.S.A.

KLUWER ACADEMIC PUBLISHERS

Distributors for North, Central and South America:
Kluwer Academic Publishers
101 Philip Drive
Assinippi Park
Norwell, Massachusetts 02061 USA
Telephone (781) 871-6600
Fax (781) 871-6528
E-Mail <kluwer@wkap.com>

Distributors for all other countries:
Kluwer Academic Publishers Group
Distribution Centre
Post Office Box 322
3300 AH Dordrecht, THE NETHERLANDS
Telephone 31 78 6392 392
Fax 31 78 6546 474
E-Mail <orderdept@wkap.nl>

Electronic Services <http://www.wkap.nl>

Library of Congress Cataloging-in-Publication Data

A C.I.P. Catalogue record for this book is available
from the Library of Congress.

ISBN 978-1-4419-5084-0

Printed on acid-free paper.

Printed in the United States of America

Contents

List of Figures

List of Tables

List of Tables

Preface

This book is the first volume of a two-book series. Here fluid flows are examined. The second book deals with the reaction, transport, and fate of chemicals and microbes. Both books are designed to address the needs of students, researchers, teachers, and practitioners. It is assumed that readers have some knowledge of matrix equations, calculus, partial differential equations, and subsurface phenomena. Through years of interaction with students and practitioners, I have found that the average student or practitioner working in subsurface hydrology is familiar with finite-difference methods; but lacks a basic background in finite element methods. Thus, finite-element methods are described in great detail, but finite-difference methods are covered briefly. To address the needs of practitioners and enable graduate students to apply finite-element modeling of subsurface hydrology to research investigations or field problems, four basic, general-purpose computer codes and their documentation are included in the attached floppy diskettes that accompany this book. Step-by-step finite-element procedures to implement these four codes are provided, with sometimes repetitious statements, so that readers can read and adapt the four models independently of each other.

A numerical subsurface model is comprised of three components: a theoretical basis to translate our understanding of subsurface phenomena into partial or partial-integral differential equations, a numerical method to approximate these governing equations, and a computer implementation to generate a generic code for research as well as for practical applications. The content of this book is arranged around this theme. Chapter 1 introduces the fundamental processes occurring in subsurface media, the concept required to integrate these processes, and the basic theories used to obtain governing equations. The Reynolds transport theorem, which is used extensively in this book, is stated. The macroscopic balance equations of mass, momentum, energy, and entropy are given based on a series of papers written by Gray and Hassanizadeh in the 1980s. These macroscopic balance equations are included in the first chapter as a matter of interest for research scientists and advanced readers, who should thoroughly read the series of papers referenced here. Chapter 2 covers the finite-difference and finite-element methods that can be used to approximate the governing equations of subsurface processes. While other numerical methods may be more elegant than these two general classes of methods, often they are applicable only to specific problems. Thus, while some of these specific methods are mentioned briefly, they are not treated extensively. The procedures of finite differences are presented almost in a verbatim fashion via a personal communication with Professor Gray of Notre Dame University in the mid-1980s. Finite-element methods are covered in much greater detail than finite-difference methods because while the former is not physically intuitive, the latter is readily comprehended with a common physical sense.

The finite-element methods presented in this book are drawn from a wide range of sources, particularly the books written by Huebner and by Huyarkorn and Pinder. I try to portray finite element methods in a manner in which readers can easily derive base functions and apply these procedures to and design a computer code of their particular problems. To achieve this goal, I have sometimes repetitiously emanated the steps. Also covered in Chapter 2 are various algorithms to solve linear matrix equations, which include direct elimination methods, basic iterative schemes, prevailing conjugate gradient algorithms, and multigrid methods. While multigrid methods have been widely applied to computational fluid dynamic problems, their applications to practical subsurface fluid flow problems are very limited. Because of the uniform convergent rate, the multigrid methods will eventually become the ultimate choice of solving linear matrix equations when they are applicable. Thus, multigrid methods are described in more detail than any other method used to solve linear matrix equations.

Chapter 3 is devoted to the finite-element modeling of subsurface fluid flows. First, governing equations of subsurface fluid flows, as well as solute and thermal transport, are derived. Then finite-element models for fluid flows in saturated and variably saturated media under isothermal conditions are given. This chapter provides details on how the finite-element approach, known as the Galerkin finite-element method, can be used to develop finite-element codes that can be used to solve a wide range of problems. Included in this chapter are the illustrative examples that show how the four attached computer codes can be used to solve various types of problems. These four computer codes are designed for generic applications, not only for research problems but also for realistic field cases. Practically all field problems ranging from the simplest regionwide groundwater flows to the most complicated saturated-unsaturated flows in three-dimensional spaces can be simulated with one of these four codes. Particular features are placed on the rigorous implementation of moving free surfaces in three-dimensional saturated flow problems. This is in contrast to the widely used finite-difference codes in which ad hoc approaches were employed to deal with moving-boundary problems. A brief discussion on the analysis of stability and convergence of a given numerical scheme is also treated in this chapter because I feel that any reader should have the capability of at least making a preliminary assessment of the stability and convergence of the numerical scheme he or she employs.

The successful completion of this manuscript has been made possible by the efforts of many colleagues and students who have made numerous suggestions and given generously their time in reading the early drafts of this book. Particular recognition is due to Dr. Jing-Ru Cheng, who conducted many simulations presented in the book using the four attached computer codes, and Dr. Ming-Hsu Li, who drafted most of the figures using VISIO and TECHPLOT software. Special thanks are extended to Teresa A. Esser of Kluwer Acadmic Publishers, who provided the most encouragement in the preparation and publication of this book. Last, but not least, I would like to thank my lovely wife, Shu-Shen, who made many sacrifices during the preparation of this text.

1 FUNDAMENTALS OF THE SUBSURFACE SYSTEM

1.1. Study of the Subsurface System

This book is concerned with the mathematical description and numerical modeling of subsurface media. It is about the subsurface media that control the movement of fluids (including water, nonaqueous liquids, and gas), the migration of chemicals, the transfer of heat, and the deformation of media. It is about the physical laws that describe the flux of fluid, heat, and chemicals, and the relationship between stress and strain. It is about the chemical reactions along with fluid flows. It is about the biological interaction within the flow and thermal domain and among chemical constituents. It is about numerical methods needed to conduct simulations of both fluid flows and advection-dominant transport. In short, the study of the subsurface system is the investigation of major processes occurring in the subsurface and the interplay of these processes with the media through which they occur. Understanding the mechanisms controlling the occurrence of these processes and their interplay is the ultimate goal of this book because it provides a method for the prediction of the occurrence of these processes in the media. To make this goal possible, accurate numerical methods to efficiently and accurately approximate mathematical descriptions are of ultimate importance. Extensive coverages of finite element methods used in fluid flows and hybrid Lagrangian-Eulerian approaches best suited to deal with advection-dominant transport are included.

1.1.1. The Processes

The major processes that will be included in this book are fluid flow, heat transport, chemical transport, and deformation. A fifth major process is the electric current, which provides the basis for geophysical measurements of subsurface parameters. Even though only four major processes are included, the number of computational models that can be conceived is very large. This is so because there are numerous application-dependent variations, depending on the type of media, the phase of fluid, and the factors causing these processes. In addition, for convenience, equations describing the same process are often cast in terms of different dependent variables. Table 1.1.1 provides a summary of major processes, dependent variables, and application-dependent variations that are commonly encountered in subsurface hydrology.

Table 1.1.1 Major Processes with Dependent Variables and Application-Dependent Variations (After Mercer and Faust, 1981).

Major Processes	Dependent Variables	Application-Dependent Variations
Fluid flow	Fluid pressure Pressure head Hydraulic head Degree of saturation	Porous media Fractured media Single-fluid phase Multi-fluid phase
Heat transport	Temperature Enthalpy or internal energy	Same as fluid flow plus convection/advection conduction or diffusion compaction/consolidation
Chemical transport	Concentration	Same as fluid flow plus physical factors: Advection/convection Dispersion/diffusion Decay/consolidation Chemical reactions: Aqueous speciation Adsorption/desorption Ion-exchange Precipitation/dissolution Oxidation/reduction Acid/base reaction Biological interactions: Microbial population dynamics Substrate utilization Biotransformation Adaption Cometabolism
Deformation	Strain or strain rate	Elastic media/plastic media Visco-elastic media Visco-plastic media Discrete fracture media

Many terms need to be defined in studying the subsurface system; the most important conceptual terms are (1) the medium itself, (2) the storage capacity of the medium, and (3) the flux through the medium. The porous medium will be described in the next section. The *storage capacity* of the medium with respect to a state variable is the amount of attributes added to a unit volume of the medium to raise the state variable by unit. For example, if the state variable is chemical concentration, then the storage capacity is the amount of chemical mass (chemical mass is the attribute to the concentration) added to a unit volume of the medium to raise the concentration by one. If there is θ volume of water in a unit volume of the medium, then storage capacity is the θ units of chemical mass. The *flux* is defined as the amount of attribute per unit time through a unit area normal to the direction of the flux. Again, if the state variable is concentration, the flux is the amount of chemical mass per unit time through a unit area normal to flux. If the state variable is pressure, then the flux is the amount of water mass per unit time through unit area with its normal direction coinciding with the direction of the flux.

1.1.2. Porous Medium

A porous medium is a multiphase medium with at least one solid phase and one fluid phase (Greenkorn, 1983), and the phases are distributed within the volume such that the thermodynamic properties at each phase can be meaningfully described using a volume averaging procedure (Fig. 1.1.1).

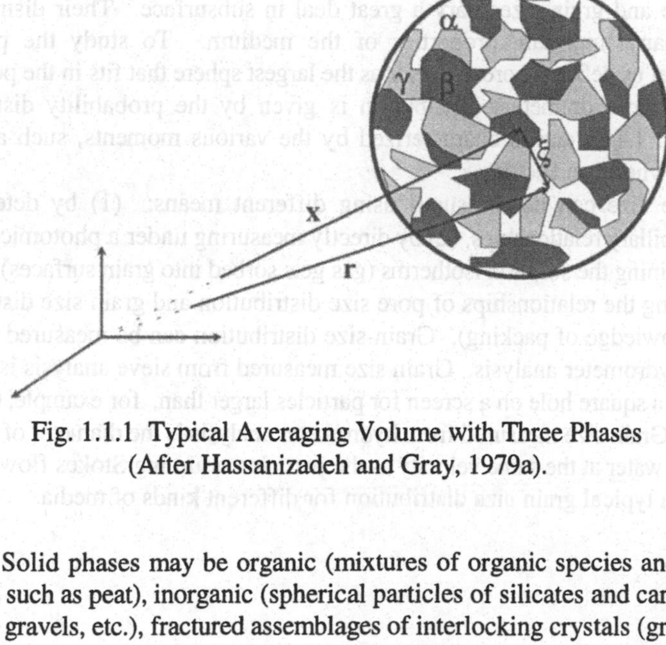

Fig. 1.1.1 Typical Averaging Volume with Three Phases
(After Hassanizadeh and Gray, 1979a).

Solid phases may be organic (mixtures of organic species and amorphous materials such as peat), inorganic (spherical particles of silicates and carbonates such as sands, gravels, etc.), fractured assemblages of interlocking crystals (granite, quartz, and other metamorphic forms), clay (an assemblage of plate-like minerals, low permeability, high porosity, active adsorption sites), gels of silica (created at evaporate sites), and precipitation of carbonates (limestone, etc.).

Fluid phases include liquids such as water, oil, and nonaqueous phase liquid (NAPL) and carbon dioxide and gases such as air, steams, natural gas, carbon dioxide, and methane.

Some important characteristics of a porous medium including the following: space not occupied by the solids is denoted as *pore*; space occupied by the solids is termed *matrix*; unconnected pores, *dead-end pores*, do not affect flow, but can be important as "storing" contaminants; pores that are formed at the time of formation of the geological unit are termed *primary pores*; pores that are formed subsequent to the formation of geological units are termed *secondary pores* (solution openings, fractures, roots, phase transformations).

Porous media may be described at three different scales (or three levels of observation). At the molecular level, individual molecules are considered. Classical mechanics can be employed to describe the behavior of individual molecules. At the microscopic level, behaviors at the pore level are considered. Either statistical mechanics or fluid mechanics can be employed to describe the behaviors at pore level. At the macroscopic level, the average behavior over a *representative elemental volume* (REV) is considered; that is, the bulk properties of the medium are defined. As we move REV through the domain, the bulk properties such as porosity, bulk density, and so on vary smoothly. The bulk properties are the *macroscopic properties*. If we cannot describe the bulk properties as average properties, then macroscopic description is inappropriate. There are three possible approaches to describe the macroscopic behaviors: intuitive engineering approach, continuum theory of mixtures, and average theory of microscopic behaviors.

Pore and grain sizes vary a great deal in subsurface. Their distributions determine many important properties of the medium. To study the pore-size distribution, let us define a pore diameter as the largest sphere that fits in the pore. The pore-size (or pore diameter) distribution is given by the probability distribution function (Fig. 1.1.2) and is characterized by the various moments, such as mean, variance, skewness, and so on.

Pore size can be measured using different means: (1) by determining saturation-capillary relationships, (2) by directly measuring under a photomicrograph, (3) by determining the sorption isotherms (gas gets sorbed into grain surfaces), and (4) by determining the relationships of pore size distribution and grain size distribution (requires knowledge of packing). Grain-size distribution can be measured by sieve analysis or hydrometer analysis. Grain size measured from sieve analysis is the side dimension of a square hole on a screen for particles larger than, for example, 0.06 mm (200 mesh). Grain size measured from hydrometer analysis is the diameter of a sphere that settles in water at the same velocity as the particles assuming Stokes flow. Figure 1.1.3 shows a typical grain-size distribution for different kinds of media.

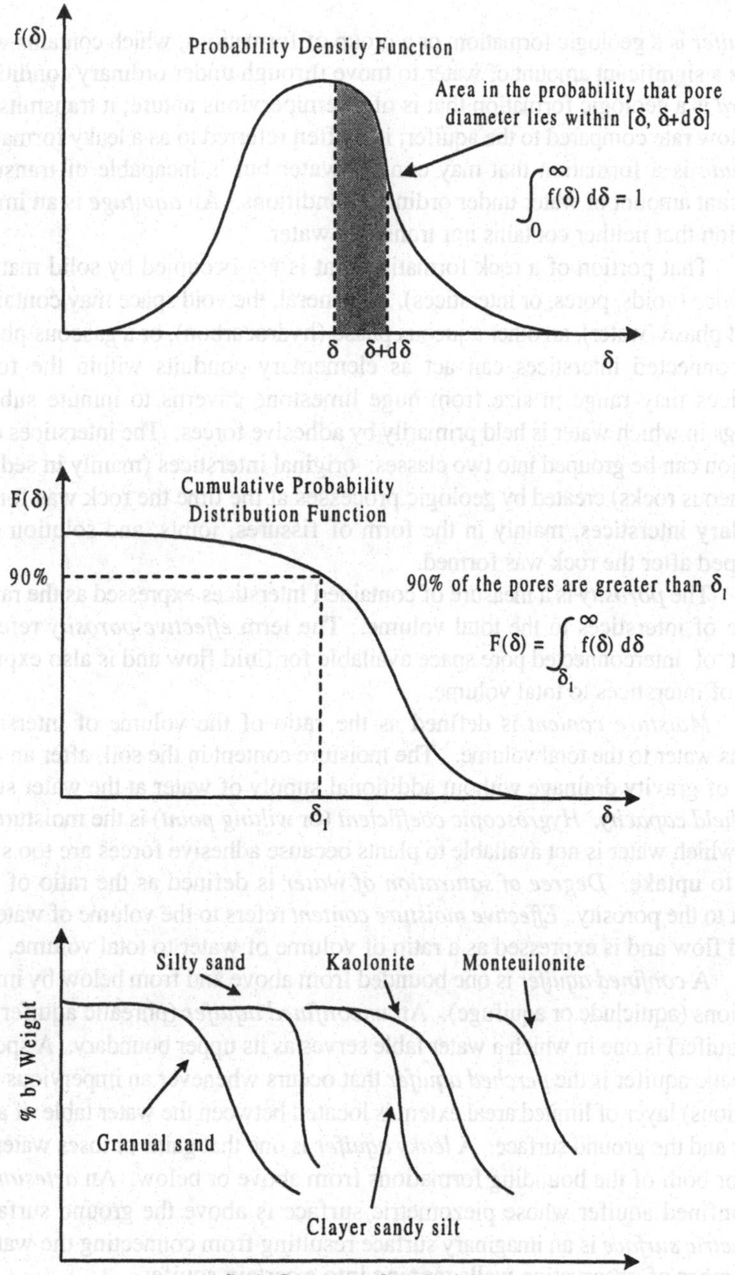

Fig. 1.1.3. A Typical Grain-Size Distribution

1.1.3. Definition of Terms (Bear, 1979; Todd, 1980)

An *aquifer* is a geologic formation, or a group of formations, which contains water and permits a significant amount of water to move through under ordinary conditions. An *aquitard* is a geologic formation that is of a semipervious nature; it transmits water at a very low rate compared to the aquifer; it is often referred to as a leaky formation. An *aquiclude* is a formation that may contain water but is incapable of transmitting a significant amount of water under ordinary conditions. An *aquifuge* is an impervious formation that neither contains nor transmits water.

That portion of a rock formation that is not occupied by solid matter is the *pore space* (voids, pores, or interstices). In general, the void space may contain in part a liquid phase (water), an other aqueous phase (hydrocarbon), or a gaseous phase (air). Only connected interstices can act as elementary conduits within the formation. Interstices may range in size from huge limestone caverns to minute subcapillary openings in which water is held primarily by adhesive forces. The interstices of a rock formation can be grouped into two classes: original interstices (mainly in sedimentary and igneous rocks) created by geologic processes at the time the rock was formed and secondary interstices, mainly in the form of fissures, joints, and solution passages developed after the rock was formed.

The *porosity* is a measure of contained interstices expressed as the ratio of the volume of interstices to the total volume. The term *effective porosity* refers to the amount of interconnected pore space available for fluid flow and is also expressed as a ratio of interstices to total volume.

Moisture content is defined as the ratio of the volume of interstices that contains water to the total volume. The moisture content in the soil, after an extended period of gravity drainage without additional supply of water at the water surface, is called *field capacity*. *Hygroscopic coefficient* (or *wilting point*) is the moisture content below which water is not available to plants because adhesive forces are too strong for plants to uptake. *Degree of saturation of water* is defined as the ratio of moisture content to the porosity. *Effective moisture content* refers to the volume of water subject to fluid flow and is expressed as a ratio of volume of water to total volume.

A *confined aquifer* is one bounded from above and from below by impervious formations (aquiclude or aquifuge). An *unconfined aquifer* (phreatic aquifer or water table aquifer) is one in which a water table serves as its upper boundary. A special case of phreatic aquifer is the *perched aquifer* that occurs whenever an impervious (or semi-impervious) layer of limited areal extent is located between the water table of a phreatic aquifer and the ground surface. A *leaky aquifer* is one that gains or loses water through either or both of the bounding formations from above or below. An *artesian aquifer* is a confined aquifer whose piezometric surface is above the ground surface. The *piezometric surface* is an imaginary surface resulting from connecting the water levels of a number of observation wells tapping into a certain aquifer.

Specific storativity (*specific storage*) is defined as the volume of water per unit volume of saturated aquifer released from or added to the storage for a unit decline or rise in head. *Aquifer storativity* is defined as the volume of water released from or taken into storage per unit surface area of aquifer per unit change in the component of head normal to that surface. *Storage coefficient* is the aquifer storativity of a confined aquifer. *Specific yield* is the aquifer storativity of the phreatic aquifer. *Specific retention* is the volume of water retained in an aquifer per unit area and unit drop of water table.

The *permeability* (*intrinsic permeability*) of a rock or soil defines its ability to transmit a fluid. This is a property of the medium only and is independent of fluid properties. *Hydraulic conductivity* is defined as the volume of water transmitted in unit time through a cross section of unit area, measured at right angles to the direction of flow, under unit hydraulic gradient. *Transmissivity* is the rate of water transmitted through a unit width of aquifer under a unit hydraulic gradient.

Components are a set of linearly independent "basis" chemical entities such that every species can be uniquely represented as a combination of those components, and no other component can be represented by other components than itself. A *species* is the product of a chemical reaction involving components as reactants.

Complexation is a chemical reaction involving two or more ions to form a solute species. *Solute* is any dissolved ion or complex. *Adsorption* (*surface complexation*) is a chemical reaction that results in accumulation of solutes at a solid-liquid interface. *Ion-exchange* is a chemical reaction involving the substitution of one ion with another ion on the surface. *Sorption* is defined as one or more of several reactions that result in the concentration of solutes at a solid-liquid interface. Thus, both adsorption and ion-exchange can be considered sorption reaction. *Oxidation* is defined as the removal of electrons by an atom or atoms, and *reduction* as the gain of electrons by an atom or atoms. An *oxidizing agent* is any material that takes on electrons, and a *reducing agent* is any material that gives up electrons. *Precipitation* is a chemical reaction involving removal of dissolved chemicals; *dissolution* is a reverse reaction of precipitation. *Acid-base* reactions are chemical reactions involving the transfer of protons.

Molar concentration (*molarity*) is the number of moles of solute contained in one liter of solution. *Molality* (*molal concentration*) is the number of moles of solute per kilogram of solvent contained in the solution. *Mole fraction* of any component in a solution is defined as the number of moles of that component, divided by the total number of moles of all components in the solution. *Activity* of a component is a fictitious quantity that, when substituted for the mole fraction of that component, will satisfy the chemical potential equation of that component for an ideal solution at constant temperature and pressure. Thus, activity can be considered an effective concentration. *Activity coefficient* is defined as a coefficient that multiplies the molal concentration (or mole fraction) of that component to yield the activity. *Fugacity* is the pressure value needed, at a given temperature, to make the properties of a real gas satisfy the equations of an ideal gas. *Fugacity coefficient* is the coefficient that multiplies the partial pressure of a component to yield the fugacity of that component.

The *stoichiometric coefficient* of a component in a species is the coefficient of that component in the chemical balance equation to form the species. The thermodynamic *equilibrium constant* of a species is the reaction quotient, defined as the activity of the species raised to a power equal to the coefficient of the species in the balance equation, divided by the product of all component species activities, each raised to the power equal to its stoichiometric coefficient in the species. *Solubility* is the quantity of a given compound that dissolves in a solution at equilibrium.

1.2. The Fundamental Concept of Continuum

The study of subsurface systems is a complex problem. To tackle the problem, we must know the most fundamental concept of continuum. Before we talk about the concept of continuum representation of the media (fluids, solids, gases, porous media), let us differentiate a physical point from a mathematical point. A physical point has a finite size while a mathematical point has a size of zero. Thus, if a mathematical point is used as a physical point, nothing can be considered as a continuum. The concept of physical point must be used to have any meaning in the study of any medium. The question is how big a volume should be used to represent a physical point. To resolve this fundamental question, we must know the level of observation we can conduct in a porous medium described above. Basically, there may be three levels of observations: (1) molecular level (individual molecules), (2) microscopic level (behavior at the pore level), and (3) macroscopic level (average behavior over a representative elemental volume (REV) of the porous medium; bulk properties). At the molecular level, one uses the theory of mechanics to describe the system of molecular behavior given the position and momentum. It is almost impossible to study the subsurface system at the molecular level because of the astronomical number of molecules (including liquid molecules, gaseous molecules, and solid molecules) involved.

At the microscopic level, one can use the principle of fluid mechanics or the theory of statistical mechanics to describe the subsurface system. A fundamental postulation in a microscopic level approach is that statistical properties of the motion of a very large number of particles may be inferred from the laws governing the motion of individual particles. Using the microscopic approach, only the average response of ensembled molecules can be determined, and the information about individual molecules cannot be retrieved. The physical structure in the microscopic approach is such that the molecular structure is overlooked and one regards the materials (fluids, solids, etc.) as a continuum with smoothly varying properties (they must be smooth if an abserver is to be able to use classical mathematics to describe). A particle is the ensemble of many molecules contained in a volume. Its size must be much larger than the mean free path of a single molecule (λ) (which implies many molecules in a particle) and must be sufficiently small that averaging still maintains values relevant to the description of the bulk fluid (or solid) properties (which implies a smaller than characteristic length L describing changes in nonhomogeneous materials). Microscopic fluid properties are identified with the centroid of the particle located at a mathematical point P(**x**). It is emphasized again that a physical point with finite size V_0 is identified with a mathematical point P(**x**) with zero size. V_0 is the volume of a particle at point P.

Consider, for example, the fluid density. Let us define ρ_i as the fluid density at a point P(**x**) determined by measuring the fluid mass Δm_i in a volume ΔV_i. As the size of ΔV_i varies, the density $\rho_i = \Delta m_i/\Delta V_i$ can vary, as in Fig. 1.2.1. If ΔV_i is smaller than λ^3, erratic density results depending on whether the ΔV_i happens to cover the fluid molecules. As the ΔV_i increases, there is a region where the density ρ_i remains fairly constant. When ΔV_i is greater than L^3, then nonhomogeneous fluid effect enters the picture. By conducting experiments for many points with P arbitrarily close to one another, one generates a fictitious, smooth medium, called a fluid, for which a continuous function of space ρ(x) is defined. Examples of other microscopic properties are chemical concentrations (mass transport by molecular diffusion due to molecular diffusivity), viscosity (momentum transfer), and heat (kinetic energy transfer due to thermal diffusivity).

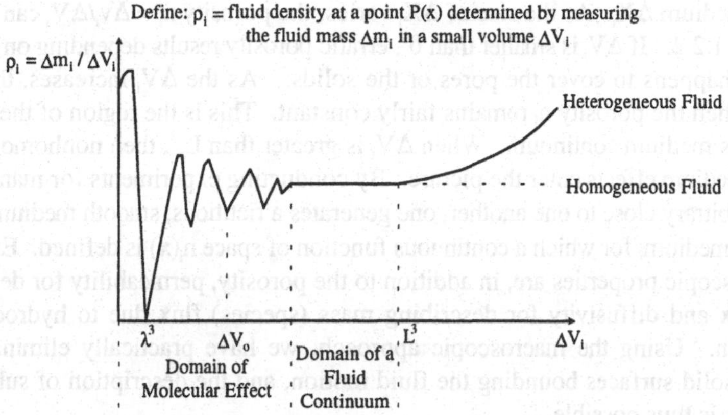

Fig. 1.2.1 Fluid Density as a Function of Measuring Size.

In principle, since we have the basic equations governing the flow (say, the Navier-Stokes equations derived based on the principle of fluid mechanics) and the boundary conditions on the global boundary and on the solid surfaces that bound the flowing fluid, a solution at the microscopic level is possible. However, it is almost impossible to describe in any exact or even approximate mathematical manner the complicated geometry of solid surfaces bounding the flowing fluid. Furthermore, under pressure, the geometry of the bounding surfaces also changes with time and space. Thus, an approach at the microscopic level is precluded in the investigation of subsurface systems.

Because of our inability to describe the subsurface processes at the microscopic level, we resort to the macroscopic level of approach. In the macroscopic level, a fundamental postulation is that statistical properties of fluid motion of a very large number of pores may be inferred from the laws governing the microscopic motion in individual pores. Using the macroscopic approach, only the average response of a representative elemental volume (REV) can be determined, and the information about fluid flow or transport in individual pores cannot be retrieved. The physical structure in the macroscopic approach is such that the pore-level behavior is overlooked, and one can regard the porous medium as continuum with smoothly varying properties (must be smooth to use classical mathematics to describe). A REV is the ensemble of many pores contained in a volume. Its size must be much larger than the diameter of pores (δ) (which implies many pores in a REV) and must be sufficiently small that averaging over a REV still maintains values relevant to bulk porous medium properties (which implies smaller than characteristic length L_p describing changes in non-homogeneous porous medium). Macroscopic porous medium properties are identified with the centroid of the REV located at a mathematical point P(**x**). It is emphasized again that a physical or material point with finite size V_o is identified with a mathematical point P(**x**) with zero size. V_o is the volume of a REV at point P.

Consider, for example, the porosity. Let us define n_i as the medium porosity at a point P(**x**) determined by measuring the volume of void space Δv_i in a volume of porous medium ΔX_i. As the size of ΔX_i varies, the porosity $n_i = \Delta v_i/\Delta V_i$ can vary, as in Figure 1.2.2. If ΔV_i is smaller than δ^3, erratic porosity results depending on whether the ΔV_i happens to cover the pores or the solids. As the ΔV_i increases, there is a region when the porosity n_i remains fairly constant. This is the region of the domain of porous medium continuum. When ΔV_i is greater than L_p^3, then nonhomogeneous porous medium effects enter the picture. By conducting experiments for many points with P arbitrary close to one another, one generates a fictitious, smooth medium, called a porous medium, for which a continuous function of space n_i(**x**) is defined. Examples of macroscopic properties are, in addition to the porosity, permeability for describing fluid flux and diffusivity for describing mass (species) flux due to hydrodynamic dispersion. Using the macroscopic approach, we have practically eliminated the internal solid surfaces bounding the fluid motion, and the description of subsurface processes is thus possible.

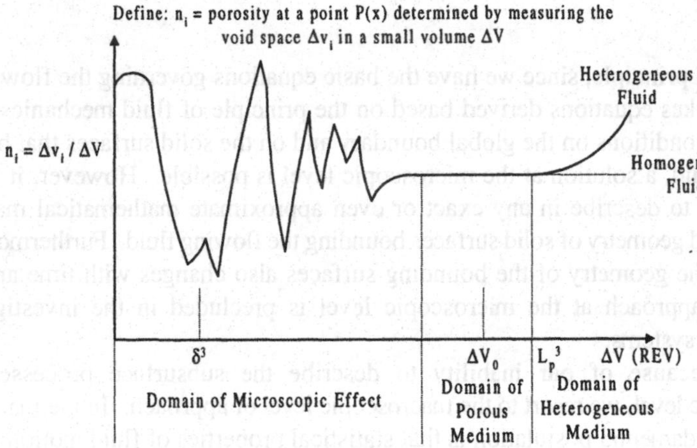

Fig. 1.2.2 Porosity as a Function of Measuring Size.

Once our physical point is greater than the REV, we regard the porous medium as a continuum with smoothly varying properties. The question is: can we have smoothly varying properties (i.e. porosity, permeability, etc.) in subsurface media. Recall that there are two groups of pores: the primary pores and the secondary pores. If the porous medium is made entirely of the primary pores, it is highly likely we will have the smoothly varying properties. However, if the porous medium is made of a mixture of primary and secondary pores, odds are against smoothly varying property. Now enters the concept of different levels of heterogeneity. As we learn from geological data, we may have three levels of heterogeneity: (1) grains of different size, (2) fractures and joints, and (3) tectonic fault (Fig. 1.2.3). The size of the physical point (or the REV) associated with the porous media may be applicable to the first level of heterogeneity.

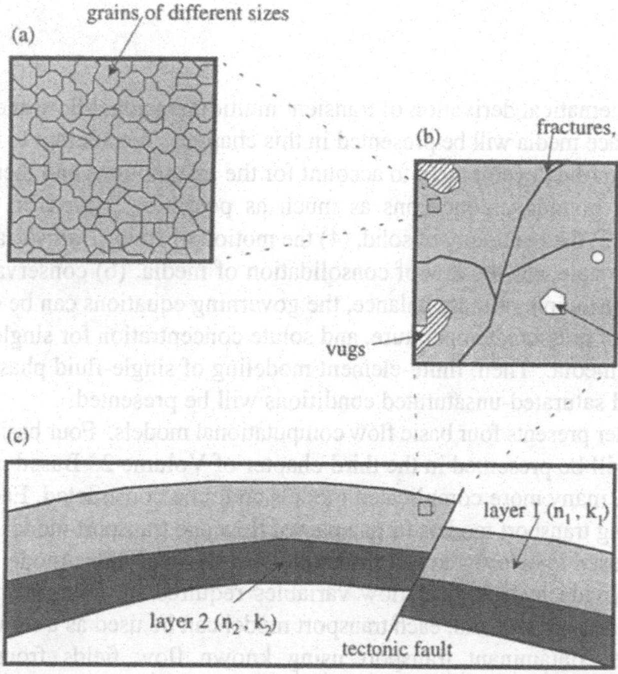

Figure 1.2.3 Three Levels of Heterogeneity of the Subsurface Medium.

For the second and third levels of heterogeneity, a different size of physical point must be determined. Following the same argument that leads us to obtain an REV for the porous media, we can obtain an REV for the fracture media (Fig. 1.2.4). In this case, we recognize that an overlapping REV for the porous media and fractures exists. This overlapping REV naturally represents the size of a physical point, and we can treat the problem as an equivalent continuum or a double continuum.

3 FINITE-ELEMENT MODELING OF SINGLE-FLUID PHASE FLOWS

First, a formal mathematical derivation of transient multidimensional flow and solute transport in subsurface media will be presented in this chapter. We attempt to include all important steps in the development to account for the assumptions and their bases and to incorporate boundary conditions as much as possible. Based on (1) the continuity of fluid, (2) the continuity of solid, (4) the motion of fluid (Darcy's law), (4) the equation of the state, (5) the law of consolidation of media, (6) conservation of energy, and (7) the principle of mass balance, the governing equations can be derived for the distribution of pressure, temperature, and solute concentration for single-phase flow in subsurface media. Then, finite-element modeling of single-fluid phase flows under saturated and saturated-unsaturated conditions will be presented.

This chapter presents four basic flow computational models. Four basic solute transport models will be presented in the third chapter of Volume 2. Based on these eight basic models, many more complicated models could be constructed. Each flow and its corresponding transport models form a pair of flow and transport models. Each flow model can be used to investigate subsurface flows as a stand-alone model. It can also be used to provide hydrological flow variables required by its corresponding transport model. On the other hand, each transport model can be used as a stand alone model to simulate contaminant transport using known flow fields from either measurements or simulations by other flow models. It can also be used in tandem with its corresponding flow model to simulate contaminant transport with the flow model creating hydrological variables.

The common features of the four models presented in this chapter are their flexibility and versatility in modeling as wide a range of real-world problems as possible. All these models are designed to (1) treat heterogeneous and anisotropic media consisting of as many geologic formations as desired; (2) consider both distributed and point sources/sinks that are spatially and temporally dependent; (3) accept the prescribed initial conditions or obtain them by simulating a steady state version of the system under consideration; (4) deal with a transient head distributed over the Dirichlet boundary; (5) handle time-dependent fluxes due to pressure

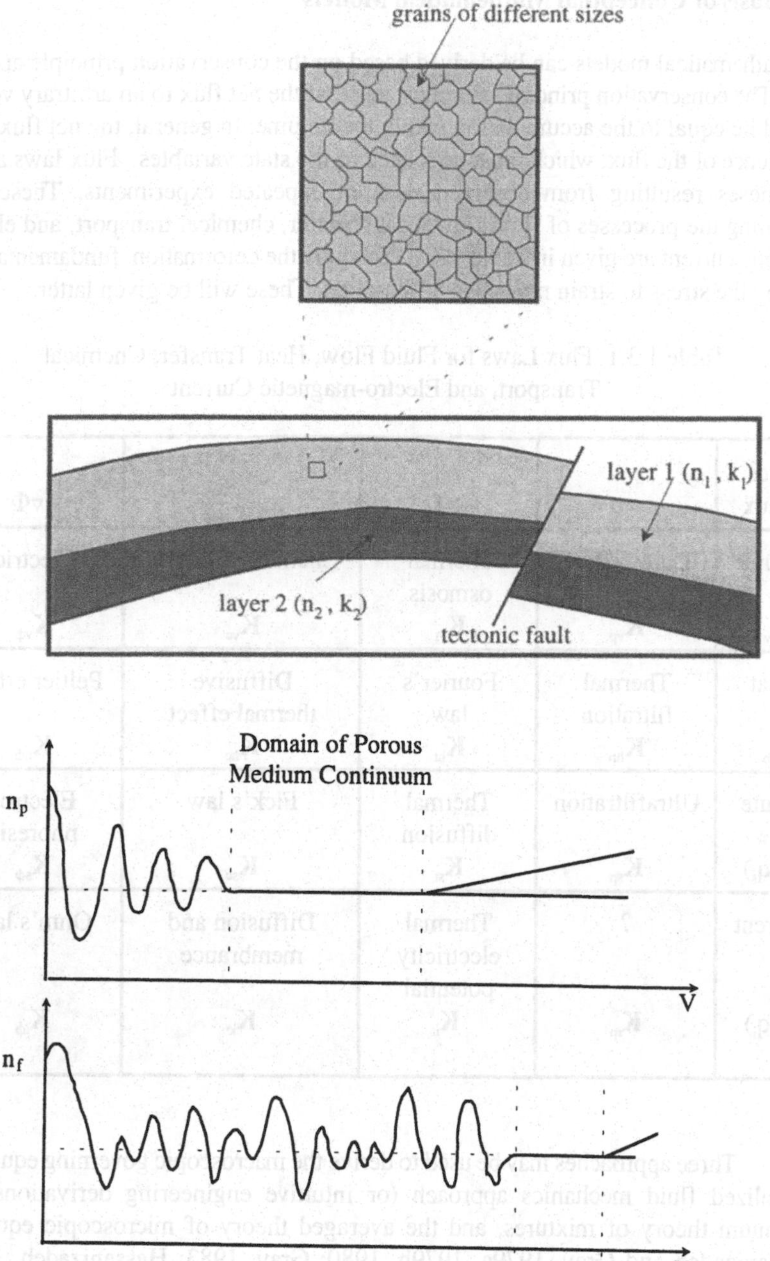

Fig. 1.2.5 REVs of Fractured Media Without Overlapping REV.

1.3. Basis of Conceptual Mathematical Models

All mathematical models can be derived based on the conservation principle and flux laws. The conservation principle simply states that the net flux to an arbitrary volume should be equal to the accumulation within the volume. In general, the net flux is the divergence of the flux, which must be related to the state variables. Flux laws are the hypotheses resulting from observed facts or repeated experiments. These laws governing the processes of fluid flow, heat transfer, chemical transport, and electro-magnetic current are given in Table 1.3.1. To study the deformation, fundamental laws relating the stress to strain must also be supplied. These will be given latter.

Table 1.3.1 Flux Laws for Fluid Flow, Heat Transfer, Chemical Transport, and Electro-magnetic Current

Forces Flux	$-\nabla p$	$-\nabla T$	$-\nabla C$	$-\nabla\Phi$
Volume J_v	Darcy's law K_{vp}	Thermal osmosis K_{vt}	Chemical osmosis K_{vc}	Electric $K_{v\phi}$
Heat J_h	Thermal filtration K_{hp}	Fourier's law K_{ht}	Diffusive thermal effect K_{hc}	Peltier effect $K_{h\phi}$
Solute J_s (q_s)	Ultrafiltration K_{sp}	Thermal diffusion K_{st}	Fick's law K_{sc}	Electro-phoresis $K_{s\phi}$
Current J_i (q_i)	? K_{ip}	Thermal electricity potential K_{it}	Diffusion and membrance K_{ic}	Ohm's law $K_{i\phi}$

Three approaches may be used to derive the macroscopic governing equations: generalized fluid mechanics approach (or intuitive engineering derivations), the continuum theory of mixtures, and the averaged theory of microscopic equations (Hassanizadeh and Gray, 1979a, 1979b, 1980; Gray, 1983; Hassanizadeh, 1986a, 1986b). Microscopic equations, on the other hand, can be derived based on fluid mechanics (Bird et al., 1960) or statistical mechanics methods (Sposito et al., 1979). To derive microscopic equations using the fluid mechanics method or the macroscopic governing equations using the generalized fluid mechanics approach, Reynolds' transport theory is often used.

1.3.1. Reynolds Transport Theory

The derivation of Reynolds transport theorem can be found elsewhere (Owczarek, 1964). Please read these pages very carefully. This theorem is written as

$$\frac{D}{Dt}\int_{v} F\, dv = \frac{\partial}{\partial t}\int_{V} F\, dV + \int_{S} F(\mathbf{V}\cdot\mathbf{n}dS), \tag{1.3.1}$$

where v = a moving volume containing a constant amount of materials, V = fixed volume in space that instantaneously coincides with the moving material volume v, S = the surface bounding the fixed volume V, **n** = outwardly normal unit vector to the surface S, and **V** = velocity of the moving surface. Equation (1.3.1) states that the rate of change of integral of function F, taken over a moving material volume v, is equal to the rate of change of this integral, taken over the fixed volume V that instantaneously coincides with v, and the flux of F out through the bounding surface. Equation (1.3.1) is known as Reynolds transport theorem.

In fact, we can rewrite Eq. (1.3.1) into the following form:

$$\frac{D}{Dt}\int_{v} F\, dv = \int_{v} \frac{\partial F}{\partial t}\, dv + \int_{\Gamma} F(\mathbf{V}\cdot\mathbf{n}d\Gamma), \tag{1.3.2}$$

where Γ = the surface bounding the moving volume v.

1.4. Microscopic Balance Equations

The balance equation for any thermodynamic property (such as specific internal energy) in any microscopic medium (fluids, solids, and so on) can be derived based on the conservation principle of continuum mechanics. The derivation will be given below. Let ρ_i = mass of the i-th species per unit medium (fluids, solids, and so on) volume, ψ_i = any microscopic thermodynamic property of the i-th species (such as specific internal energy), $\mathbf{i}_i$ = microscopic surface flux vector of the i-th species (such as molecular diffusion theory, heat flux at a radiating body), f_i = specific external supply of the i-th species, g_i = specific internal generation of the i-th species within the volume, t = time, and ∇ = del operator. Applying the conservation principle in integral form, we obtain

$$\frac{D}{Dt}\int_{v}\rho_i\psi_i dv = -\int_{a}\mathbf{n}\cdot\mathbf{i}_i da + \int_{v}\rho_i g_i dv + \int_{v}\rho_i f_i dv, \tag{1.4.1}$$

where v = material volume containing a constant amount of the fluid medium including all species (L^3), a = surface enclosing the material volume v (L^2).

By Reynolds transport theorem (Owczarek, 1964), Eq. (1.4.1) can be written

$$\int_{v}\frac{\partial\rho_i\psi_i}{\partial t}dv + \int_{a}\mathbf{n}\cdot(\rho_i\psi_i\mathbf{v}_i)da + \int_{a}\mathbf{n}\cdot\mathbf{i}_i da = \int_{v}\rho_i g_i dv + \int_{v}\rho_i f_i dv, \tag{1.4.2}$$

where $\mathbf{v}_i$ = microscopic mass-weighted mean velocity of molecules for the i-th species.

Applying the Gaussian divergence theorem to Eq. (1.4.2) and using the fact that v is arbitrary, one can obtain the following microscopic balance equation for the thermodynamic property ψ_i associated with the i-th species in differential form:

$$\frac{\partial \rho_i \psi_i}{\partial t} + \nabla \cdot (\rho_i \mathbf{v}_i \psi_i) + \nabla \cdot \mathbf{i}_i = \rho_i (f_i + g_i). \qquad (1.4.3)$$

The first term in Eq. (1.4.3) is the rate of accumulation within the REV, the second term is the net outward flux via advection, the third term is the net outward flux via surface flux, the fourth term is the net external supply rate, and the fifth term is the net internal production rate. Individual balance laws for mass, momentum, energy, and entropy can be obtained by appropriate choice of ψ_i in Table 1.4.1.

Table 1.4.1 Variables Selected for the Microscopic Balance Equation.

Quantity	ψ_i	$\mathbf{i}_i$	f_i	g_i
Mass	1	$\mathbf{j}_i$	q_i	r_i
Momentum	$\mathbf{v}_i$	σ_i	$\mathbf{g}_i$	$r_i \mathbf{v}_i + \mathbf{t}_i$
Energy	$e_i + v_i^2/2$	$\sigma_i \cdot \mathbf{v}_i + \mathbf{q}_i$	$\mathbf{g}_i \cdot \mathbf{v}_i + h_i$	$r_i(e_i + v_i^2/2) + \mathbf{t}_i \cdot \mathbf{v}_i + u_i$
Entropy	s_i	$\mathbf{f}_i$	b_i	$r_i s_i + w_i + \Gamma_i$

The variables in Table 1.4.1 are defined as follows: $\mathbf{j}_i$ = surface mass flux for the i-th species, q_i = the externally supplied mass of the i-th species, r_i = rate of production of the i-th species mass due to chemical reaction, $\mathbf{v}_i$ = specific momentum of the i-th species, σ_i = microscopic partial stress tensor for the species exterior to a diaphragm, on the species interior to that diaphragm, through the diaphragm surface, $\mathbf{g}_i$ = external body force (e.g. due to gravity or ionic attraction), $\mathbf{t}_i$ = internal body force exerted on the species i by all other species coexisting with i within the diaphragm, e_i = internal energy density function for the i-th species, $\mathbf{q}_i$ = surface heat flux of the i-th species, h_i = external supply of energy to the i-th species, u_i = exchange of energy between species i and all other species, s_i = internal entropy density function for species i, $\mathbf{f}_i$ = surface entropy flux for species i, b_i = external supply of entropy to species i,w_i = exchange of entropy between species i and all other species,and Γ_i = rate of net production of entropy of the species i.

The exchanges of mass, momentum, energy, and entropy thate place among the species are all internal to the fluid phase. That is, they are actions and counter actions and thus do not give rise to a net production of corresponding properties of the fluid phase. Therefore, they must satisfy the following restrictions:

$$\sum_i \rho_i r_i = 0 \tag{1.4.4}$$

$$\sum_i \rho_i (r_i v_i + t_i) = 0 \tag{1.4.5}$$

$$\sum_i \rho_i \left[r_i \left(e_i + v_i^2/2 \right) + t_i \cdot v_i + u_i \right] = 0 \tag{1.4.6}$$

$$\sum_i \rho_i (r_i s_i + w_i) = 0. \tag{1.4.7}$$

Theoretically, after making appropriate constitutive assumptions on surface fluxes and internal production rates, one should be able to obtain a solution for equations (1.4.3) constrained by (1.4.4), subject to appropriate global boundary and initial conditions and the interfacial conditions with other phases (such as solids). However, the highly complicated geometry of interfaces makes it virtually impossible to apply interfacial boundary conditions. Furthermore, the microscopic quantities, in most cases, are not practically measurable, and any constitutive postulates will not be able to be subjected to experimental verification. In fact, in practice, only average values of microscopic quantities are measured and are of interest. Therefore, we must resort to a macroscopic approach.

1.5. Macroscopic Balance Equations

The derivation of the general macroscopic balance equation based on the microscopic balance equation (Bird et al., 1960; Bowen, 1976) was extensively covered in a series of papers (Fick, 1855; Gray and Lee, 1977; Hassanizadeh and Gray, 1979a, 1979b, 1980; Hassanizadeh, 1986a, 1986b). To gain an understanding of macroscopic balance equations, these papers must be critically evaluated by serious readers. The theory of averaging procedures and assumptions involved in the derivation of the general macroscopic balance equation will not be repeated here.

Applying the principle of average procedures to the general microscopic balance equation, Eq. (1.4.3), one can obtain the following general macroscopic balance law for the thermodynamic property associated with the i-th species in the α-th phase (Hassanizadeh, 1986a):

$$\frac{\partial \rho_i^\alpha \theta^\alpha \Psi_i^\alpha}{\partial t} + \nabla \cdot \left(\rho_i^\alpha \theta^\alpha \mathbf{V}_i^\alpha \Psi_i^\alpha \right) + \nabla \cdot \mathbf{I}_i^\alpha = \rho_i^\alpha \theta^\alpha \left(F_i^\alpha + G_i^\alpha + e^\alpha(\rho_i \psi_i) + D_i^\alpha \right) \tag{1.5.1}$$

subject to

$$\sum_\alpha \rho_i^\alpha \theta^\alpha \left[e^\alpha(\rho_i \psi_i) + D_i^\alpha \right] = 0 \tag{1.5.2}$$

in which

$$\theta^\alpha = \frac{1}{\Delta V}\int_{\Delta V}\gamma_\alpha\, dV \tag{1.5.3}$$

$$\rho_i^\alpha = \frac{1}{\theta^\alpha \Delta V}\int_{\Delta V}\rho_i\gamma_\alpha\, dV \tag{1.5.4}$$

$$\mathbf{V}_i^\alpha = \frac{1}{\rho_i^\alpha\theta^\alpha \Delta V}\int_{\Delta V}\rho_i\mathbf{v}_i\gamma_\alpha\, dV \tag{1.5.5}$$

$$\Psi_i^\alpha = \frac{1}{\rho_i^\alpha\theta^\alpha \Delta V}\int_{\Delta V}\rho_i\psi_i\gamma_\alpha\, dV \tag{1.5.6}$$

$$\mathbf{N}\cdot\mathbf{I}_i^\alpha = \frac{1}{\Delta A}\int_{\Delta A}\left(\mathbf{i}_i - \rho_i\tilde{\mathbf{v}}_i\tilde{\psi}_i\right)\cdot\mathbf{n}\,\gamma_\alpha dA \tag{1.5.7}$$

$$F_i^\alpha = \frac{1}{\rho_i^\alpha\theta^\alpha \Delta V}\int_{\Delta V}\rho_i f_i\gamma_\alpha\, dV \tag{1.5.8}$$

$$G_i^\alpha = \frac{1}{\rho_i^\alpha\theta^\alpha \Delta v}\int_{\Delta V}\rho_i g_i\gamma_\alpha\, dV \tag{1.5.9}$$

$$e^\alpha(\rho_i\psi_i) = \frac{1}{\rho_i^\alpha\theta^\alpha \Delta V}\sum_{\beta\neq\alpha}\int_{\Delta A_{\alpha\beta}}\rho_i\psi_i(\mathbf{w} - \mathbf{v}_i)\cdot\mathbf{n}^{\alpha\beta}dA \tag{1.5.10}$$

$$D_i^\alpha = \frac{1}{\rho_i^\alpha\theta^\alpha \Delta V}\sum_{\beta\neq\alpha}\int_{\Delta A_{\alpha\beta}}\mathbf{i}_i\cdot\mathbf{n}^{\alpha\beta}dA, \tag{1.5.11}$$

where ρ_i^α = macroscopic intrinsic volume averaged mass density of the i-th species in the α-phase, θ^α = macroscopic fraction of voids occupied by the α phase, Ψ_i^α = any macroscopic mass averaged thermodynamic property of the i-th species in the α phase, $\mathbf{V}_i^\alpha$ = macroscopic mass averaged velocity of the particles for the i-th species in the α phase, $\mathbf{I}_i^a$ = macroscopically averaged surface flux of Ψ_i^α, F_i^α = macroscopic source/sink term representing external supply of Ψ_i^α to the α-th phase, G_i^α = macroscopic source/sink term representing the production of Ψ_i^α by internal reaction within the α-th phase, $e^\alpha(\rho_i\psi_i)$ = macroscopic source/sink terms representing the transfer of ψ_i to the α-th phase due to phase change at the interfaces, D_i^α = macroscopic source/sink terms representing the transfer of Ψ_i^α to the α-th phase due to interphase diffusion, ΔV = representative elemental volume (REV), γ_α = α phase distribution function, $\tilde{\mathbf{v}}_i$ = difference between microscopic and macroscopic velocity of the i-th species, $\tilde{\psi}_i$ = difference between microscopic and macroscopic thermodynamic property of the i-th species, $\mathbf{w}$ = velocity of the interface $\alpha\beta$, $\Delta A^{\alpha\beta}$ = interfacial area of the interface $\alpha\beta$

with the REV, and ΔA = surface area enclosing the REV ΔV.

Using Eqs. (1.5.3) through (1.5.11), one can obtain the macroscopic balance equations for mass, momentum, energy, and entropy from Eq. (1.5.1) by assigning the appropriate variables to ψ_i, $\mathbf{i}_i$, f_i, and g_i as given in Table 1.3.1. The detail procedure to obtain these equations can be found elsewhere (Hassanizadeh, 1986a).

1.5.1. Macroscopic Mass Balance Equation

Substituting ψ_i equal to 1, $\mathbf{i}_i$ equal to $\mathbf{j}_i$, f_i equal to q_i, and g_i equal to internal mass production of the i-th species r_i into Eqs. (1.5.3) through (1.5.11), we have the following balance equation for the mass of the i-th species in the α-th phase (Hassanizadeh, 1986a):

$$\frac{\partial \rho_i^\alpha \theta^\alpha}{\partial t} + \nabla\cdot\left(\rho_i^\alpha \theta^\alpha \mathbf{v}_i^\alpha\right) - \nabla\cdot\mathbf{J}_i^\alpha = \rho_i^\alpha \theta^\alpha\left[Q_i + R_i^\alpha + e^\alpha(\rho_i) + J_i^\alpha\right] \tag{1.5.12}$$

in which

$$\mathbf{N}\cdot\mathbf{J}_i^\alpha = \frac{1}{\Delta A}\int_{\Delta A}\left(\mathbf{j}_i - \rho_i\tilde{\mathbf{v}}_i\right)\cdot\mathbf{n}\gamma_\alpha dA \tag{1.5.13}$$

$$Q_i^\alpha = \frac{1}{\rho_i^\alpha \theta^\alpha \Delta V}\int_{\Delta V}\rho_i q_i \gamma_\alpha \, dV \tag{1.5.14}$$

$$R_i^\alpha = \frac{1}{\rho_i^\alpha \theta^\alpha \Delta V}\int_{\Delta V}\rho_i r_i \gamma_\alpha \, dV \tag{1.5.15}$$

$$e^\alpha(\rho_i) = \frac{1}{\rho_i^\alpha \theta^\alpha \Delta V}\sum_\beta \int_{\Delta A^{\alpha\beta}} \rho_i(\mathbf{w} - \mathbf{v}_i)\cdot\mathbf{n}^{\alpha\beta} dA \tag{1.5.16}$$

$$J_i^\alpha = \frac{1}{\rho_i^\alpha \theta^\alpha \Delta V}\sum_\beta \int_{\Delta A^{\alpha\beta}} \mathbf{j}_i\cdot\mathbf{n}^{\alpha\beta} dA. \tag{1.5.17}$$

The first term on the right-hand side of Eq. (1.5.12) represents the external source added to the i-th species, the second the internal production of mass of the i-th species due to homogeneous chemical reactions with other species or due to biological processes and/or decay within the phase, the third term the production of mass of the i-th species due to phase change through the interfaces, and the fourth term the production of mass of the i-th species due to interphase diffusion through interfaces.

1.5.2. Macroscopic Momentum Balance Equation

Taking ψ_i equal to the momentum density $\mathbf{v}_i$, $\mathbf{i}_i$ equal to the partial stress tensor for the i-th species σ_i, f_i equal to the external body force (such as gravity or ionic attractions) exerted on the i-th species $\mathbf{g}_i$, and g_i equal to $r_i \cdot \mathbf{v}_i$ plus the internal body force exerted on the i-th species by all other species $\mathbf{t}_i$ in Eqs. (1.5.3) through (1.5.11), we have the following balance equation for the momentum of the i-th species (Hassanizadeh, 1986a):

$$\frac{\partial \rho_i^\alpha \theta^\alpha \mathbf{V}_i^\alpha}{\partial t} + \nabla \cdot \left(\rho_i^\alpha \theta^\alpha \mathbf{V}_i^\alpha \mathbf{V}_i^\alpha \right) - \nabla \cdot \boldsymbol{\Sigma}_i^\alpha = \tag{1.5.18}$$

$$\rho_i^\alpha \theta^\alpha \mathbf{G}_i^\alpha + \rho_i^\alpha \theta^\alpha \left(\mathrm{R}_i^\alpha \mathbf{V}_i^\alpha + \hat{\mathbf{T}}_i^\alpha \right) + \rho_i^\alpha \theta^\alpha \left\{ e^\alpha(\rho_i) \mathbf{V}_i^\alpha + \left[e^\alpha(\rho_i \tilde{\mathbf{v}}_i) + \boldsymbol{\Upsilon}_i^\alpha \right] \right\}$$

in which

$$\mathbf{N} \cdot \boldsymbol{\Sigma}_i^\alpha = \frac{1}{\Delta A} \int_{\Delta A} \left(\sigma_i - \rho_i \tilde{\mathbf{v}}_i \tilde{\mathbf{v}}_i \right) \cdot \mathbf{n} \gamma_\alpha dA \tag{1.5.19}$$

$$\mathbf{G}_i^\alpha = \frac{1}{\rho_i^\alpha \theta^\alpha \Delta V} \int_{\Delta V} \rho_i \mathbf{g}_i \gamma_\alpha dV \tag{1.5.20}$$

$$\hat{\mathbf{T}}_i^\alpha = \mathbf{T}_i^\alpha + \frac{1}{\rho_i^\alpha \theta^\alpha \Delta V} \int_{\Delta V} \rho_i \left(\mathrm{r}_i \tilde{\mathbf{v}}_i \right) \gamma_\alpha dV \tag{1.5.21a}$$

$$\mathbf{T}_i^\alpha = \frac{1}{\rho_i^\alpha \theta^\alpha \Delta V} \int_{\Delta V} \rho_i \mathbf{t}_i \gamma_\alpha dV \tag{1.5.21b}$$

$$e^\alpha(\rho_i \tilde{\mathbf{v}}_i) + \frac{1}{\rho_i^\alpha \theta^\alpha \Delta V} \sum_{\beta \neq \alpha} \int_{\Delta A^{\alpha\beta}} \rho_i \tilde{\mathbf{v}}_i (\mathbf{w} - \mathbf{v}_i) \cdot \mathbf{n}^{\alpha\beta} dA \tag{1.5.22}$$

$$\boldsymbol{\Upsilon}_i^\alpha = \frac{1}{\rho_i^\alpha \theta^\alpha \Delta V} \sum_{\beta \neq \alpha} \int_{\Delta A^{\alpha\beta}} \sigma_i \cdot \mathbf{n}^{\alpha\beta} dA. \tag{1.5.23}$$

The first term on the right-hand side of Eq. (1.5.18) represents the effect of external body force on the transport of momentum; the second and third terms are the effect of internal mass production and internal force exerted on the i-th species by all other species, respectively; and the fourth, fifth, and sixth terms reflect the effect of interphase mass exchange, interphase momentum exchange, and interphase momentum diffusion, respectively.

1.5.3. Macroscopic Energy Balance Equation

Taking ψ_i equal to the density of internal energy and kinetic energy $(e_i + v_i^2/2)$, $\mathbf{i}_i$ equal to $\sigma_i \cdot v_i$ plus the heat flux of the i-th species $\mathbf{q}_i$, f_i equal to $\mathbf{g}_i \cdot v_i$ plus the external supply of energy to the i-th species h_i, and g_i equal to $[r_i \cdot (e_i + v_i^2/2) + \mathbf{t}_i \cdot v_i]$ plus the internal exchange of energy between the i-th species and all other species u_i in Eqs. (1.5.3) through (1.5.11), we have the following balance equation for the energy of the i-th species (Hassanizadeh, 1986a):

$$\frac{\partial \rho_i^\alpha \theta^\alpha \left(\hat{E}_i^\alpha + \mathbf{V}_i^{\alpha^2}/2\right)}{\partial t} + \nabla \cdot \left[\rho_i^\alpha \theta^\alpha \left(\hat{E}_i^\alpha + \mathbf{V}_i^{\alpha^2}/2\right) \mathbf{V}_i^\alpha\right] - \nabla \cdot \left(\boldsymbol{\Sigma}_i^\alpha \cdot \mathbf{V}_i^\alpha + \hat{\mathbf{Q}}_i^\alpha\right) =$$
$$- \rho_i^\alpha \theta^\alpha \left(\mathbf{G}_i^\alpha \cdot \mathbf{V}_i^\alpha - \hat{H}_i^\alpha\right) + \rho_i^\alpha \theta^\alpha \left[R_i^\alpha \left(\hat{E}_i^\alpha + \mathbf{V}_i^{\alpha^2}/2\right) + \hat{\mathbf{T}}_i^\alpha \cdot \mathbf{V}_i^\alpha + \hat{U}_i^\alpha\right] + \tag{1.5.24}$$
$$\rho_i^\alpha \theta^\alpha \left[e^\alpha(\rho_i)\left(\hat{E}_i^\alpha + \mathbf{V}_i^{\alpha^2}/2\right) + e^\alpha\left(\rho_i \tilde{\mathbf{v}}_i\right) \cdot \mathbf{V}_i^\alpha + e^\alpha\left(\rho_i \tilde{e}_i + \rho_i \tilde{\mathbf{v}}_i^2/2\right) + \hat{Q}_i^\alpha\right]$$

in which

$$\hat{E}_i^\alpha = E_i^\alpha + \frac{1}{\rho_i^\alpha \theta^\alpha \Delta V} \int_{\Delta V} \rho_i \left(\tilde{\mathbf{v}}_i^2/2\right) \gamma_\alpha dV \tag{1.5.25a}$$

$$E_i^a = \frac{1}{\rho_i^\alpha \Delta V} \int_{\Delta V} \rho_i e_i \, \gamma_\alpha \, dV \tag{1.5.25b}$$

$$\mathbf{N} \cdot \hat{\mathbf{Q}}_i^\alpha = \mathbf{N} \cdot \mathbf{Q}_i^\alpha + \frac{1}{\Delta A} \int_{\Delta A} \left[\sigma_i \cdot \tilde{\mathbf{v}}_i - \rho_i \tilde{\mathbf{v}}_i \left(e_i + \tilde{\mathbf{v}}_i^2/2\right)\right] \cdot \mathbf{n} \gamma_\alpha dA \tag{1.5.26a}$$

$$\mathbf{N} \cdot \mathbf{Q}_i^\alpha = \frac{1}{\Delta A} \int_{\Delta A} \mathbf{q}_i \cdot \mathbf{n} \, \gamma_\alpha dA \tag{1.5.26b}$$

$$\hat{H}_i^\alpha = H_i^\alpha + \frac{1}{\rho_i^\alpha \theta^\alpha \Delta V} \int_{\Delta V} \rho_i \left(\mathbf{g}_i \tilde{\mathbf{v}}_i\right) \gamma_\alpha dV \tag{1.5.27a}$$

$$H_i^\alpha = \frac{1}{\rho_i^\alpha \theta^\alpha \Delta V} \int_{\Delta V} \rho_i h_i \gamma_\alpha dV \tag{1.5.27b}$$

$$\hat{U}_i^\alpha = U_i^\alpha + \frac{1}{\rho_i^\alpha \theta^\alpha \Delta V} \int_{\Delta V} \rho_i \left[\tilde{\mathbf{t}}_i \cdot \tilde{\mathbf{v}}_i + \tilde{r}_i \left(\tilde{e}_i + \tilde{\mathbf{v}}_i^2/2\right)\right] \gamma_\alpha dV \tag{1.5.28a}$$

$$U_i^\alpha = \frac{1}{\rho_i^\alpha \theta^\alpha \Delta V} \int_{\Delta V} \rho_i u_i \gamma_\alpha dV \qquad (1.5.28b)$$

$$e^\alpha(\rho_i \tilde{e}_i + \rho_i \tilde{v}_i^2/2) + \frac{1}{\rho_i^\alpha \theta^\alpha \Delta V} \sum_{\beta \neq \alpha} \int_{\Delta A^{\alpha\beta}} \rho_i(\tilde{e}_i + \tilde{v}_i^2/2)(\mathbf{w}_i - \mathbf{v}) \cdot \mathbf{n}^{\alpha\beta} dA \qquad (1.5.29)$$

$$\hat{Q}_i^\alpha = Q_i^\alpha + \frac{1}{\rho_i^\alpha \theta^\alpha \Delta V} \sum_{\beta \neq \alpha} \int_{\Delta A^{\alpha\beta}} (\sigma_i \cdot \tilde{\mathbf{v}}_i) \cdot \mathbf{n}^{\alpha\beta} dA + \Upsilon_i^\alpha \cdot \mathbf{V}_i^\alpha \qquad (1.5.30a)$$

$$Q_i^\alpha = \frac{1}{\rho_i^\alpha \theta^\alpha \Delta V} \sum_{\beta} \int_{\Delta A^{\alpha\beta}} \mathbf{q}_i \cdot \mathbf{n}_i^{\alpha\beta} dA \qquad (1.5.30b)$$

$$\tilde{e}_i = e_i - \hat{E}_i^\alpha . \qquad (1.5.31)$$

The first and second terms on the right-hand side of Eq. (1.5.24) reflect the effect of external body force and external heat supply, respectively, on the energy transport; the third, fourth, and fifth terms are the effect of internal mass production, internal forces exerted on the i-th species by all other species, and internal transfer of energy to the i-th species by all other species, respectively; and the sixth, seventh, eighth, and ninth terms are the effect of interphase mass exchange, interphase momentum exchange and interphase momentum diffusion, the interphase energy exchange, and interphase energy diffusion, respectively.

1.5.4. Macroscopic Entropy Balance Equation

Taking ψ_i equal to entropy density s_i, $\mathbf{i}_i$ equal to the surface flux of entropy associated with the i-th species $\mathbf{f}_i$, f_i equal to the external supply of entropy to the i-th species b_i, and g_i equal to $r_i \cdot s_i$ plus the internal exchange of entropy to i-th species by all other species w_i and the rate of net production of entropy of the i-th species in Eqs. (1.5.3) through (1.5.11), we have the following balance equation for the entropy of the i-th species (Hassanizadeh, 1986a):

$$\frac{\partial \rho_i^\alpha \theta^\alpha S_i^\alpha}{\partial t} + \nabla \cdot (\rho_i^\alpha \theta^\alpha \mathbf{V}_i^\alpha S_i^\alpha) - \nabla \cdot \mathbf{F}_i^\alpha = -\rho_i^\alpha \theta^\alpha B_i^\alpha + \rho_i^a \theta^a (R_i^a S_i^a + \hat{W}_i^\alpha) + \qquad (1.5.32)$$
$$\rho_i^\alpha \theta^\alpha \{ e^\alpha(\rho_i) S_i^\alpha + [e^\alpha(\rho_i \tilde{s}_i) + F_i^\alpha] \} + \rho_i^\alpha \theta^\alpha \Gamma_i^\alpha$$

in which

$$S_i^\alpha = \frac{1}{\rho_i^\alpha \theta^\alpha \Delta V} \int_{\Delta V} \rho_i s_i \gamma_\alpha dV \tag{1.5.33}$$

$$\mathbf{N} \cdot \mathbf{F}_i^\alpha = \frac{1}{\Delta A} \int_{\Delta A} (\mathbf{f}_i - \rho_i \tilde{\mathbf{v}}_i \tilde{s}_i) \cdot \mathbf{n} \gamma_\alpha dA \tag{1.5.34}$$

$$B_i^\alpha = \frac{1}{\rho_i^\alpha \theta^\alpha \Delta V} \int_{\Delta V} \rho_i b_i \gamma_\alpha dV \tag{1.5.35}$$

$$\hat{W}_i^\alpha = W_i^\alpha + \frac{1}{\rho_i^\alpha \theta_i^\alpha \Delta V} \int_{\Delta V} \rho_i (\tilde{r}_i \tilde{s}_i) \gamma_\alpha dV \tag{1.5.36a}$$

$$W_i^\alpha = \frac{1}{\rho_i^\alpha \theta^\alpha \Delta V} \int_{\Delta V} \rho_i w_i \gamma_\alpha dV \tag{1.5.36b}$$

$$e^\alpha(\rho_i \tilde{s}_i) = \frac{1}{\rho_i^\alpha \theta^\alpha \Delta V} \sum_{\beta \neq \alpha} \int_{\delta A^{\alpha\beta}} \rho_i \tilde{s}_i (\mathbf{w} - \mathbf{v}_i) \cdot \mathbf{n}^{\alpha\beta} dA \tag{1.5.37}$$

$$F_i^\alpha = \frac{1}{\rho_i^\alpha \theta^\alpha \Delta V} \sum_{\beta \neq \alpha} \int_{\Delta A^{\alpha\beta}} \mathbf{f}_i \cdot \mathbf{n}^{\alpha\beta} dA \tag{1.5.38}$$

$$\Gamma_i^\alpha = \frac{1}{\rho_i^\alpha \theta^\alpha \Delta V} \int_{\Delta V} \rho_i \Gamma_i \gamma_\alpha dV. \tag{1.5.39}$$

The first term on the right-hand side of Eq. (1.5.32) represents the effect of external supply on the transport of entropy; the second and third terms are the effect of internal mass production and internal transfer of entropy to the i-th species by all other species within the phase, respectively; the fourth, fifth, and sixth terms reflect the effect of interphase mass exchange, interphase entropy exchange, and interphase entropy diffusion, respectively; and the last term is the rate of net production of entropy.

The macroscopic balance equations for mass, momentum, energy, and entropy are summaried in Table 1.5.1. The variables in Table 1.5.1 are defined by Eqs. (1.5.12) through (1.5.39). All macroscopic variables are expressed in terms of microscopic variables. If innovative experimental instrumentations are developed to measure all the microscopic variables, then all macroscopic variables in Eqs. (1.5.12) through (1.5.39) can be computed via simple integration. The macroscopic balance equations may be validated with these innovative experiments.

Table 1.5.1 Variables Selected for the Macroscopic Balance Equation.

Quantity	Ψ_i^α	I_i^α	F_i^α	G_i^α	$e^\alpha(\rho_i\psi_i)$	D_i^α
Mass	1	J_i^α	Q_i^α	R_i^α	$e^\alpha(\rho_i)$	J_i^α
Momentum	V_i^α	Σ_i^α	G_i^α	$R_i^\alpha V_i^\alpha + \hat{\mathbf{T}}_i^\alpha$	$e^\alpha(\rho_i)V_i^\alpha + e^\alpha(\rho_i v_i)$	Y_i^α
Energy	$\hat{E}_i^\alpha + V_i^{\alpha 2}/2$	$\Sigma_i^\alpha \cdot V_i^\alpha + Q_i^\alpha$	$G_i^\alpha \cdot V_i^\alpha - \hat{H}_i^\alpha$	$R_i^\alpha(\hat{E}_i^\alpha + V_i^{\alpha 2}/2) + \hat{\mathbf{T}}_i^\alpha \cdot V_i^\alpha + \hat{U}_i^\alpha$	$e^\alpha(\rho_i)(\hat{E}_i^\alpha + V_i^{\alpha 2}/2) + e^\alpha(\rho_i v_i)V_i + e^\alpha(\rho_i e_i + \rho_i v_i^2/2)$	$\hat{Q}_i^\alpha$
Entropy	S_i^α	F_i^α	B_i^α	$R_i^\alpha S_i^\alpha + \hat{W}_i^\alpha + \Gamma_i^\alpha$	$e^\alpha(\rho_i)S_i^\alpha + e^\alpha(\rho_i s_i)$	F_i^α

1.6. References

Bear, J. 1979. Hydraulics of Groundwater. New York: McGraw-Hill.

Bird, R. B., W. E. Stewart, and E. N. Lightfoot. 1960. Transport Phenomena. New York: Wiley.

Bowen, R. M. 1976. Theory of mixtures: In A. C. Eringen ed., Continuum Physics. Vol. 3, Pt. I. New York: Academic Press.

Fick, A. E. 1855. Philosophical Magazine. 10:30.

Gray, W. G. 1983. General conservation equations for multi-phase systems: 4. Constitutive theory including phase change. Advances in Water Resources 6:130-140.

Gray, W. G. and P. C. Y. Lee. 1977. On the theorems for local volume averaging of multiphase systems. Int. J. Multiphase Flow. 3:333.

Greenkorn, Robert A. 1983. Flow Phenomena in Porous Media. New York: Marcel Dekker.

Hassanizadeh, M. and W. G. Gray. 1979a. General conservation equations for multi-phase systems: 1. Averaging procedure. Advances in Water Resources. 2:131-144.

Hassanizadeh, M. and W. G. Gray. 1979b. General conservation equations for multi-phase systems: 2. Mass, momentum, energy, and entropy equations. Advances in Water Resources. 2:191-2034.

Hassanizadeh, M. and W. G. Gray. 1980. General conservation equations for multi-phase systems: 3. Constitutive theory for porous media. Advances in Water Resources. 3:25-40.

Hassanizadeh, M. 1986a. Derivation of basic equations of mass transport in porous media. Part 1. Macroscopic balance laws. Advances in Water Resources. 9:196-206.

Hassanizadeh, M. 1986b. Derivation of basic equations of mass transport in porous media. Part 2. Generalized Darcy's law and Fick's laws. Advances in Water Resources. 9:207-222.

Mercer, J. W. and C. R. Faust. 1981. Ground-Water Modeling. National Water Well Association.

Owczarek, J. A. 1964. Fundamental of Gas Dynamics. Scranton, PA: International Textbook.

Sposito, G., V. K. Gupta, and R. N. Bhattacharya. 1979. Foundation theories of solute transport in porous media: a critical review. Advances in Water Resources. 2:59-68.

Todd, D. K. 1980. Groundwater Hydrology. 2nd Ed. New York: Wiley.

2 NUMERICAL METHODS APPLIED TO SUBSURFACE HYDROLOGY

Numerical methods, as shall be described in this book, are merely tools used to enable one to replace differential equations governing the subsurface processes with approximation sets of algebraic equations or matrix equations, which are subsequently solved using the methods of linear algebra and requiring the manipulation of computers (Fig. 2.1). If the differential equations were solved exactly by analytic procedures, the solution would appear as some combination of mathematical functions. Subsequent interest in the value of the solution at various locations within a domain of interest would require that the functions be evaluated. Often, when the functions are of a complex form, the computer must be used to determine the values of the function at the points of interest. In many cases the analytical solution will be in terms of an infinite series or some transcendental functions that can be evaluated only approximately. Nevertheless, it is often possible to control the accuracy of the evaluation by careful use of the computer. The steps outlined above do require some facility with number manipulation on the computer and do yield an approximate value of the solution at points of interest. However, the actual steps involve numerical evaluation of an analytical solution to a differential equation rather than numerical solution to the differential equation. The differences between these concepts is the presence of an exact analytical expression as an intermediate step in the former case and the use of an approximation to the differential equation in the latter case.

Numerical methods are used because they allow for solution of a broad range of problems that otherwise could not be treated. Problems involving irregularly shaped boundaries, materials with high variation in properties, nonlinear differential equations, and complex boundary conditions that may be unsolvable analytically often reduce to a straightforward exercise with numerical methods. Furthermore, changes in problem geometry or physical parameters are often easily accommodated within a numerical model where the use of analytical methods would require a complete reanalysis of the problem. Nevertheless, analytical solutions are still useful in that they often furnish information on the functional dependence of one property on other properties. Analytical solutions to idealized systems also serve as standards for comparison of numerically computed results - a verification procedure required by any numerical solution. Perhaps the greatest advantage of analytical solutions over numerical solution is that convergency, stability, and accuracy often are not serious problems whereas these problems often are the biggest headache and challenge of numerical methods.

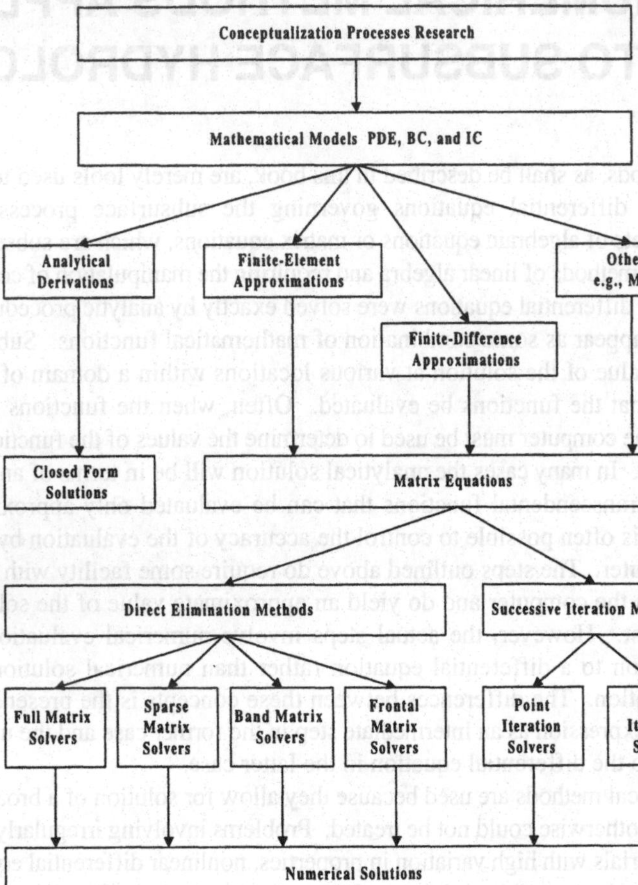

Fig. 2.1 Basic Ingredients in Subsurface Hydrology

The most common numerical techniques are finite-difference methods (FDM) and finite element methods (FEM). Although the spectral element methods (SEM) are widely used in computational fluid dynamics (Hussaini et al., 1983; Gottlieb et al., 1984; Orssag, 1980; Zang et al., 1989), they are rarely employed in subsurface hydrology. The fundamental distinction between FEMs and FDMs is that the former is based directly on approximating the function (Pinder and Gray, 1977), whereas the latter is based on approximation of the derivatives (Forsythe and Wasow, 1960;

Hilderbrand, 1968). A variety of other numerical methods exists, such as the method of characteristics (MOC) (Konikow and Bredehoeft, 1978), the integrated finite difference method (IFDM) (Narasimahn and Witherspoon, 1976), the integrated compartment method (ICM) (Yeh, 1981a), the finite volume method (FVM) (Jameson et al., 1981; Peric, 1985), and the boundary element methods (BEM) (Brebbia et al., 1984). Only FEMS and FDMs can deal with generic problems in computational subsurface hydrology. MOC is particular useful in dealing with transport problems with advection-dominant transport. The IFDM was evolved from the polygonal finite difference method (Hassan, 1974), which in turn resulted from the refinement of distributive parameter methods developed in the 1960s and adopted by several authors (Orlob and Woods, 1967; Tanji, 1970). This series of methods was obtained by directly applying the mass balance calculations to a number of well-defined spatial cells to yield a set of ordinary differential equations. The most serious problems with IFDM are that it cannot be applied to cases when cross-derivatives (such as anisotropic problems) appear in the differential equations and it is very difficult to extend the method for higher-order approximations. ICM based on three integral theorems of vectors recently was developed and theorized (Yeh, 1981b). While ICM can deal with anisotropic problems, it creates additional degrees of freedom at the cell. FVMs, very popular in the community of computational fluid dynamics in the last ten years, are essentially similar to IFDMs and ICMs. While BEMs can reduce the problems to one dimension less than the original problems, its applications completely depend on the ability to derive a fundamental solution of the partial differential equation. Thus, it finds most of its applications to Laplace equations with constant coefficients, to which the fundamental solution is known.

2.1. Desired Properties of Numerical Approximations

Numerical methods allow the scientist to replace a complex problem involving differential equations with a similar but more tractable problem. In the absence of any guidelines for performing the replacement, the procedure could be hazardous and ineffective. Thus, it is important to ensure that the replacement or approximation satisfies certain important properties of the exact solution: (1) conservativeness, (2) boundedness, and (3) transportiveness.

The conservative property means that the net transport across the transport domain boundaries should be equal to the total production or consumption by internal sources. It is very important from the physical point of view that physically conserved quantities (such as mass, momentum, energy, and chemical species) are conserved by numerical approximation; otherwise, unrealistic results may occur (such negative density or turbulent kinetic energy). Conservation of transported quantities is ensured if the flux across any cell face is uniquely represented so that all inner fluxes, if summed, should cancel in pairs (Fig. 2.1.1). Any scheme that processes this property is said to be conservative. For example, consider the simplest advection transport equation written in conservation form:

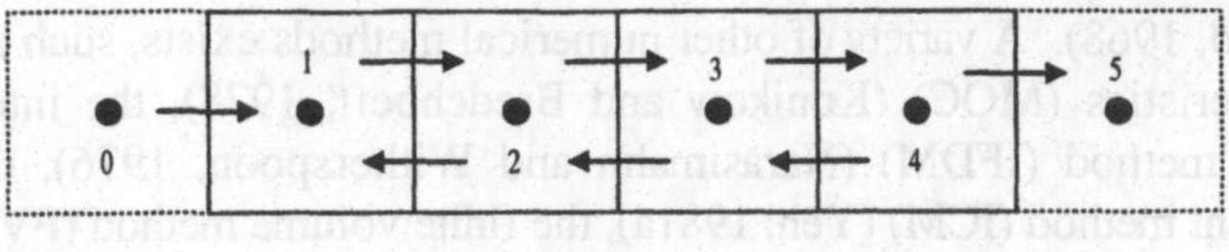

Fig. 2.1.1 Conservative Properties of a Numerical Scheme.

$$\frac{\partial C}{\partial t} + \frac{\partial VC}{\partial x} = 0, \tag{2.1.1}$$

where C is the concentration, t is the time, x is the spatial coordinate, and V is the velocity. Equation (2.1.1) can be discretized by the following finite volume approximation:

$$\begin{aligned}
\Delta C_1 \Delta x &= \left[0.25(V_0+V_1)(C_0+C_1) - 0.25(V_1+V_2)(C_1+C_2)\right]\Delta t \\
\Delta C_2 \Delta x &= \left[0.25(V_1+V_2)(C_1+C_2) - 0.25(V_2+V_3)(C_2+C_3)\right]\Delta t \\
\Delta C_3 \Delta x &= \left(0.25(V_2+V_3)(C_2+C_3) - 0.25(V_3+V_4)(C_3+C_4)\right]\Delta t \\
\Delta C_4 \Delta x &= \left[0.25(V_2+V_3)(C_2+C_3) - 0.25(V_4+V_5)(C_4+C_5)\right]\Delta t,
\end{aligned} \tag{2.1.2}$$

which is obviously a mass-conserved scheme. It can also be discretized by the following central and upstream finite difference approximations, respectively, as

$$\begin{aligned}
\Delta C_1 \Delta x &= \left[V_0 C_0 - V_2 C_2\right]\Delta t/2 \\
\Delta C_2 \Delta x &= \left[V_1 C_1 - V_3 C_3\right]\Delta t/2 \\
\Delta C_3 \Delta x &= \left[V_2 C_2 - V_4 C_4\right]\Delta t/2 \\
\Delta C_4 \Delta x &= \left[V_3 C_3 - V_5 C_5\right]\Delta t/2
\end{aligned} \tag{2.1.3}$$

$$\begin{aligned}
\Delta C_1 \Delta x &= \left[V_0 C_0 - V_1 C_1\right]\Delta t \\
\Delta C_2 \Delta x &= \left[V_1 C_1 - V_2 C_2\right]\Delta t \\
\Delta C_3 \Delta x &= \left[V_2 C_2 - V_3 C_3\right]\Delta t \\
\Delta C_4 \Delta x &= \left[V_3 C_3 - V_4 C_4\right]\Delta t.
\end{aligned} \tag{2.1.4}$$

It is seen that the central difference scheme is obviously not a mass-conserved scheme while the upstream scheme is.

The advection dispersion equation could have been written in a non-conservation form as

$$\frac{\partial C}{\partial t} + V\frac{\partial C}{\partial x} + C\frac{\partial V}{\partial x} = 0. \tag{2.1.5}$$

The discretization of Eq. (2.1.5) with central and upstream finite differences would yield

$$\begin{aligned}\Delta C_1 \Delta x &= \left[V_1(C_0 - C_2) + C_1(V_0 - V_2)\right]\Delta t/2\\ \Delta C_2 \Delta x &= \left[V_2(C_1 - C_3) + C_2(V_1 - V_3)\right]\Delta t/2\\ \Delta C_3 \Delta x &= \left[V_3(C_2 - C_4) + C_3(V_2 - V_4)\right]\Delta t/2\\ \Delta C_4 \Delta x &= \left[V_4(C_3 - C_5) + C_4(V_3 - V_5)\right]\Delta t/2\end{aligned} \qquad (2.1.6)$$

$$\begin{aligned}\Delta C_1 \Delta x &= \left[V_1(C_0 - C_1) + C_1(V_0 - V_1)\right]\Delta t\\ \Delta C_2 \Delta x &= \left[V_2(C_1 - C_2) + C_2(V_1 - V_2)\right]\Delta t\\ \Delta C_3 \Delta x &= \left[V_3(C_2 - C_3) + C_3(V_2 - V_3)\right]\Delta t\\ \Delta C_4 \Delta x &= \left[V_4(C_3 - C_4) + C_4(V_3 - V_4)\right]\Delta t.\end{aligned} \qquad (2.1.7)$$

Neither the central difference nor the upstream scheme yields mass-conserved discretization for this case.

Ensuring conservation does not guarantee that all the other important properties of transport processes will be maintained within the solution domain. For example, in the absence of sources/sinks and under steady-state conditions, the values of transported quantity ψ in the interior of the solution domain should be bounded by the values of ψ_b on the domain boundaries, that is, no resulting value should be larger than the maximum nor smaller than the minimum boundary value (Gosman and Lai, 1982):

$$\min(\psi_b) \le \psi \le \max(\psi_b). \qquad (2.1.8)$$

If a matrix resulting from a numerical approximation is diagonally dominant and of the positive type (please do not be confused with the matrix that is the positive definite), then the condition of boundedness is satisfied. A matrix is diagonally dominant if

$$\sum_{j \neq i} \frac{|a_{ij}|}{|a_{ii}|} \le 1 \text{ for all i } \textit{and } \sum_{j \neq i} \frac{|a_{ij}|}{|a_{ii}|} < 1 \text{ for at least i.} \qquad (2.1.9)$$

A matrix is of the positive type if

$$\frac{a_{ij}}{a_{ii}} < 0 \text{ for all } j \neq i \text{ and for all } i. \qquad (2.1.10)$$

These are sufficient, but not always necessary, conditions for the boundedness criterion to be satisfied. Whether the presence of a nonpositive type element will drive ψ_p out of bounds depends on the smoothness of ψ field around ψ_p. However, this is the only unconditional guarantee that a bounded solution will be obtained. Failure to satisfy the boundedness requirement will be prone to produce "overshoots," "undershoot," or "wiggles," as will be demonstrated later.

Figure 2.1.2 illustrates what is often termed the *transportive* properties of a fluid flow (Roache, 1976). If there is a constant source of ψ at point P in the flow field with uniform velocity and diffusivity, then the shape of contours of constant ψ will be influenced by the ratio of advection to diffusion, that is, by the Peclet number (P_e = uL/D, where L is a characteristic length, u is the magnitude of the velocity, and D is the diffusion coefficient) as shown in Fig. 2.1.2. When the fluid is stagnant ($P_e = 0$), the contours will be circles with P as the center; as P_e increases, the contours become elliptical, shifted in the direction of flow, until, for $P_e = \infty$ (no diffusion), they collapse into the streamline from P downstream. The implication is that high P_e flows events at node P will have a weak or no influence on upstream regions, while downstream regions will be strongly affected. Failure to account for these features of fluid flow can give rise to unrealistic results or even unstable solutions.

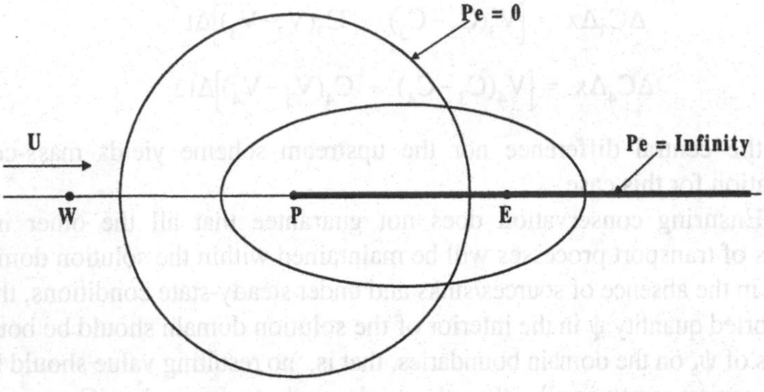

Fig. 2.1.2 Illustration of Transport Property of a Fluid Flow.

2.2. Finite-Difference Approximations

In elementary calculus, the derivative is defined as the difference between values of a function at two points divided by the distance between the points taken in the limit as the distance approaches zero. For example,

$$\frac{df}{dx} = \lim_{\Delta x \to 0} \frac{f(x + \Delta x) - f(x)}{\Delta x} . \tag{2.2.1}$$

Finite difference procedures involve the inverse of the above process such that a derivative based on a differential distance dx is approximated as a difference based on some distance Δx. Of course, in making the approximation an error is introduced. Finite-difference theory is concerned with ensuring that the error is of a magnitude that is tolerable within the constraints of the problem under study, the size of the computer, and the budget of the scientist.

The finite-difference philosophy is based on replacing the differential equation that holds for the entire domain under study with a number of discrete approximations at selected points. The approximations are chosen according to the judgement of the user and in such a way that the approximation at one point depends to some degree on the nearby points. A simple approximation at a point may depend only on the immediate neighboring points. A complex approximation conceivably could depend on all other points. Thus, although the finite-difference procedure utilizes local approximations, we see that it is possible for points throughout the entire domain to contribute to the local approximation.

The finite-difference method consists of developing an approximation to a differential equation at many locations throughout the domain of interest. The modeler discretizes the domain by selecting nodal points at which the approximations are to be used. Although, in theory, one may select points anywhere in the domain, Shoup (1978) pointed out that regular mesh patterns are usually required and economical mesh grading is difficult to obtain. It is suggested that the advantages of the finite-difference procedures lie in the use of a simple, regular grid that thus allows the user to easily develop many alternative approximations to a differential equation. If one wishes to use an irregular grid and develop all the bookkeeping necessary to keep track of nodal locations and derivatives approximation, then the finite element procedure might as well be used. There are two common types of grids to locate regularly patterned points: mesh centered and block centered. These are shown in Figure 2.2.1. Associated with the grids are node points that represent the position at which the solution of the unknown values such as head, velocity, temperature, and concentration are obtained. In the meshed-centered grid, the nodes are located on the intersection of grid lines, whereas in the block-centered grid the nodes are centered between grid lines. The choice of the type of grid to use depends largely on the boundary conditions. The mesh-centered grid is convenient for problems where values of the unknown are specified, whereas the block-centered grid has an advantage in problems where the flux is specified across the boundary. From a practical point of view, the differences in the two types of grids are minor.

Two methods are commonly used for the development of finite-difference approximations: the curve-fitting technique and the Taylor series expansion technique. We will examine the techniques in detail for one-dimensional problems and indicate how they are extended to problems of higher dimensionality. Alternatively, a more intuitive technique can be used to obtain the final difference equations by considering the fluxes into and out of a finite-difference cell. This intuitive technique is usually limited to a lower order of approximations and is easier to understand.

2.2.1. Curve Fitting for Finite-Difference Approximations

Consider the grid in Figure 2.2.2 which lies in a portion of the one-dimensional x-domain. Assume that this domain is discretized and some of the nodes selected are indicated as x_0 through x_5. The distance between nodes is not necessarily constant.

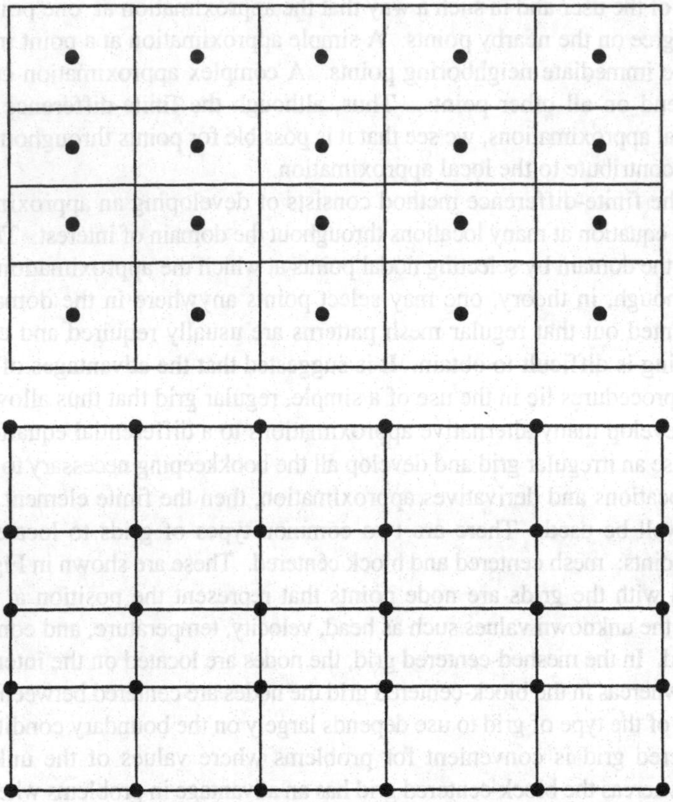

Fig. 2.2.1 Grids Showing Block- and Grid-Centered Nodes.

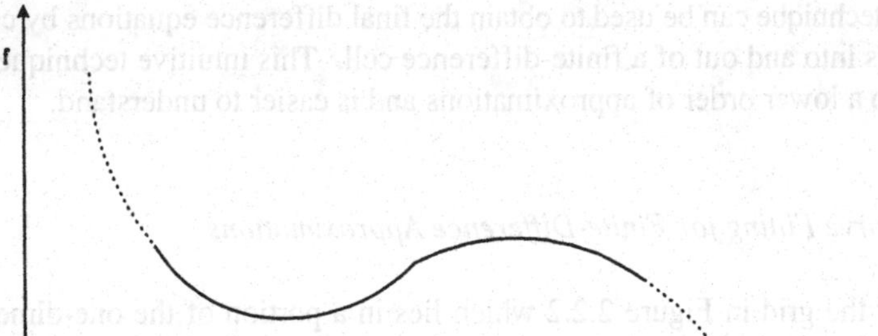

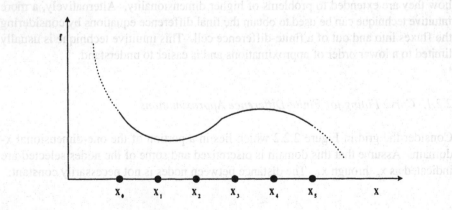

Figure 2.2.2 Plot of f as a Function of x.

We are interested in determining approximations to the derivatives of some function f(x). By the finite-difference procedure we will approximate these derivatives in terms of f evaluated at the locations where $f(x_i) = f_i$. If we are interested in the n-th derivative of f, we must use at least (n + 1) nodal locations to obtain a nontrivial approximation. We wish to approximate f(x) with a simpler function that would give exact values of f(x) at these (n + 1) points. The most commonly used and the simplest approximating functions are polynomials. One type of polynomial is the Lagrange polynomial, defined as

$$L_k^{(n)}(x) = \prod_{m \neq k}^{n} \frac{x - x_m}{x_k - x_m} = \frac{(x-x_o)\ldots(x-x_{k-1})(x-x_{k+1})\ldots(x-x_n)}{(x_k-x_o)\ldots(x_k-x_{k-1})(x_k-x_{k+1})\ldots(x_k-x_n)}. \tag{2.2.2}$$

Since $L_k^{(n)}(x)$ is the product of n linear factors, it is clearly a polynomial of degree n. We note that when $x = x_k$, the numerator and denominator of $L_k(x_k)$ are identical and the polynomial has unit value. But when $x = x_m$ and $m \neq k$, the polynomial vanishes. This fact can be used to represent an arbitrary function f(x) over an interval on the x-axis. For example, we may approximate f as

$$\hat{f} = \sum_{i=0}^{3} f_i L_i^{(3)}(x) \qquad \text{on the interval } x_o \leq x \leq x_3. \tag{2.2.3}$$

Because we have used four nodes, we can approximate up to the third derivative of $\hat{f}$:

$$\frac{d\hat{f}}{dx} = \sum_{i=0}^{3} f_i \frac{dL_i^{(3)}(x)}{dx} \tag{2.2.4}$$

$$\frac{d^2\hat{f}}{dx^2} = \sum_{i=0}^{3} f_i \frac{d^2L_i^{(3)}(x)}{dx^2} \tag{2.2.5}$$

$$\frac{d^3\hat{f}}{dx^3} = \sum_{i=0}^{3} f_i \frac{d^3L_i^{(3)}(x)}{dx^3}. \tag{2.2.6}$$

Note that the approximations to $d\hat{f}/dx$ and $d^2\hat{f}/dx^2$ given in Eqs. (2.2.4) and (2.2.5) depend on space x. This means that different approximations will be obtained at different locations. However, the approximation given by Eq. (2.2.6) for the third derivative does not depend on space and hence we obtain a constant value for the third derivative when using a four-node approximation. The accuracy of the approximation seems to decrease as we go to higher derivatives with an approximation using a finite number of points. Thus, to maintain a similar order of accuracy for all derivatives, we would need to use a greater number of points for higher derivatives.

If we consider the case where the distance between nodes is equal with a spacing of h, and if we evaluate these derivatives at, for example, x_1, we obtain

$$\frac{d\hat{f}}{dx}\bigg|_{x=x_1} = \frac{-2f_o - 3f_1 + 6f_2 - f_3}{6h} \tag{2.2.7}$$

$$\frac{d^2\hat{f}}{dx^2}\bigg|_{x=x_1} = \frac{f_o - 2f_1 + f_2}{h^2} \tag{2.2.8}$$

$$\frac{d^3\hat{f}}{dx^3}\bigg|_{x=x_1} = \frac{-f_o + 3f_1 - 3f_2 + f_3}{h^3}. \tag{2.2.9}$$

The above approximations depend on (1) the location where the derivative is evaluated, (2) the number of points used in the approximation, and (3) the locations of the points used in the approximation. In developing a derivative expression at a node, one is free to determine (2) and (3) arbitrarily except for the constraint that an efficient finite-difference formulation will have a regular mesh pattern. For example, at node x_1, the following alternative approximations for the first derivative may be obtained using polynomial interpolation, and the indicated nodes where consecutively numbered nodes are assumed to be separated by a distance h:

For three nodes located at x_o, x_1, and x_2, we have

$$\frac{d\hat{f}}{dx}\bigg|_{x=x_1} = \frac{-f_o + f_2}{2h}. \tag{2.2.10}$$

For two nodes located at x_o and x_1,

$$\frac{d\hat{f}}{dx}\bigg|_{x=x_1} = \frac{-f_o + f_1}{h}. \tag{2.2.11}$$

For two nodes located at x_1 and x_2,

$$\frac{d\hat{f}}{dx}\bigg|_{x=x_1} = \frac{-f_1 + f_2}{h}. \tag{2.2.12}$$

For three nodes located at x_1, x_2, and x_3, we have

$$\frac{d\hat{f}}{dx}\bigg|_{x=x_1} = \frac{-3f_1 + 4f_2 - f_3}{2h}. \tag{2.2.13}$$

In a computer code it is therefore possible to select any of these or the many other possibilities that exist for the approximation to the derivative. The actual selection is usually made to optimize the accuracy and efficiency of the code.

In the above discussion, nothing has been said about the size of the error generated when using discrete approximations to the derivative. Indeed, one drawback to using polynomial interpolation to generate the difference approximations is that no error estimate is obtained directly. It is possible to determine the error by applying the difference expression successively to polynomials of higher degree until an error is detected. Then the dependence of this error on the grid size and the derivatives of the polynomial may be obtained. For example, to determine the error of (2.2.13), consider the case where $x_1 = h$, $x_2 = 2h$, and $x_3 = 3h$. Equation (2.2.13) is the expression for df/dx. Since we are interested in df/dx, there will be an error such that

$$\left.\frac{df}{dx}\right|_{x=x_1} = \frac{-3f_1 + 4f_2 - f_3}{2h} + E. \tag{2.2.14}$$

When f = 1, df/dx = 0. Equation (2.2.14) reproduces this result exactly with E = 0. When f = x, df/dx = 1 and Eq. (2.2.14) becomes

$$1 = \frac{-3(h) + 4(2h) - (3h)}{2h} + E \quad or \quad 1 = 1 + E. \tag{2.2.15}$$

Thus, Eq. (2.2.13) is exact for first degree of polynomials. When $f = x^2$, $df/dx = 2x$; thus df/dx evaluated at x = h is 2h. Equation (2.2.14) becomes

$$2h = \frac{-3(h^2) + 4(4h^2) - (9h^2)}{2h} + E \quad or \quad 2h = 2h + E. \tag{2.2.16}$$

This shows that Eq. (2.2.13) is exact for the second degree polynomials. For the third degree function $f = x^3$, $df/dx = 3x^2$; hence Eq. (2.2.14) becomes

$$3h^2 = \frac{-3(h^3) + 4(8h^3) - (27h^3)}{2h} + E, \tag{2.2.17}$$

which implies

$$3h^2 = h^2 + E, \qquad E = 2h^2. \tag{2.2.18}$$

Because the error is, for the first time, nonzero for the third-degree polynomial, it is clear that the error depends on the third derivative or

$$E = 2h^2 = C\,\frac{d^3f}{dx^3}, \tag{2.2.19}$$

where C is a constant. Because the third derivative of x^3 is 6, $C = h^2/3$. The error given by Eq. (2.2.19) is the leading error term, and Equation (2.2.14) becomes

$$\left.\frac{df}{dx}\right|_{x=x_1} = \frac{-3f_1 + 4f_2 - f_3}{2h} + \frac{h^2}{3}\frac{d^3f}{dx^3} + E_1, \tag{2.2.20}$$

where E_1 accounts for additional error that depends on derivatives higher than the third. Of course, the components of E_1 may be calculated by continuing to examine polynomials of higher degree. For practical purposes, the leading term is usually the term of interest and E_1 would not be computed. For Eq. (2.2.9), the leading error term is proportional to d^4f/dx^4, that is,

$$\left.\frac{df}{dx}\right|_{x=x_1} = \frac{-2f_o - 3f_1 + 6f_2 - f_3}{6h} + \frac{h^3}{12}\frac{d^4f}{dx^4} + E_1. \qquad (2.2.21)$$

Although the above method for determining the error of an approximation may seem clumsy, in fact it may be the simplest method for determining the truncation error of a given complex expression. When an expression is being derived by the Taylor series method, as discussed below, the error term is obtained directly as a byproduct.

In summary, we may obtain one-dimensional approximations to the p-th derivative of a function from interpolation polynomials of degree p or higher. The leading error term of the approximation may also be obtained by using the method of successive substitution of polynomials into the difference expressions. The difference expressions obtained may be substituted into differential equations as a basis for a numerical finite difference solution computer code. This curve-fitting method of obtaining a finite-difference approximation to the differential equations can be extended to two and higher dimensions without much difficulty.

2.2.2. *Taylor Series Expansion for Finite-Difference Approximations*

The Taylor series method does not provide any additional information about difference expressions than can be obtained by the interpolation procedure. This method is slightly more elegant and may provide difference expressions on an irregular multi-dimensional grid more easily than they could be obtained using interpolation. Furthermore, the Taylor series approach allows one to proceed directly from the nodal locations to the difference expression without obtaining an intermediate curve-fit polynomial.

Difference approximations to derivatives are obtained from Taylor series expansions around the point where the derivative is to be evaluated. For example, with reference to Fig. 2.2.2, we may derive an approximation to df/dx evaluated at x_1 that makes use of information at x_1, x_2, and x_3. If

$$x_2 - x_1 = h \quad \textit{and} \quad x_3 - x_1 = \beta h, \qquad (2.2.22)$$

we may perform Taylor series expansion to obtain f_2 and f_3 in terms of f and its derivatives at x_1.

$$f_2 = f_1 + \left.\frac{df}{dx}\right|_1 h + \left.\frac{d^2f}{dx^2}\right|_1 \frac{h^2}{2} + \left.\frac{d^3f}{dx^3}\right|_1 \frac{h^3}{6} + \ldots \qquad (2.2.23)$$

$$f_3 = f_1 + \left.\frac{df}{dx}\right|_1 \beta h + \left.\frac{d^2f}{dx^2}\right|_1 \frac{\beta^2h^2}{2} + \left.\frac{d^3f}{dx^3}\right|_1 \frac{\beta^3h^3}{6} + \ldots \qquad (2.2.24)$$

To obtain an expression for df/dx evaluated at x_1 with the greatest accuracy possible using this three-point difference expression, we combine Eqs. (2.2.23) and (2.2.24) so that the lowest-order term in which we have no interest is eliminated. In the current instance, we eliminate d^2f/dx^2 evaluated at x_1 between Eqs. (2.2.23) and (2.2.24) to obtain

$$\beta^2 f_2 - f_3 = (\beta^2 - 1)f_1 + \beta(\beta - 1)h\left.\frac{df}{dx}\right|_1 + \beta^2(1 - \beta)\frac{h^3}{6}\left.\frac{d^3f}{dx^3}\right|_1 + \cdots \tag{2.2.25}$$

Solution of Eq. (2.2.25) yields

$$\left.\frac{df}{dx}\right|_1 = \frac{(1 - \beta^2)f_1 + \beta^2 f_2 - f_3}{\beta(\beta - 1)h} + \frac{\beta h^2}{6}\left.\frac{d^3f}{dx^3}\right|_1 + \cdots \tag{2.2.26}$$

where the last term is the leading error term of the expression. If we consider a grid that has evenly spaced nodes such that $x_j = jh$, then β is equal to 2 and Eq. (2.2.26) becomes

$$\left.\frac{df}{dx}\right|_1 = \frac{-3f_1 + 4f_2 - f_3}{2h} + \frac{h^2}{3}\left.\frac{d^3f}{dx^3}\right|_1 + \cdots \tag{2.2.27}$$

Equation (2.2.27) is identical to Eq. (2.2.20), which was developed using the method of polynomial interpolation.

If we wish to obtain d^2f/dx^2 from knowledge of f at x_1, x_2, and x_3, we may again use Eqs. (2.2.23) and (2.2.24). In this instance the lowest-order term that is not of interest is the first derivative term. Elimination of the first derivative between Eqs. (2.2.23) and (2.2.24) yields

$$\beta f_2 - f_3 = (\beta - 1)f_1 + \beta(\beta - 1)\frac{h^2}{2}\left.\frac{d^2f}{dx^2}\right|_1 + \beta(1 - \beta^2)\frac{h^3}{6}\left.\frac{d^3f}{dx^3}\right|_1 + \cdots \tag{2.2.28}$$

Rearrangement of Eq. (2.2.28) gives us

$$\left.\frac{d^2f}{dx^2}\right|_1 = \frac{2(1 - \beta)f_1 + 2\beta f_2 - 2f_3}{\beta(1 - \beta)h^2} - (1 + \beta)\frac{h}{3}\left.\frac{d^3f}{dx^3}\right|_1 + \cdots \tag{2.2.29}$$

When the grid is regular, $x_j = jh$ and $\beta = 2$, so Eq. (2.2.29) becomes

$$\left.\frac{d^2f}{dx^2}\right|_1 = \frac{f_1 - 2f_2 + f_3}{h^2} + h\left.\frac{d^3f}{dx^3}\right|_1 + \cdots \tag{2.2.30}$$

It is noted that in both Eqs. (2.2.27) and (2.2.30) the error term is proportional to d^3x/dx^3. Thus the expressions for the first and second derivatives are exact for polynomials of zero, first, and second degrees. The error of the first derivative is proportional to h^2 while that of the second derivative is only proportional to h. Thus, as the grid spacing is decreased, the error in the first derivative expression will decrease at a faster rate than the error in the second derivative. Furthermore, this indicates that to obtain difference expressions of a certain order of accuracy, we need more grid points as the derivative becomes higher. One final observation that is important is that the expressions in Eqs. (2.2.27) and (2.2.30) are one-sided in that all nodes are located on one side of the point where the derivative is evaluated. Certainly when nodes are drawn from both sides of the point of interest, a higher degree of accuracy may be obtained using a fixed number of points. Even spacing of the nodes used also contributes to improved accuracy of the approximation.

2.2.3. *Finite-Difference Approximations on Arbitrary Nodes (2-D)*

The use of interpolating polynomials and of the Taylor series expansion concept can be adapted for development of finite difference expression in multiple dimensions. For discussion purposes, only two-dimensional expressions will be considered. Extrapolation to higher dimensions is straightforward.

In two dimensions, interpolation polynomials are most easily applied when nodes lie on the boundaries of elements that have sides parallel with the coordinate axes. In other cases it may become difficult to determine the appropriate form for the polynomials. Figure 2.2.3 contains a number of examples that help to clarify this point. In Figures 2.2.3(a) to 2.2.3(c), we see nodes that lie on the sides of rectangles whose sides are parallel with the x- and y-axes. The interpolation polynomial is determined from the values of the function at the node locations and by deciding the degree of the independent variable to be used in the interpolation. In Figures 2.2.3(a) and 2.2.3(b), we have built up the degree of the polynomial by taking enough factors that are in terms of the lowest orders of the independent variable.

For Figure 2.2.3(c), we have done the same thing but elected to use x^2y^2 rather than x^3 or y^3 because this keeps the interpolation symmetric. The interpolation formula for Figure 2.2.3(c) is thus

$$\begin{aligned} \hat{f} &= A_1 + A_2x + A_2y + A_4xy + \\ &A_5x^2 + A_6y^2 + A_7x^2y + A_8xy^2 + A_9x^2y^2. \end{aligned} \tag{2.2.31}$$

The nine parameters in this expression may be determined by requiring this expression to be exact at the nine nodes where f is specified. Approximations to

$$\frac{\partial f}{\partial x}, \frac{\partial f}{\partial y}, \frac{\partial^2 f}{\partial x \partial y}, \frac{\partial^2 f}{\partial x^2}, \frac{\partial^2 f}{\partial y^2}, \frac{\partial^3 f}{\partial x^2 \partial y}, \frac{\partial^3 f}{\partial x \partial y^2}, \frac{\partial^4 f}{\partial x^2 \partial y^2} \tag{2.2.32}$$

may be obtained from the interpolant. It is noted that even though one fourth derivative in f is derivable from Eq. (2.2.31), not all the third derives can be deduced.

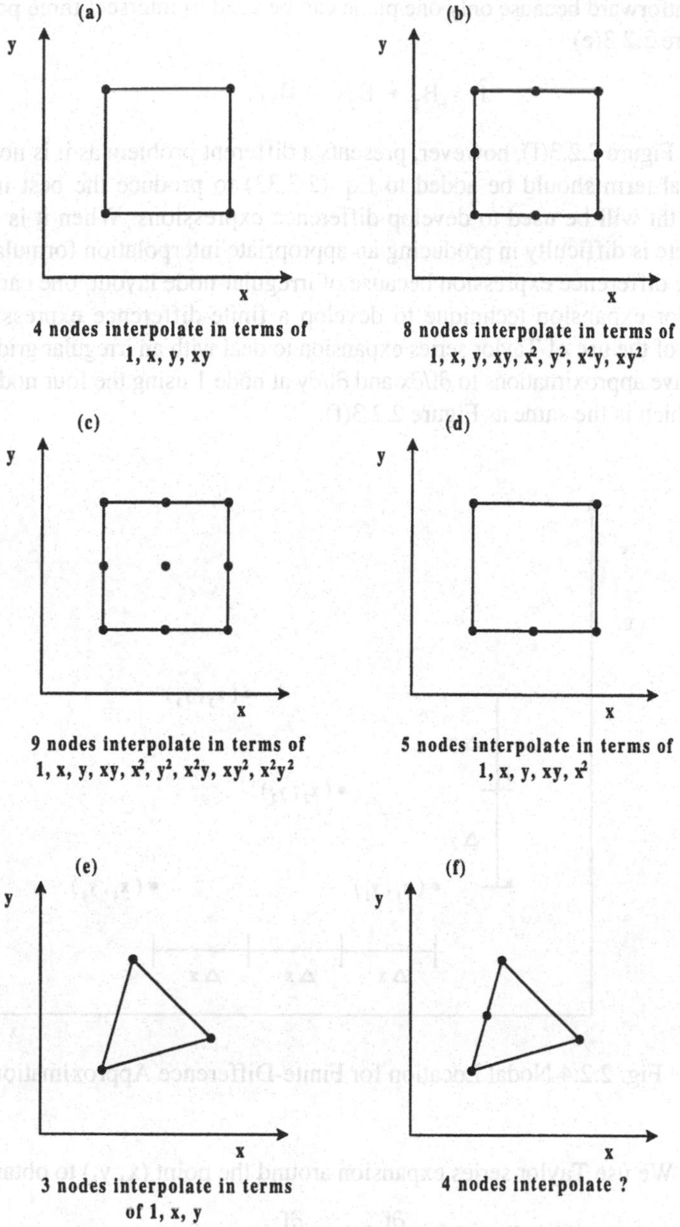

Fig. 2.2.3 Various Locations of Nodes in Two Dimensions.

Figure 2.2.3(d) is a rectangle, which begins to point out some of the difficulties that arise in generating approximating polynomials. Here we have elected to add a term in x^2 to the interpolation used in Figure 2.2.3(a). As the number of nodes increases or if this rectangle were not aligned with the coordinate system, the choice of interpolation polynomial would be complicated. In Figure 2.2.3(e), the interpolation is straightforward because only one plane can be used to intersect three points. Thus, for Figure 2.2.3(e)

$$\hat{f} = B_1 + B_2x + B_3y. \tag{2.2.33}$$

Figure 2.2.3(f), however, presents a different problem as it is not clear what additional term should be added to Eq. (2.2.33) to produce the best interpolation formula tht will be used to develop difference expressions. When it is not clear or when there is difficulty in producing an appropriate interpolation formula to develop the finite difference expression because of irregular node layout, one can simply use the Taylor expansion technique to develop a finite-difference expression. As an example of the use of Taylor series expansion to deal with an irregular grid system, we shall derive approximations to $\partial f/\partial x$ and $\partial f/\partial y$ at node 1 using the four nodes in Figure 2.2.4, which is the same as Figure 2.2.3(f).

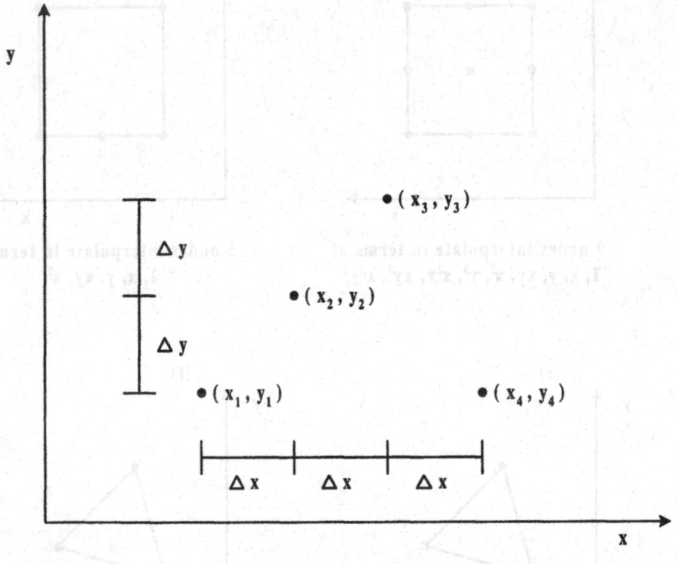

Fig. 2.2.4 Nodal Location for Finite-Difference Approximations.

We use Taylor series expansion around the point (x_1,y_1) to obtain

$$f_2 = f_1 + \frac{\partial f}{\partial x}\Delta x + \frac{\partial f}{\partial y}\Delta y + \frac{\partial^2 f}{\partial x^2}\frac{\Delta x^2}{2} + \frac{\partial^2 f}{\partial y^2}\frac{\Delta y^2}{2} + \frac{\partial^2 f}{\partial x \partial y}\Delta x \Delta y + \ldots \tag{2.2.34}$$

$$f_3 = f_1 + 2\frac{\partial f}{\partial x}\Delta x + 2\frac{\partial f}{\partial y}\Delta y + 4\frac{\partial^2 f}{\partial x^2}\frac{\Delta x^2}{2} + 4\frac{\partial^2 f}{\partial y^2}\frac{\Delta y^2}{2} + 4\frac{\partial^2 f}{\partial x \partial y}\Delta x \Delta y + \ldots \tag{2.2.35}$$

$$f_4 = f_1 + 3\frac{\partial f}{\partial x}\Delta x + 9\frac{\partial^2 f}{\partial x^2}\frac{\Delta x^2}{2} + \ldots \tag{2.2.36}$$

To obtain a good approximation of $\partial f/\partial x$, we would eliminate the lowest-order terms that represent error. It can be seen that if we eliminate $\partial f/\partial y$ between Eqs. (2.2.34) and (2.2.35), we will also eliminate $\partial f/\partial x$ and obtain

$$2f_2 - f_3 = f_1 - \frac{\partial^2 f}{\partial x^2}\Delta x^2 - \frac{\partial^2 f}{\partial y^2}\Delta y^2 - 2\frac{\partial^2 f}{\partial x \partial y}\Delta x \Delta y - \ldots \tag{2.2.37}$$

We could use this expression in Eq. (2.2.36) to eliminate the error term in Δx^2, but doing so would introduce second-order errors in Δy^2 and $\Delta x \Delta y$. Thus, second-order term approximations to $\partial f/\partial x$ are obtained either from Eq. (2.2.36) directly as

$$\frac{\partial f}{\partial x} = \frac{f_4 - f_1}{3\Delta x} - \frac{3}{2}\frac{\partial^2 f}{\partial x^2}\Delta x + \ldots \tag{2.2.38}$$

or by combining Eqs. (2.2.37) with Eq. (2.2.36) to eliminate the $\partial^2 f/\partial x^2$ term, which will result in

$$\frac{\partial f}{\partial x} = \frac{18f_2 + 2f_4 - 9f_3 - 11f_1}{6\Delta x} + \frac{3}{2}\frac{\partial^2 f}{\partial y^2}\frac{\Delta y^2}{\Delta x} + 3\frac{\partial^2 f}{\partial x \partial y}\Delta y + \ldots \tag{2.2.39}$$

Equation (2.2.39) contains the undesirable feature that if $\Delta x \to 0$, the first error term becomes large. Finite-difference expressions that are useful typically exhibit the property that as the grid size is decreased the difference expression approaches the differential expression and the error approaches zero. For this reason, Eq. (2.2.38) is probably preferable to Eq. (2.2.39).

To obtain an expression for $\partial f/\partial y$, we first eliminate $\partial f/\partial x$ from Eqs. (2.2.34) and (2.2.36); the substraction of 3 x Eq. (2.3.34) from Eq. (2.2.36) yields the following expression:

$$f_4 - 3f_2 = -2f_1 - 3\frac{\partial f}{\partial y}\Delta y + 6\frac{\partial^2 f}{\partial x^2}\frac{\Delta x^2}{2} - 3\frac{\partial^2 f}{\partial y^2}\frac{\Delta y^2}{2} - 3\frac{\partial^2 f}{\partial x \partial y}\Delta x \Delta y + a\frac{\partial^3 f}{\partial x^3}\Delta x^3 + \ldots \tag{2.2.40}$$

We then eliminate $\partial f/\partial x$ from Eqs. (2.2.34) and (2.2.36); the substraction of 3 x Eq.

(2.3.35) from 2 x Eq. (2.3.36) gives

$$2f_4 - 3f_3 = -f_1 - 6\frac{\partial f}{\partial y}\Delta y + \\ 6\frac{\partial^2 f}{\partial x^2}\frac{\Delta x^2}{2} - 12\frac{\partial^2 f}{\partial y^2}\frac{\Delta y^2}{2} - 12\frac{\partial^2 f}{\partial x \partial y}\Delta x \Delta y + 2a\frac{\partial^3 f}{\partial x^3}\Delta x^3 + \ldots \tag{2.2.41}$$

Finally, we can subtract Eq. (2.2.40) from Eq. (2.2.41) to eliminate the $\partial^2 f/\partial x^2$,

$$\frac{\partial f}{\partial y} = \frac{-f_4 + 3f_3 - 3f_2 + f_1}{3\Delta y} - 3\frac{\partial^2 f}{\partial y^2}\frac{\Delta y}{2} - 3\frac{\partial^2 f}{\partial x \partial y}\Delta x + a\frac{\partial^3 f}{\partial x^3}\frac{\Delta x^3}{\Delta y} + \ldots \tag{2.2.42}$$

Besides the two-leading error terms in Eq. (2.2.42), additional error terms that contain $1/\Delta y$ exist. Thus, if Δy is very small, the approximation given by Eq. (2.2.42) could contain a large error. An alternative approximation to $\partial f/\partial y$ may be derived by eliminating $\partial^2 f/\partial y^2$ and $\partial^2 f/\partial x \partial y$ between Eqs. (2.2.40) and (2.2.41) to obtain

$$\frac{\partial f}{\partial y} = \frac{-2f_4 - 3f_3 + 12f_2 - 7f_1}{6\Delta y} - \frac{3}{2}\frac{\partial^2 f}{\partial x^2}\frac{\Delta x^2}{\Delta y} + \ldots \tag{2.2.43}$$

Again the error term in Eq. (2.2.43) contains $1/\Delta y$. Thus, it is seen that even with a powerful tool such as the Taylor expansion, it is very difficult to obtain a finite difference expression with good accuracy when irregular nodes are encountered. Fortunately, in the finite difference approximation, we are primarily concerned with orthogonal grids of some regularity. Under such circumstances, the extension of Taylor series expansion and interpolating formulation from one-dimensional problems to two- and three-dimensional problems is as easy as eating rice (a Chinese proverb). This point will be addressed later.

2.2.4. Flux-Based Finite Difference Approximation

Fundamental to any numerical approach to solving partial differential equations is the concept of discretization, wherein a compound region is represented as a number of continuous subregions. While regular-shaped subregions of one kind are normally used in the finite-difference discretization, irregular-grid finite-difference discretization has been developed (Thacker, 1977). Irregular shapes of more than one kind have been employed in finite element discretization. Practical considerations, however, dictate that a very limited number of shapes, for example, triangular and quadrilateral (two-dimensional cases) or tetrahedral and hexahedral (three-dimensional cases), be used in a particular problem. Besides the difficulty of deriving base functions for shapes other than these simple ones, programming limitations prohibit the consideration of a large number of various shapes simultaneously. In the flux-based approach of finite difference approximation, one can divide the region of interest into various shapes and sizes (Yeh, 1981a, 1981b).

Thus, the first step of the intuitive approach is to divide a compound region into any number of subregions of various shapes and sizes. In this book, each subregion is termed a compartment rather than a discrete element so as not to be confused with finite-element methods. Nor it is termed a box or a cell since these generally refer to regular shapes such as rectangular or hexahedral prisms. The centroid of a compartment is defined as a node, and the line that connects two nodes is called a *connector*. The connector represents the interface of two neighboring compartments. Thus, a connector is characterized by its two end nodes, the interfacial area, directional cosines of a unit vector normal to the interface, and two length scales representing the distances from two nodes to the interface, respectively. A compartment is characterized by its volume and the node representing it. With these definitions, the region of interest is ready for discretization. For example, Figure 2.2.5 shows the region R divided into 10 compartments, 14 interior-to-interior connectors, and 11 interior-to-boundary compartment connectors. For convenience, one may wish to create 11 imaginary boundary compartments and thus, in this particular discretization, 11 interior-to-boundary node connectors to represent the outside region shown in Figure 2.2.5. The perpendicular distances from boundary nodes to the boundary are zero. In other words, the boundary compartment nodes are located right on the boundary (Fig. 2.2.5). Of course, any other discretization is possible.

The major task in employing irregular compartments of various shapes and sizes is, however, the attention and effort demanded for the discretization. Fortunately, methods for automating the discretization of complex regions have been reported elsewhere (Thacker et al., 1980). The only limitation on the irregularity and size of all compartments will be the consideration of computational stability and convergence. Instabilities associated with the irregular compartment discretization are quite like those associated with variable coefficients on uniform compartments (Thacker, 1977). Therefore, for numerical stability, it is important that the sizes and shapes of compartments be smoothly distributed through the region. In other words, transition of the compartment sizes and shapes from one subregion to another should be gradual in consideration with the coefficients of the differential equations. Properly designed, this should lead to the creation of a well-behaved matrix for the resulting algebraic equations.

To use the intuitive approach, differential equations first must be rearranged and regrouped into the combinations of the gradient, divergence, and curl of some quantities. The following integral theorems of vectors are then used to transform the integration of any property over the region to that over the boundary:

$$\int \Delta F \, dV = \int \mathbf{n} F \, dS \tag{2.2.44}$$

$$\int \nabla \cdot \mathbf{U} \, dV = \int \mathbf{n} \cdot \mathbf{U} \, dS \tag{2.2.45}$$

$$\int \nabla \times \mathbf{U} \, dV = \int \mathbf{n} \times \mathbf{U} \, dS, \tag{2.2.46}$$

where F and **U** are any scalar and vector functions, respectively, dV is the differential volume in the compartment, dS is the differential surface area on the compartment interfaces, and **n** is an outward unit vector normal to the enclosing surface. The right-hand side of Eqs. (2.2.44) through (2.2.46) can be approximated as follows:

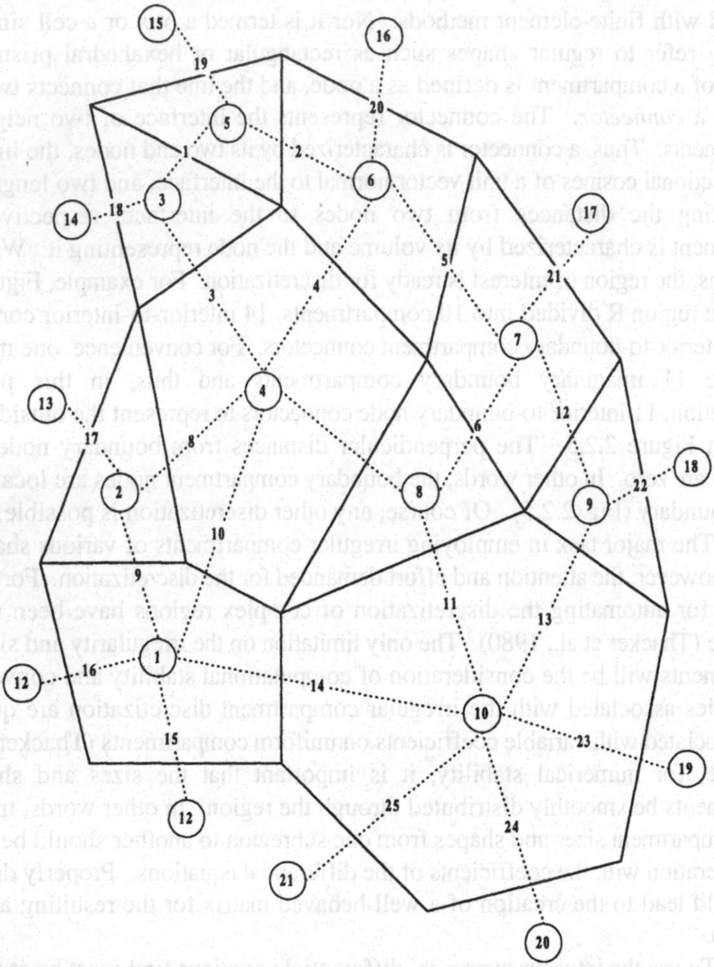

Fig. 2.2.5 Example of Arbitrary Spatial Discretization of a Region.

$$\int \mathbf{n} F dS = \sum_{j \in N_i} \mathbf{n_{ij}} F_{ij} S_{ij} \tag{2.2.47}$$

$$\int \mathbf{n} \cdot \mathbf{U} dS = \sum_{j \in N_i} \mathbf{n_{ij}} \cdot \mathbf{U_{ij}} S_{ij} \tag{2.2.48}$$

$$\int \mathbf{n} \times \mathbf{U} dS = \sum_{j \in N_i} \mathbf{n_{ij}} \times \mathbf{U_{ij}} S_{ij} \tag{2.2.49}$$

where the subscript ij indicates that the values are to be evaluated at the interface of compartments i and j as shown in Figure 2.2.6 and N_i is the set of node numbers surrounding the compartment i. It should be noted that the summation in Eqs. (2.2.47) through (2.2.49) is to be performed over all the interfaces enclosing compartment i. For example, there are four interfaces, ij, ik, il, and im, surrounding compartment i in Figure 2.2.6.

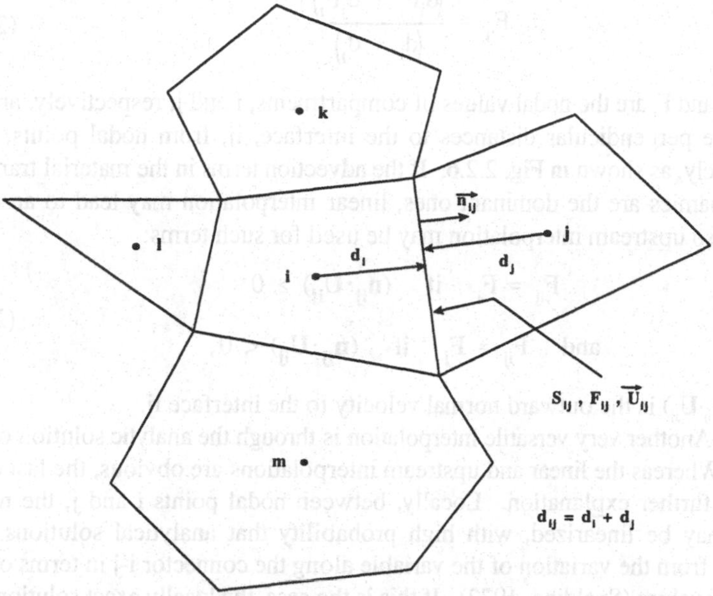

Fig. 2.2.6 Definition of Compartment and Interfacial Values.

A particularly attractive feature of Eqs. (2.2.44) through (2.2.49) is their ability to conserve the properties, F and **U**. For example, Eq. (2.2.48) applied to the continuity equation would guarantee that the mass is conserved numerically. In addition, the novel feature of applying Eqs. (2.2.44) through (2.2.46) to any spatial derivative is to reduce the problem to the approximation of $(n-1)^{th}$ order derivatives on the compartment interfaces rather than the approximation of nth order derivatives at the nodal points as used in the irregular-grid finite-difference method (Thacker, 1977). This is particularly significant with regard to first- and second-order derivative terms in any partial differential equations. Instead of having to approximate both the first and second derivatives by the finite-difference method, one simply has to approximate the function itself and the first-order derivative with simple finite difference interpolation at the interface. Because the first-order derivatives usually define the flux of the property under consideration, the reduction in the order of spatial derivatives makes it simple to build the physical representation at the interface, in particular the interface of different media.

Equations (2.2.47) through (2.2.49) involve the interfacial values. To close the system such that a set of algebraic equations can be obtained in which the number of unknowns equals the number of equations, interpolation methods must be employed to express the interfacial values in terms of the compartment values or the nodal values. The key to the success of the intuitive approach lies in the use of appropriate interpolations because they are intimately related to the stability of the numerical solution. There may be infinite varieties of interpolations; the intuitive and simplest is the linear interpolation, given by

$$F_{ij} = \frac{(d_i F_j + d_j F_i)}{(d_i + d_j)}, \tag{2.2.50}$$

where F_i and F_j are the nodal values of compartments, i and j, respectively, and d_i and d_j are the perpendicular distances to the interface, ij, from nodal points, i and j, respectively, as shown in Fig. 2.2.6. If the advection terms in the material transport or flow dynamics are the dominant ones, linear interpolation may lead to an unstable solution so upstream interpolation may be used for such terms:

$$\begin{aligned} F_{ij} &= F_i \quad \text{if} \quad (\mathbf{n}_{ij} \cdot \mathbf{U}_{ij}) \geq 0 \\ \text{and} \quad F_{ij} &= F_j \quad \text{if} \quad (\mathbf{n}_{ij} \cdot \mathbf{U}_{ij}) < 0, \end{aligned} \tag{2.2.51}$$

where $(\mathbf{n}_{ij} \cdot \mathbf{U}_{ij})$ is the outward normal velocity to the interface ij.

Another very versatile interpolation is through the analytic solution of a local region. Whereas the linear and upstream interpolations are obvious, the last category requires further explanation. Locally, between nodal points i and j, the nonlinear terms, may be linearized, with high probability that analytical solutions may be obtained from the variation of the variable along the connector i-j in terms of its two boundary values (Spalding, 1972). If this is the case, the locally exact solution may be used as the basis for interpolation. Using the exact solution to obtain the interfacial values is naturally superior to any other method. However, it should be noted that exact solutions are seldom obtainable because of the high nonlinearity in the flow dynamic equations. Nevertheless, this exact interpolation usually is obtainable when one is dealing with the transport of materials having locally constant velocity.

The most serious deficiency that may be encountered in the intuitive approach is that the interpolated value is only a representative value of the interface. The value cannot be identified with a definite point on the interface. This deficiency can be overcome by first interpolating the value of each of the corner points of the interface in terms of the node values whose compartments join at the corner and then interpolating the value at any point on the interface in terms of the values at the corner points. This two-step interpolation procedure would greatly complicate the problem. However, it is worth pursuing for the improvement of the intuitive approach.

2.3. Finite-Element Approximations

To apply the finite-element method, a region in space is subdivided into a set of smaller regions or finite elements. Nodes are located at the vertices of the elements, along the sides of the elements, and within the elements. Interpolation polynomials, or base functions, are defined over the elements that are piecewise continuous and, in some instances, have derivatives that are continuous over element boundaries. The economy of the finite-element procedure lies in the definition of base functions to be nonzero only over a limited portion of the grid. Typically each base function is associated with a node and is nonzero only in elements that include that node.

The finite-element procedure allows for the easy incorporation of irregular node placement and element size in a computer simulation. In fact, once the code has been written with all the necessary bookkeeping for tracking node locations and element membership, alteration of the grid is quite simple. In some instances a finite-element code simply may be altered to allow a much different problem to be solved than the one originally considered. Additionally, spatial variation of physical properties such as aquifer thickness or permeability is easily accounted for in finite-element codes. The option to cluster nodes in a region of the grid where large gradients exist and yet use a coarse grid where gradients are expected to be small is a primary advantage of the finite-element method. The price for this flexibility is higher overhead costs for the finite-element method than for the finite-difference method. Thus, selection of one method as opposed to another should be made by the modeler after consideration of the physical features of the problem to be solved and the available computer as well as the computer code.

Finite-element approximations can be made using either the variational procedure or the method of weighted residuals. In order to use the variational procedure, one must be able to find the functional for extremization. This is not always easy.

2.3.1. The Method of Weighted Residuals

The method of weighted residuals is a general mathematical procedure whose applicability is not limited to finite-element numerical methods. To demonstrate this method, we will consider the time-space differential operator L, which operates on f such that

$$L(f) - b = 0 \quad \text{in} \quad V, \tag{2.3.1}$$

where V is the domain of interest and b is a specified function. We now assume that f may be approximated by $\hat{f}$, which is made up a linear combination of trial functions and satisfies the principal boundary conditions of the problem. Thus we select

$$f \approx \hat{f} = \sum_{j=1}^{N} f_j \phi_j(x,t) \tag{2.3.2}$$

or

$$f \approx \hat{f} = \sum_{j=1}^{N} f_j(t)N_j(x), \qquad (2.3.3)$$

where **x** refers to spatial dependence, t refers to temporal dependence, f_j's are coefficients to be determined, and ϕ_j's are a set of selected interpolations or base functions that depend on time and space while N_j's are a set of selected base functions that are only spatially dependent. In most of the finite-element applications, Eq. (2.3.3) is used and finite differencing in time of $f_j(t)$ is performed while the spatial discretization is performed using finite element concepts. The index N refers to the number of base functions selected for the approximation. Also it should be noted that whether or not f_j is the value of $\hat{f}$ at node j depends on the set of base functions selected. In general f_j's are merely undetermined coefficients. Because $\hat{f}$ is only an approximation to f, substitution of $\hat{f}$ into Eq. (2.3.1) may produce an error or residual R such that

$$L(\hat{f}) - b = R(x,t). \qquad (2.3.4)$$

If $\hat{f}$ were the exact solution of Eq. (2.3.1), the residual would be zero. By the method of weighted residuals, R is forced to zero in some average sense by appropriate selection of the parameters f_j. The constraints that allow for f_j to be determined are the weighted residual integral of the form

$$\int_V R(x,t)\,W_i(x,t)\;dV \qquad i = 1, 2, \ldots, N. \qquad (2.3.5)$$

where $W_i(\mathbf{x},t)$ is the set of weighting functions. If the approximation Eq. (2.3.3) is used for $\hat{f}$, the weighting functions will depend only on space and the integration is performed only over the spatial domain or

$$\int_{V_x} R(x,t)W_i(x)\;dV \qquad i = 1, 2, \ldots, N. \qquad (2.3.6)$$

The application of Eq. (2.3.5) or (2.3.6) to finite elements, rather than the entire domain, is accomplished by selecting base functions, ϕ_j or N_j, and weighting functions, W_i, which are nonzero only on a limited portion of the grid. The form of weighting in the spatial domain is of greater concern here as the finite-element discretization depends on the form of the weighting function selected. When element sides are not parallel or perpendicular to the spatial axes, as will be the case when the finite-element concept is being exploited to maximum advantage, there will be no uncoupling of spatial directions as was obtained with finite differences. Nevertheless, the finite-element discretization in space depends on the node locations, the interpolation functions selected, the weighting functions used, and the way in which the integrations are carried out. For illustrative purposes we will now consider some of the most commonly used spatial weighting functions.

2.3.1.1. Subdomain Method. Here the spatial domain is divided into N subdomains, S_i, and the weighting functions are

$$W_i(x) = \begin{cases} 1 & \text{if } x \text{ in } V_i \\ 0 & \text{if } x \text{ not in } V_i \end{cases} \tag{2.3.7}$$

Thus, Eq. (2.3.6) becomes

$$\int_{V_{ix}} R(x,t)\, dV = 0, \tag{2.3.8}$$

where V_{ix} is the region where $W_i = 1$. The error of the approximation averaged over the region is thus zero.

2.3.1.2. Collocation Method. Here N points, known as collocation points, are selected at which the weighting function is a dirac delta function or

$$W_i = \delta(x - x_i). \tag{2.3.9}$$

Thus, Eq. (2.3.6) becomes

$$\int_{V_x} R(x,t)\delta(x - x_i)dV = R(x_i,t) = 0,\ i = 1, 2, \ldots, N. \tag{2.3.10}$$

This method is equivalent to requiring the residual or error to be zero at certain points. The integrations are obviously easy to perform with this method but other disadvantages arise. It is usually necessary to select base functions that have at least continuous first derivatives across element boundaries and to solve for the derivatives of $\hat{f}$ as well as $\hat{f}$ at the collocation points. This method is still in the development stage for field applications, but some interesting preliminary results have been reported by Shapiro and Pinder (1981) and Hayes et al. (1981).

2.3.1.3. Least Square Method. In this approach, the weighting function is

$$W = p(x) \frac{\partial R(x,t)}{\partial f_i}, \tag{2.3.11}$$

where p(x) is an arbitrary positive function. Thus, Eq. (2.3.6) becomes

$$\int_{V_x} p(x)R(x,t) \frac{\partial R(x,t)}{\partial f_i}\, dV = 0. \tag{2.3.12}$$

This expression is equivalent to minimizing the integral

$$I = \int_{V_x} p(x)R^2(x,t)dV \tag{2.3.13}$$

with respect to each coefficient f_i. This method is not generally used because it leads to a relatively cumbersome and complicated integral that must be evaluated.

2.3.1.4. Galerkin Method. For this procedure the base functions in Eqs. (2.3.2) or

(2.3.3) are also used as the weighting functions. Thus, Eq. (2.3.6) becomes

$$\int_{V_x} R(x,t)N_i(x)dV = 0, \quad i = 1, 2, \ldots, N. \tag{2.3.14}$$

This procedure is the most commonly used of the weighted residual procedures for hydrological modeling.

2.3.1.5. Petrov-Galerkin Method. For this procedure the weighting functions are one or more orders higher than the base function. Included in this class of weighting functions are the upstream weighting functions (Huyakorn and Pinder, 1977), the orthogonal weighting function (Yeh, 1983), and the orthogonal upstream weighting functions (Yeh, 1985). The application of upstream weighting functions to practical problems has been extensive. However, the effectiveness of the upstream weighting function in circumventing numerical problems has proved elusive. The applications of orthogonal and orthogonal-upstream weighting functions to field problems have not been extensive. Another class of Petrov-Galerkin methods is the N+2 upstream weighting functions (Westerink et al., 1988), which has yielded some promising results for a simple test case (Carrano and Yeh, 1995) but has not been extensively tested.

2.3.2. An Example to Illustrate the Method of Weighted Residuals

In practice, nine steps are employed to apply the finite-element method to differential equations: (1) divide the region into elements and nodes, (2) define base functions for each node, (3) define weighting functions for each node, (4) approximate the function in terms of base function and node values, (5) define the residual as the difference between true solution and approximate solution, (6) set weighted residual to zero, (7) derive matrix equations, (8) incorporate boundary conditions, and (9) use initial conditions to advance the solution through time.

The above steps are best explained by showing how they are applied to a simple example, as follows. The differential equation is

$$\frac{\partial f}{\partial t} - \frac{\partial^2 f}{\partial x^2} = 0. \tag{2.3.15}$$

The initial condition is

$$f = 0 \quad \text{at} \quad t = 0. \tag{2.3.16}$$

The boundary conditions are

$$f = 1 \quad \text{at} \quad x = x_1 \tag{2.3.17}$$

and

$$-\frac{\partial f}{\partial x} = f \quad \text{at} \quad x = x_4. \tag{2.3.18}$$

The above initial and boundary value problem governed by Eqs. (2.3.15) through (2.3.18) will be used to illustrate the procedures to obtain a matrix equation with the weighted residual finite-element method as follows.

2.3.2.1. Step 1. Let us assume that the region of interest can be divided into three elements and four nodes (Fig. 2.3.1).

Fig. 2.3.1 Division of a Region into Three Elements and Four Nodes.

2.3.2.2. Step 2. The second step in the finite-element formulation is to define the base and weighting functions in both global and local coordinates. The simplest function associated with a node, which satisfies the conditions that the values of the function are one at the node and zeros at all other nodes, is a linear function over the elements connecting to the node. Therefore, we define base functions both globally and locally (Fig. 2.3.2) as follows:

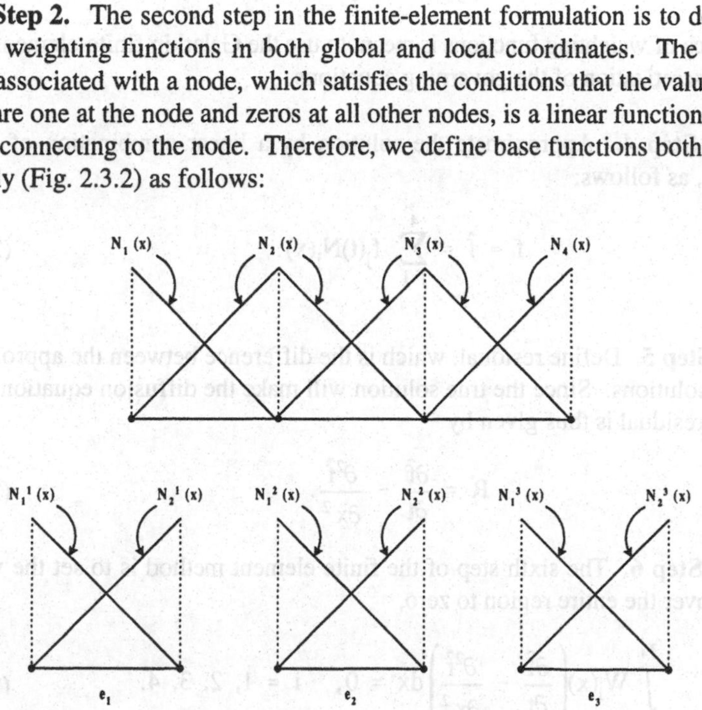

Fig. 2.3.2 Definition of Global and Local Base Functions.

where $N_1(x)$, $N_2(x)$, $N_3(x)$, and $N_4(x)$ are the base functions for nodes 1, 2, 3, and 4, respectively; $N_1^1(x)$ and $N_2^1(x)$ are the first and second base functions for element 1, respectively; $N_1^2(x)$ and $N_2^2(x)$ are the first and second base functions for element 2, respectively; and $N_1^3(x)$ and $N_2^3(x)$ are the first and second base functions for element 3, respectively. It is seen that

$$N_1^1(x) = N_1(x), \; N_2^1(x) = N_2(x) \quad \text{over element } e_1 \tag{2.3.19}$$

$$N_1^2(x) = N_2(x), \quad N_2^2(x) = N_3(x) \quad \text{over element } e_2 \tag{2.3.20}$$

and

$$N_1^3(x) = N_3(x), \quad N_2^3(x) = N_4(x) \quad \text{over element } e_3. \tag{2.3.21}$$

2.3.2.3. Step 3. For simplicity, assume that the weighting functions are identical to base functions, as follows:

$$\begin{aligned} W_1(x) &= N_1(x), \quad W_2(x) = N_2(x) \\ W_3(x) &= N_3(x), \quad W_4(x) = N_4(x). \end{aligned} \tag{2.3.22}$$

This choice of weighting functions is meant to use the Galerkin finite-element method for the discretization of the governing equations.

2.3.2.4. Step 4. Approximate the solution by a linear combination of the base functions, as follows:

$$f \simeq \hat{f} = \sum_{j=1}^{4} f_j(t) N_j(x). \tag{2.3.23}$$

2.3.2.5. Step 5. Define residual, which is the difference between the approximation and true solutions. Since the true solution will make the diffusion equation equal to zero, the residual is thus given by

$$R = \frac{\partial \hat{f}}{\partial t} - \frac{\partial^2 \hat{f}}{\partial x^2}. \tag{2.3.24}$$

2.3.2.6. Step 6. The sixth step of the finite element method is to set the weighted residual over the entire region to zero,

$$\int_{x_1}^{x_4} W_i(x) \left(\frac{\partial \hat{f}}{\partial t} - \frac{\partial^2 \hat{f}}{\partial x^2} \right) dx = 0, \quad i = 1, 2, 3, 4. \tag{2.3.25}$$

2.3.2.7. Step 7. The derivation of the matrix equation is the centerpiece of the finite-element method. While in finite-difference methods the matrix equation is derived based on the differentiation, in finite-element methods the matrix equation is derived based on integration of base and weighting functions as demonstrated below. Let us integrate Eq. (2.3.25) over the entire region of interest by part

$$\int_{x_1}^{x_4} W_i(x) \frac{\partial \hat{f}}{\partial t} dx + \int_{x_1}^{x_4} \frac{dW_i(x)}{dx} \frac{\partial \hat{f}}{\partial x} dx = W_i(x) \frac{\partial \hat{f}}{\partial x} \bigg|_{x_1}^{x_4}. \tag{2.3.26}$$

Substituting the approximate solution, Eq. (2.3.3), into Eq. (2.3.26), we obtain

$$\sum_{j=1}^{4}\left(\int_{x_1}^{x_4} W_i N_j dx\right)\frac{df_j}{dt} + \sum_{j=1}^{4}\left(\int_{x_1}^{x_4} \frac{dW_i}{dx}\frac{dN_j}{dx}dx\right) f_j = W\left.\frac{\partial \hat{f}}{\partial x}\right|_{x_1}^{x_4}. \tag{2.3.27}$$

Denoting a_{ij}, b_{ij}, and B_i as

$$a_{ij} = \int_{x_1}^{x_4} W_i N_j dx,\ b_{ij} = \int_{x_1}^{x_4} \frac{dW_i}{dx}\frac{dN_j}{dx}\, dx,\ \textit{and}\ B_i = W_i\left.\frac{\partial \hat{f}}{\partial x}\right|_{x_1}^{x_4}, \tag{2.3.28}$$

we can write Eq. (2.3.27) as the matrix equation

$$[a]\left\{\frac{df}{dt}\right\} + [b]\{f\} = \{B\} \tag{2.3.29}$$

or as a system of equations

$$\begin{aligned} &a_{11}\frac{df_1}{dt} + a_{12}\frac{df_2}{dt} + a_{13}\frac{df_3}{dt} + a_{14}\frac{dt_4}{dt} + \\ &b_{11}f_1 + b_{12}f_2 + b_{13}f_3 + b_{14}f_4 = B_1 \end{aligned} \tag{2.3.30}$$

$$\begin{aligned} &a_{21}\frac{df_1}{dt} + a_{22}\frac{df_2}{dt} + a_{23}\frac{df_3}{dt} + a_{24}\frac{dt_4}{dt} + \\ &b_{21}f_1 + b_{22}f_2 + b_{23}f_3 + b_{24}f_4 = B_2 \end{aligned} \tag{2.3.31}$$

$$\begin{aligned} &a_{31}\frac{df_1}{dt} + a_{32}\frac{df_2}{dt} + a_{33}\frac{df_3}{dt} + a_{34}\frac{dt_4}{dt} + \\ &b_{31}f_1 + b_{32}f_2 + b_{33}f_3 + b_{34}f_4 = B_3 \end{aligned} \tag{2.3.32}$$

$$\begin{aligned} &a_{41}\frac{df_1}{dt} + a_{42}\frac{df_2}{dt} + a_{43}\frac{df_3}{dt} + a_{44}\frac{dt_4}{dt} + \\ &b_{41}f_1 + b_{42}f_2 + b_{43}f_3 + b_{44}f_4 = B_4, \end{aligned} \tag{2.3.33}$$

where [a] is the mass matrix, [b] is the stiff matrix, and {B} is the boundary load vector.

The remaining task in the finite-element method is the evaluation of a_{ij}, b_{ij}, and B_i. These integrations can be obtained element by element. First the integration over the entire domain is decomposed into that over each element, then the globally defined functions are made equivalent to locally defined functions as demonstrated below:

$$a_{11} = \int_{x_1}^{x_4} N_1N_1dx = \int_{x_1}^{x_2} N_1N_1dx + \int_{x_2}^{x_3} N_1N_1dx + \int_{x_3}^{x_4} N_1N_1dx = \tag{2.3.34}$$

$$\underbrace{\int_{e_1} N_1^1N_1^1dx}_{a_{11}^1} + \int_{e_2} 0\ 0\ dx + \int_{e_3} 0\ 0\ dx$$

$$a_{12} = \int_{x_1}^{x_4} N_1N_2dx = \int_{x_1}^{x_2} N_1N_2dx + \int_{x_2}^{x_3} N_1N_2dx + \int_{x_3}^{x_4} N_1N_2dx = \tag{2.3.35}$$

$$\underbrace{\int_{e_1} N_1^1N_2^1dx}_{a_{12}^1} + \int_{e_2} 0\ N_1^2dx + \int_{e_3} 0\ 0\ dx$$

$$a_{13} = \int_{x_1}^{x_4} N_1N_3dx = \int_{x_1}^{x_2} N_1N_3dx + \int_{x_2}^{x_3} N_1N_3dx + \int_{x_3}^{x_4} N_1N_3dx = \tag{2.3.36}$$

$$\int_{e_1} N_1^1 0\ dx + \int_{e_2} 0\ N_2^2dx + \int_{e_3} 0\ N_1^3\ dx$$

$$a_{14} = \int_{x_1}^{x_4} N_1N_4dx = \int_{x_1}^{x_2} N_1N_4dx + \int_{x_2}^{x_3} N_1N_4dx + \int_{x_3}^{x_4} N_1N_4dx = \tag{2.3.37}$$

$$\int_{e_1} N_1^1 0\ dx + \int_{e_2} 0\ 0\ dx + \int_{e_3} 0\ N_2^3\ dx$$

$$a_{21} = \int_{x_1}^{x_4} N_2N_1dx = \int_{x_1}^{x_2} N_2N_1dx + \int_{x_2}^{x_3} N_2N_1dx + \int_{x_3}^{x_4} N_2N_1dx = \tag{2.3.38}$$

$$\underbrace{\int_{e_1} N_2^1N_1^1dx}_{a_{21}^1} + \int_{e_2} N_1^2\ 0\ dx + \int_{e_3} 0\ 0\ dx$$

$$a_{22} = \int_{x_1}^{x_4} N_2 N_2 dx = \int_{x_1}^{x_2} N_2 N_2 dx + \int_{x_2}^{x_3} N_2 N_2 dx + \int_{x_3}^{x_4} N_2 N_2 dx = \underbrace{\int_{e_1} N_2^1 N_2^1 dx}_{a_{22}^1} + \underbrace{\int_{e_2} N_1^2 N_1^2 dx}_{a_{11}^2} + \int_{e_3} 0\ 0\ dx \tag{2.3.39}$$

$$a_{23} = \int_{x_1}^{x_4} N_2 N_3 dx = \int_{x_1}^{x_2} N_2 N_3 dx + \int_{x_2}^{x_3} N_2 N_3 dx + \int_{x_3}^{x_4} N_2 N_3 dx = \int_{e_1} N_2^1\ 0\ dx + \underbrace{\int_{e_2} N_1^2 N_2^2 dx}_{a_{12}^2} + \int_{e_3} 0\ N_1^3\ dx \tag{2.3.40}$$

$$a_{24} = \int_{x_1}^{x_4} N_2 N_4 dx = \int_{x_1}^{x_2} N_2 N_4 dx + \int_{x_2}^{x_3} N_2 N_4 dx + \int_{x_3}^{x_4} N_2 N_4 dx = \int_{e_1} N_2^1\ 0\ dx + \int_{e_2} N_1^2 0 dx + \int_{e_3} 0\ N_1^3\ dx \tag{2.3.41}$$

$$a_{31} = \int_{x_1}^{x_4} N_3 N_1 dx = \int_{x_1}^{x_2} N_3 N_1 dx + \int_{x_2}^{x_3} N_3 N_1 dx + \int_{x_3}^{x_4} N_3 N_1 dx = \int_{e_1} 0\ N_1^1 dx + \int_{e_2} N_2^2 0 dx + \int_{e_3} N_1^3\ 0\ dx \tag{2.3.42}$$

$$a_{32} = \int_{x_1}^{x_4} N_3 N_2 dx = \int_{x_1}^{x_2} N_2 N_2 dx + \int_{x_2}^{x_3} N_3 N_2 dx + \int_{x_3}^{x_4} N_3 N_2 dx = \int_{e_1} 0\ N_1^1 dx + \underbrace{\int_{e_2} N_2^2\ N_1^2\ dx}_{a_{21}^2} + \int_{e_3} 0\ 0\ dx \tag{2.3.43}$$

$$a_{33} = \int_{x_1}^{x_4} N_3N_3dx = \int_{x_1}^{x_2} N_3N_3dx + \int_{x_2}^{x_3} N_3N_3dx + \int_{x_3}^{x_4} N_3N_3dx =$$

$$\int_{e_1} 0\ 0\ dx + \underbrace{\int_{e_2} N_2^2\ N_2^2 dx}_{a_{22}^3} + \underbrace{\int_{e_3} N_1^3 N_1^3 dx}_{a_{11}^3} \tag{2.3.44}$$

$$a_{34} = \int_{x_1}^{x_4} N_3N_4dx = \int_{x_1}^{x_2} N_3N_4dx + \int_{x_2}^{x_3} N_3N_4dx + \int_{x_3}^{x_4} N_3N_4dx =$$

$$\int_{e_1} 0\ 0\ dx + \int_{e_2} N_2^2 0\ dx + \underbrace{\int_{e_3} N_1^3 N_2^3 dx}_{a_{12}^3} \tag{2.3.45}$$

$$a_{41} = \int_{x_1}^{x_4} N_4N_1dx = \int_{x_1}^{x_2} N_4N_1dx + \int_{x_2}^{x_3} N_4N_1dx + \int_{x_3}^{x_4} N_4N_1dx =$$

$$\int_{e_1} 0\ N_1^1 dx + \int_{e_2} 0\ 0\ dx + \int_{e_3} N_2^3\ 0\ dx \tag{2.3.46}$$

$$a_{42} = \int_{x_1}^{x_4} N_4N_2dx = \int_{x_1}^{x_2} N_4N_2dx + \int_{x_2}^{x_3} N_4N_2dx + \int_{x_3}^{x_4} N_4N_2dx =$$

$$\int_{e_1} 0\ N_2^1 dx + \int_{e_2} 0\ N_1^2 dx + \int_{e_3} N_2^3\ 0\ dx \tag{2.3.47}$$

$$a_{43} = \int_{x_1}^{x_4} N_4N_3dx = \int_{x_1}^{x_2} N_4N_3dx + \int_{x_2}^{x_3} N_4N_3dx + \int_{x_3}^{x_4} N_4N_3dx =$$

$$\int_{e_1} 0\ 0\ dx + \int_{e_2} 0\ N_2^2 dx + \underbrace{\int_{e_3} N_2^3 N_1^3 dx}_{a_{21}^3} \tag{2.3.48}$$

$$a_{44} = \int_{x_1}^{x_4} N_4 N_4 dx = \int_{x_1}^{x_2} N_4 N_4 dx + \int_{x_2}^{x_3} N_4 N_4 dx + \int_{x_3}^{x_4} N_4 N_4 dx =$$

$$\int_{e_1} 0\ 0\ dx + \int_{e_2} 0\ 0\ dx + \underbrace{\int_{e_3} N_2^3 N_2^3 dx}_{a_{22}^3}. \tag{2.3.49}$$

From Eqs. (2.3.34) through (2.3.49), it is seen that the global matrix [a] is made of the element matrices of all three elements, that is,

$$[a] = \begin{bmatrix} a_{11} & a_{12} & a_{13} & a_{14} \\ a_{21} & a_{22} & a_{23} & a_{24} \\ a_{31} & a_{32} & a_{33} & a_{34} \\ a_{41} & a_{42} & a_{43} & a_{44} \end{bmatrix} \tag{2.3.50}$$

or

$$[a] = \begin{bmatrix} a_{11}^1 & a_{12}^1 & 0 & 0 \\ a_{21}^1 & a_{22}^1 + a_{11}^2 & a_{12}^2 & 0 \\ 0 & a_{21}^2 & a_{22}^2 + a_{11}^3 & a_{12}^3 \\ 0 & 0 & a_{21}^3 & a_{22}^3 \end{bmatrix}. \tag{2.3.51}$$

Similarly, the matrix [b] can be obtained as

$$[b] = \begin{bmatrix} b_{11}^1 & b_{12}^1 & 0 & 0 \\ b_{21}^1 & b_{22}^1 + b_{11}^2 & b_{12}^2 & 0 \\ 0 & b_{21}^2 & b_{22}^2 + b_{11}^3 & b_{12}^3 \\ 0 & 0 & b_{21}^3 & b_{22}^3 \end{bmatrix}. \tag{2.3.52}$$

The boundary load vector {B} can be computed as follows

$$B_1 = W_1 \frac{\partial \hat{f}}{\partial x}\bigg|_{x_1}^{x_4} = N_1 \frac{\partial \hat{f}}{\partial x}\bigg|_{x_4} - N_1 \frac{\partial \hat{f}}{\partial x}\bigg|_{x_1} = * \tag{2.3.53}$$

$$B_2 = W_2 \frac{\partial \hat{f}}{\partial x}\Bigg|_{x_1}^{x_4} = N_2 \frac{\partial \hat{f}}{\partial x}\Bigg|_{x_4} - N_2 \frac{\partial \hat{f}}{\partial x}\Bigg|_{x_1} = 0 \tag{2.3.54}$$

$$B_3 = W_3 \frac{\partial \hat{f}}{\partial x}\Bigg|_{x_1}^{x_4} = N_3 \frac{\partial \hat{f}}{\partial x}\Bigg|_{x_4} - N_3 \frac{\partial \hat{f}}{\partial x}\Bigg|_{x_1} = 0 \tag{2.3.55}$$

$$B_4 = W_4 \frac{\partial \hat{f}}{\partial x}\Bigg|_{x_1}^{x_4} = N_4 \frac{\partial \hat{f}}{\partial x}\Bigg|_{x_4} - N_4 \frac{\partial \hat{f}}{\partial x}\Bigg|_{x_1} = -f_4. \tag{2.3.56}$$

Thus,

$$\{B\} = \begin{Bmatrix} * \\ 0 \\ 0 \\ -f_4 \end{Bmatrix}. \tag{2.3.57}$$

The matrix equation can be written as

$$\begin{bmatrix} a_{11} & a_{12} & 0 & 0 \\ a_{21} & a_{22} & a_{23} & 0 \\ 0 & a_{32} & a_{33} & a_{34} \\ 0 & 0 & a_{43} & a_{44} \end{bmatrix} \begin{Bmatrix} \frac{df_1}{dt} \\ \frac{df_2}{dt} \\ \frac{df_3}{dt} \\ \frac{df_4}{dt} \end{Bmatrix} + \begin{bmatrix} b_{11} & b_{12} & 0 & 0 \\ b_{21} & b_{22} & b_{23} & 0 \\ 0 & b_{32} & b_{33} & b_{34} \\ 0 & 0 & b_{43} & b_{44} \end{bmatrix} \begin{Bmatrix} f_1 \\ f_2 \\ f_3 \\ f_4 \end{Bmatrix} = \begin{Bmatrix} * \\ 0 \\ 0 \\ -f_4 \end{Bmatrix}. \tag{2.3.58}$$

2.3.2.8. Step 8. The eighth step in finite-element modeling is to incorporate the boundary conditions into the matrix equation

$$a_{11} = 0,\ a_{12} = 0,\ b_{11} = 1,\ b_{12} = 0 \tag{2.3.59}$$

and

$$* = 1. \tag{2.3.60}$$

In the meantime, move f_4 on the right-hand side of Eq. (2.3.58) to the left-hand side. The final matrix equation would become

$$\begin{bmatrix} 0 & 0 & 0 & 0 \\ a_{21} & a_{22} & a_{23} & 0 \\ 0 & a_{32} & a_{33} & a_{34} \\ 0 & a_{32} & a_{33} & a_{34} \\ 0 & 0 & a_{43} & a_{44} \end{bmatrix} \begin{Bmatrix} \frac{df_1}{dt} \\ \frac{df_2}{dt} \\ \frac{df_3}{dt} \\ \frac{df_4}{dt} \end{Bmatrix} = \begin{bmatrix} 1 & 0 & 0 & 0 \\ b_{21} & b_{22} & b_{23} & 0 \\ 0 & b_{32} & b_{33} & b_{34} \\ 0 & 0 & b_{43} & b_{44}+1 \end{bmatrix} \begin{Bmatrix} f_1 \\ f_2 \\ f_3 \\ f_4 \end{Bmatrix} = \begin{Bmatrix} 1 \\ 0 \\ 0 \\ 0 \end{Bmatrix} . \quad (2.3.61)$$

2.3.2.9. Step 9. The final step in finite-element modeling is the solution of Eq. (2.3.61) as an initial value problem. The simulation depends on what time discretization method is used.

2.3.3. Variational Method

The continuum problems have different but equivalent formulations - a differential formulation and a variational formulation. In the differential formulation, the problem is to integrate a differential or a system of differential equations subject to given boundary conditions. In the classical variational formulation, the problem is to find the unknown function or functions that extremize (maximize or minimize) or make stationary a functional or system of functionals subjected to the same boundary conditions. The two problem formulations are equivalent because the functions that satisfy the differential equations and their boundary conditions also extremize or make stationary the functional. This equivalence is apparent from the calculus of variations, which shows that the functionals are extremized or made stationary only when one or more Euler equations and their boundary conditions are satisfied.

Once the variational formulation is given, the differential formulation can easily be obtained by the Euler equation. First, a definition: a *variational formulation* specifies a scalar quantity (functional) I, which is defined by an integral form

$$I = \int_V f\left(h, x, y, z, \frac{\partial h}{\partial x}, \frac{\partial h}{\partial y}, \frac{\partial h}{\partial z} \right) dV + \int_B B.C.dB \quad (2.3.62)$$

in which h is the unknown function of x, y, and z and f is the specified operator. The solution to the problem is a function h, which makes I stationary with respect to a small change δh. The differential formulation of the above variational formulation is given by the following Euler equation (Sagon, 1961):

$$\frac{\partial f}{\partial h} - \frac{\partial}{\partial x}\left(\frac{\partial f}{\partial h_x} \right) - \frac{\partial}{\partial y}\left(\frac{\partial f}{\partial h_y} \right) - \frac{\partial}{\partial z}\left(\frac{\partial f}{\partial h_z} \right) = 0 \quad (2.3.63)$$

B.C.

The classical variational formulation of a continuum problem often has advantages over the differential formulation from the standard point of obtaining an approximate solution.

Since subsurface problems are usually stated in differential formulation, we face the question of how to obtain the classical variational formulation from differential formulation. Three approaches can be used to decide whether a classical variational principle exists for the problem: (1) attempt to derive a variational principle by using mathematical manipulation (Huebner, 1975), (2) consult mathematical text books on variational methods and try to find the problem in question and its corresponding variational principle (Mikhlin, 1964, 1965; Mikhlin and Smolitsky, 1967), or (3) use Frechet derivatives to test for the existence of the variational principle (Vainberg, 1964; Tonti, 1969; Finlayson, 1972; Finlayson and Scriven, 1967). In any event, if a variational principle can be found, then means are established for obtaining approximate solutions in the standard, integral form suitable for finite-element analysis. A simple example is used to illustrate this point. The steady-state ground water flow equation is

$$K_{xx}\frac{\partial^2 h}{\partial x^2} + K_{yy}\frac{\partial^2 h}{\partial y^2} = 0; \quad \text{B.C.} \tag{2.3.64}$$

It can be verified that the variational principle for the above partial differential equation is to extremize the following functional:

$$I = \frac{1}{2}\int\left[K_{xx}\left(\frac{\partial h}{\partial x}\right)^2 + K_{yy}\left(\frac{\partial h}{\partial y}\right)^2\right]dV + \int_B g(\text{B.C.})dB. \tag{2.3.65}$$

Now, if h is approximate by

$$h \approx \hat{h} = \sum_{j=1}^{N} h_j N_j(x), \tag{2.3.66}$$

the extremization of Eq. (2.3.66) will result in

$$\sum_{j=1}^{N}\int\left[K_{xx}\frac{\partial N_i}{\partial x}\frac{\partial N_j}{\partial x}h_j + K_{yy}\frac{\partial N_i}{\partial y}\frac{\partial N_j}{\partial y}h_j\right]dV = \int_B g'(\text{B.C.})dB, \; i \in N, \tag{2.3.67}$$

which is the same as the result from the method of weighted residual.

2.4. Base and Weighting Functions

The foundation of finite element approximation has been laid. The remaining task is to define the base (interpolating, shape, approximating) functions and weighting (testing) functions. The interpolation functions cannot be chosen arbitrarily. Rather, certain continuity requirements must be met to ensure that the convergence criteria (in the sense that our approximation solution converges to the correct solution when we use an increasing number of smaller elements, that is, when we refine the element mesh) are satisfied. It is helpful at this point to introduce a standard definition and notation to express the degree of continuity of a state variable. If the state variable is continuous, we say that we have C^0 continuity; if, in addition, first derivatives are continuous, we have C^1 continuity; if second derivatives are also continuous, we have C^2 continuity; and so on. Suppose the functions appearing under the integrals in the weighted residual equations contain derivatives up to (r+1)-th order; then, to have rigorous assurance of convergence as element size decreases, the base functions must satisfy two requirements: compatibility and completeness. A base function is said to be compatible if it poses the C^r continuity at element interfaces. A base function is complete if it has the C^{r+1} continuity within an element. For example, if

$$a_{ij}^{e} = \int_{e} N_i F N_j dx, \quad b_{ij}^{e} = \int_{e} K \frac{dN_i}{dx}\frac{dN_j}{dx}dx, \tag{2.4.1}$$

then the base function $N_i(x)$'s must have C^1 continuity within the element and C^0 continuity at the element interface. The simplest function to satisfy these compatibility and completeness requirements is the linear function.

One question practitioners often ask each other is what type of element they are using in their problems. When we speak of the type of elements, we are really speaking of four distinct pieces of information: (1) the shape of elements, (2) number and type of nodes, (3) type of nodal variables, and (4) type of base function (interpolating function, shape function, approximating function).

The shape of elements depends on the spatial dimensionality. For the one-dimensional problem, the shape of elements is usually a line segment (Fig. 2.4.1).

Fig. 2.4.1 Shape of Elements in One-Dimensional Problems.

For two-dimensional problems, the shape of the element can be a triangular element or a quadrilateral element (Fig. 2.4.2).

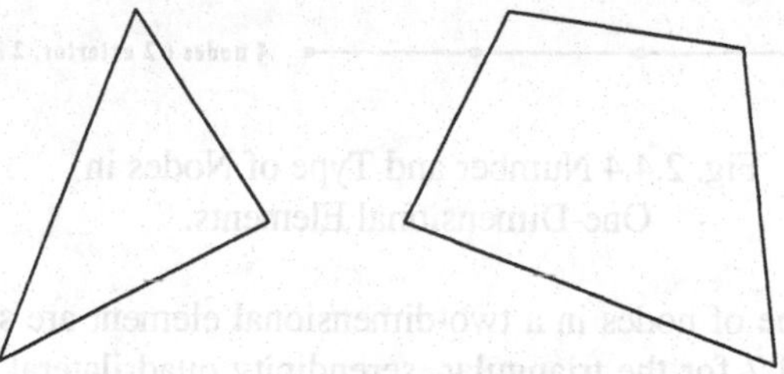

Fig. 2.4.2 Shapes of Elements in Two-Dimensional Problems.

For three-dimensional problems, there are a variety of element shapes. These include tetrahedron, hexahedron, triangular prism, pyramid, etc. (Fig. 2.4.3).

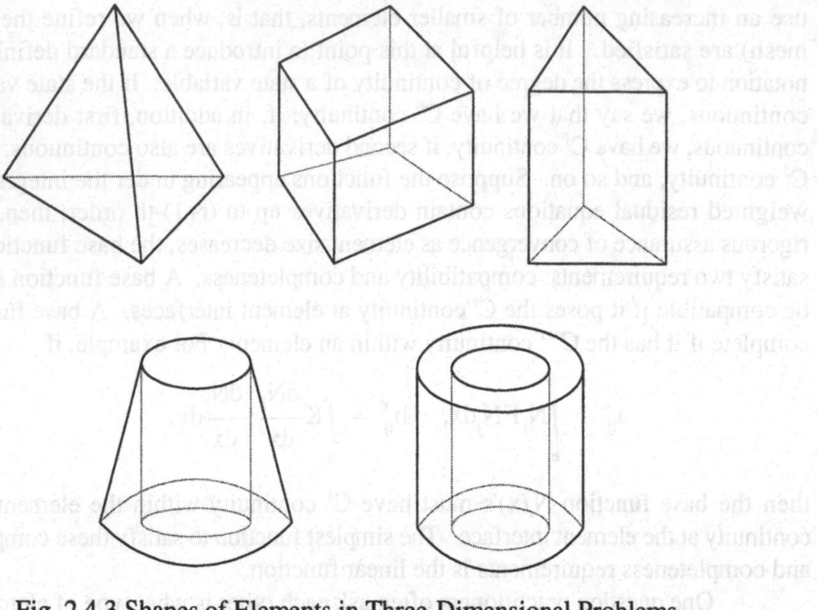

Fig. 2.4.3 Shapes of Elements in Three-Dimensional Problems.

The number of nodes in an element depends on the order of the element, and the type of nodes generally refers to the interior and exterior nodes. For a one-dimensional element, the number of nodes can be 2 (linear elements), 3 (quadratic elements), 4 (cubic elements), etc. (Fig. 2.4.4). The type of nodes is either interior or exterior. There are always two exterior nodes in a one-dimensional element, and the remaining nodes are the interior nodes (Fig. 2.4.4).

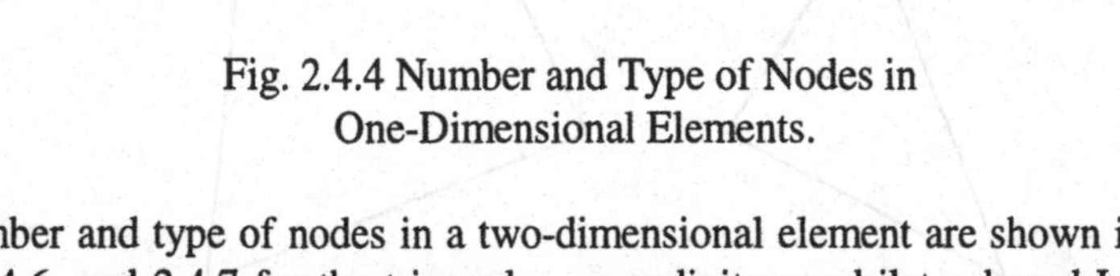

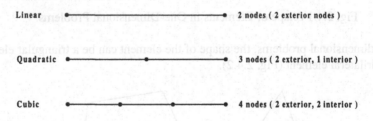

Fig. 2.4.4 Number and Type of Nodes in One-Dimensional Elements.

The number and type of nodes in a two-dimensional element are shown in Figures 2.4.5, 2.4.6, and 2.4.7 for the triangular, serendipity quadrilateral, and Lagrangian quadrilateral elements, respectively. For a triangular element, the number of nodes is given by (n+1)(n+2)/2 where n is the order of the element; for example, there are three, six, and 10 nodes for the linear, quadratic, and cubic triangular elements, respectively (Fig. 2.4.5). For a linear triangular element, all three nodes are exterior points. For a quadratic triangular element, all six nodes are exterior points, and for a cubic triangular element, nine nodes are exterior points and one point is the interior node (Fig. 2.4.5).

For a linear quadrilateral element, all four nodes are exterior points (Fig. 2.4.6). A serendipity quadratic quadrilateral element has eight nodes, all of which are exterior nodes, and a serendipity cubic quadrilateral element has all 12 exterior nodes (Fig. 2.4.6). On the other hand, for a Lagrange quadratic quadrilateral element, there are nine nodes, eight of which are exterior nodes and one is the interior nodes (Fig. 2.4.7). For a Lagrange cubic quadrilateral element, there are 16 nodes: 12 are exterior and four are interior.

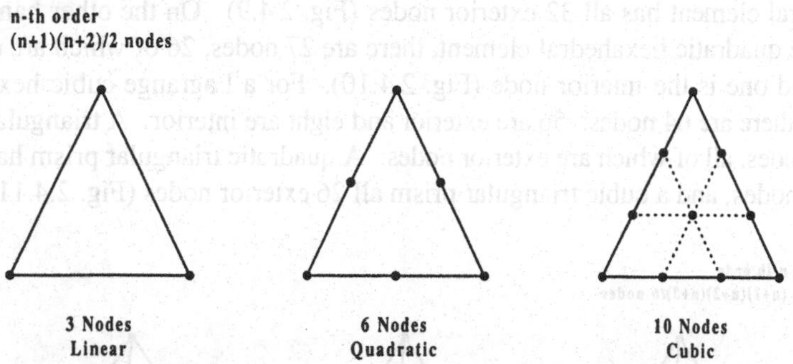

Fig. 2.4.5 Number and Type of Nodes in Triangular Elements.

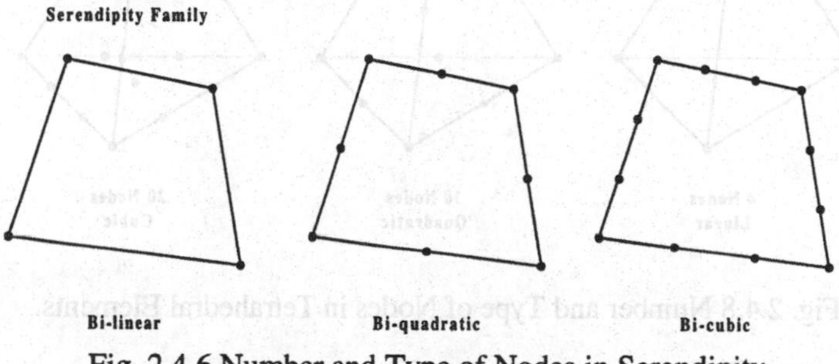

Fig. 2.4.6 Number and Type of Nodes in Serendipity Two-Dimensional Elements.

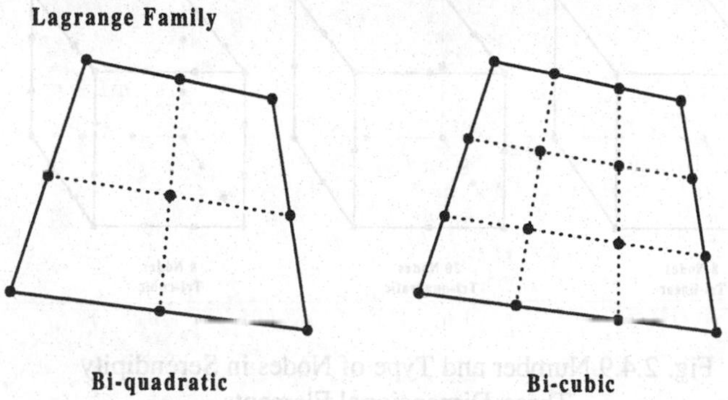

Fig. 2.4.7 Number and Type of Nodes in Lagrangian Two-Dimensional Elements.

The number and type of nodes in a three-dimensional element are shown in Figures 2.4.8 through 2.4.11for the tetrahedral, serendipity hexahedral, Lagrangian hexahedral elements, and triangular prism, respectively. For a tetrahedral element, the number of nodes is given by (n+1)(n+2)(n+3)/6, where n is the order of the element; for example, there are four, 10, and 20 nodes for linear, quadratic, and cubic tetrahedral elements, respectively, all which are exterior nodes (Fig. 2.4.8). For a linear hexahedral element, all eight nodes are exterior points (Fig. 2.4.9). A serendipity quadratic hexahedral element has 20 nodes all of which are exterior nodes, and a serendipity cubic hexahedral element has all 32 exterior nodes (Fig. 2.4.9). On the other hand, for a Lagrange quadratic hexahedral element, there are 27 nodes, 26 of which are exterior nodes and one is the interior node (Fig. 2.4.10). For a Lagrange cubic hexahedral element, there are 64 nodes: 56 are exterior and eight are interior. A triangular prism has six nodes, all of which are exterior nodes. A quadratic triangular prism has all 15 exterior nodes, and a cubic triangular prism all 26 exterior nodes (Fig. 2.4.11.)

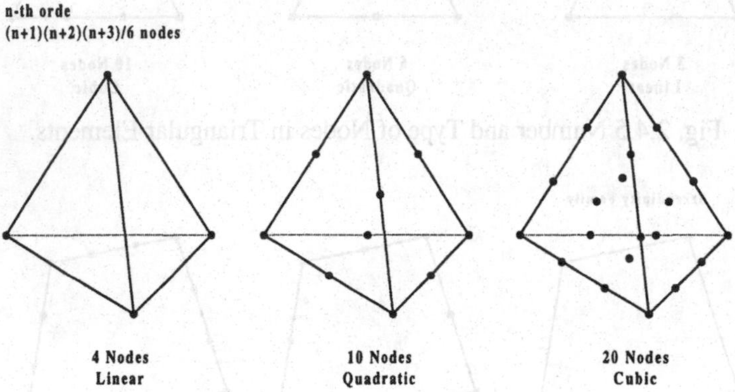

Fig. 2.4.8 Number and Type of Nodes in Tetrahedral Elements.

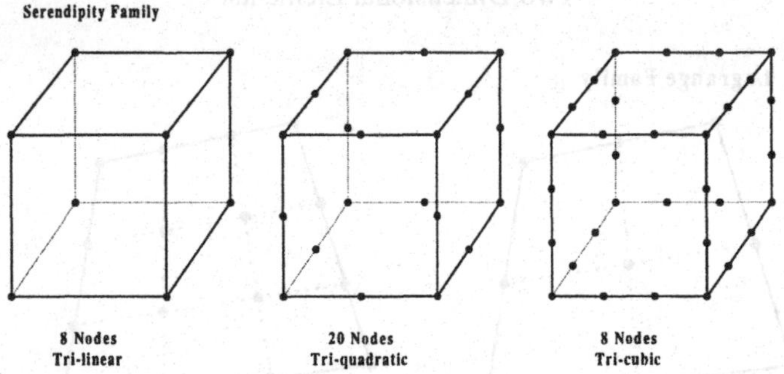

Fig. 2.4.9 Number and Type of Nodes in Serendipity
Three-Dimensional Elements.

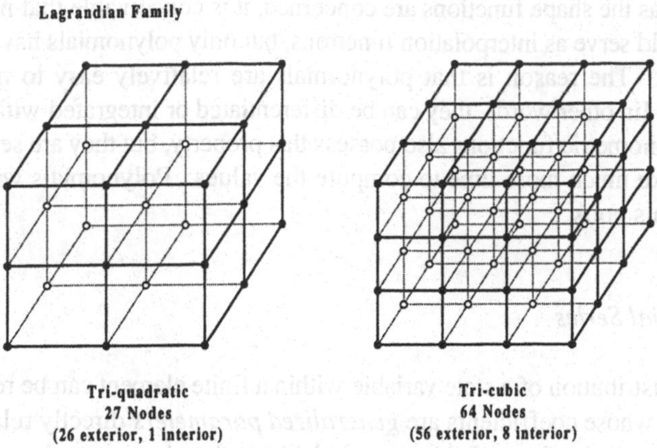

Fig. 2.4.10 Number and Type of Nodes in Lagrangian Three-Dimensional Elements

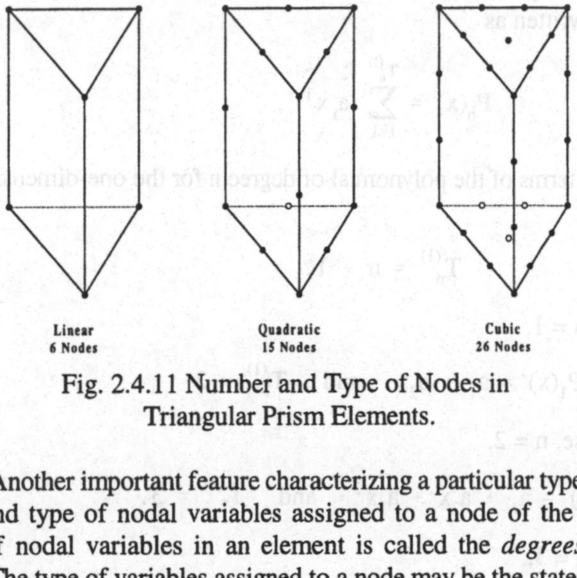

Fig. 2.4.11 Number and Type of Nodes in Triangular Prism Elements.

Another important feature characterizing a particular type of an element is the number and type of nodal variables assigned to a node of the element. The total number of nodal variables in an element is called the *degrees of freedom* of the element. The type of variables assigned to a node may be the state variable (unknown) and/or the derivative of the state variable (derivatives of the unknown). For example, one may assign just the state variable h to each of the two nodes in a linear line element; then the degrees of freedom of the element is two (Fig. 2.4.12). Alternatively, one may assign both the state variable h and its derivative to each of the two nodes. Under such circumstances, degrees of freedom of the element is four (Fig. 2.4.12).

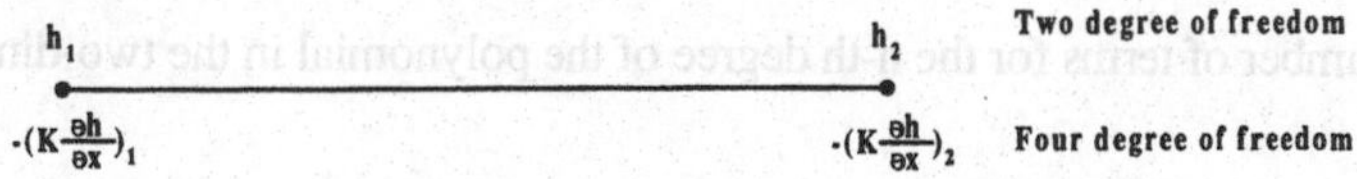

Fig. 2.4.12 Number and Type of Nodal Variables.

As far as the shape functions are concerned, it is conceivable that many types of functions could serve as interpolation functions, but only polynomials have received widespread use. The reason is that polynomials are relatively easy to manipulate mathematically. Iin other words, they can be differentiated or integrated without much difficulty. Trigonometric functions also possess this property, but they are seldom used because they take much more time to compute the values. Polynomials will be used exclusively in this book.

2.4.1. Polynomial Series

In general, the distribution of a state variable within a finite element can be represented by a polynomial whose coefficients are *generalized parameters* directly related to the nodal coordinates and values of the state variable at the element nodes. Once this polynomial is obtained, the expression for the base functions can be derived easily.

2.4.1.1. Complete Polynomial Series. In one dimension, a complete nth-order polynomial may be written as

$$P_n(x) = \sum_{i=1}^{T_n^{(1)}} a_i x^{i-1}, \tag{2.4.2}$$

where the number of terms of the polynomial of degree n for the one-dimensional case $T_n^{(1)}$ is

$$T_n^{(1)} = n + 1. \tag{2.4.3}$$

For the linear case, n = 1,

$$P_1(x) = a_1 + a_2 x \quad \text{and} \quad T_1^{(1)} = 2. \tag{2.4.4}$$

For the quadratic case, n = 2,

$$P_2(x) = a_1 + a_2 x + a_3 x^2 \quad \text{and} \quad T_2^{(1)} = 3. \tag{2.4.5}$$

For the cubic case, n = 3,

$$P_3(x) = a_1 + a_2 x + a_3 x^2 + a_4 x^2 \quad \text{and} \quad T_3^{(1)} = 4. \tag{2.4.6}$$

In two dimensions, a complete nth-order polynomial may be written as

$$P_n(x,y) = \sum_{m=1}^{T_n^{(2)}} a_m x^i y^j, \; i + j \le n, \; m = \frac{1}{2}(i+j)(i+j+1) + (j+1), \tag{2.4.7}$$

where the number of terms for the n-th degree of the polynomial in the two-dimensional case $T_n^{(2)}$ is

$$T_n^{(2)} = \frac{1}{2}(n+1)(n+2). \qquad (2.4.8)$$

For the linear case, n = 1,

$$P_1(x,y) = a_1 + a_2x + a_3y \quad \text{and} \quad T_1^{(2)} = 3. \qquad (2.4.9)$$

For the quadratic case, n = 2,

$$P_2(x,y) = a_1 + a_2x + a_3y + a_4x^2 + a_5xy + a_6y^2 \text{ and } T_2^{(2)} = 6. \qquad (2.4.10)$$

For the cubic case, n = 3,

$$P_3(x,y) = a_1 + a_2x + a_3y + a_4x^2 + a_5yx + a_6y^2 + a_7x^3 + a_8x^2y + a_9xy^2 + a_{10}y^3$$
$$\text{and} \quad T_3^{(2)} = 10. \qquad (2.4.11)$$

A convenient way of illustrating a complete two-dimensional polynomial is by means of the so-called *Pascal triangle* scheme, wherein the terms in the series are placed in a triangular array in ascending order (Fig. 2.4.13).

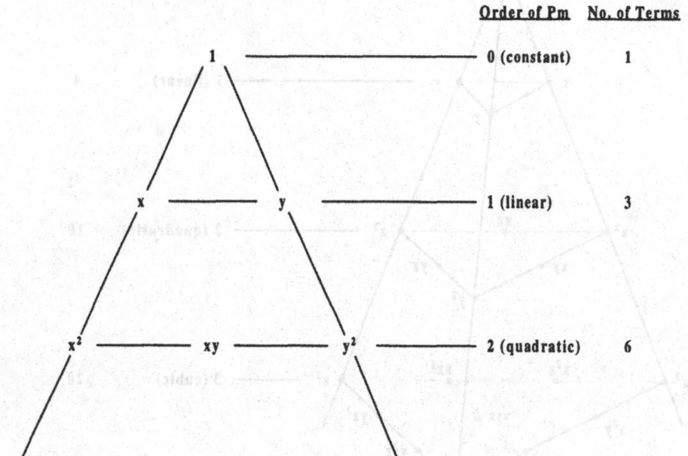

Fig. 2.4.13 A Pascal Triangle to Show a Polynomial Series.

In three dimensions, a complete n^{th}-order polynomial may be written as

$$P_n(x,y,z) = \sum_{m=1}^{T_n^{(3)}} a_m x^i y^j z^k, \quad i + j + k \leq n, \qquad (2.4.12)$$

where the number of terms for the n-th degree polynomial in the three-dimensional cases is given by

$$T_n^{(3)} = \frac{1}{6}(n + 1)(n + 2)(n + 3). \tag{2.4.13}$$

For the linear case, n = 1,

$$P_1(x,y) = a_1 + a_2x + a_2y + a_3z \quad \text{and} \quad T_1^{(3)} = 4. \tag{2.4.14}$$

For the quadratic case, n = 2,

$$P_2(x,y) = a_1 + a_2x + a_3y + a_4z + a_5x^2 + a_6xy + a_6y + a_7xz + a_8y^2 + a_9yz + a_{10}z^2 \text{and} \quad T_2^{(3)} = 6. \tag{2.4.15}$$

In a manner similar to the two-dimensional case, we can illustrate a complete three-dimensional polynomial by placing the terms at different planar levels of a tetrahedron, as illustrated in Figure 2.4.14.

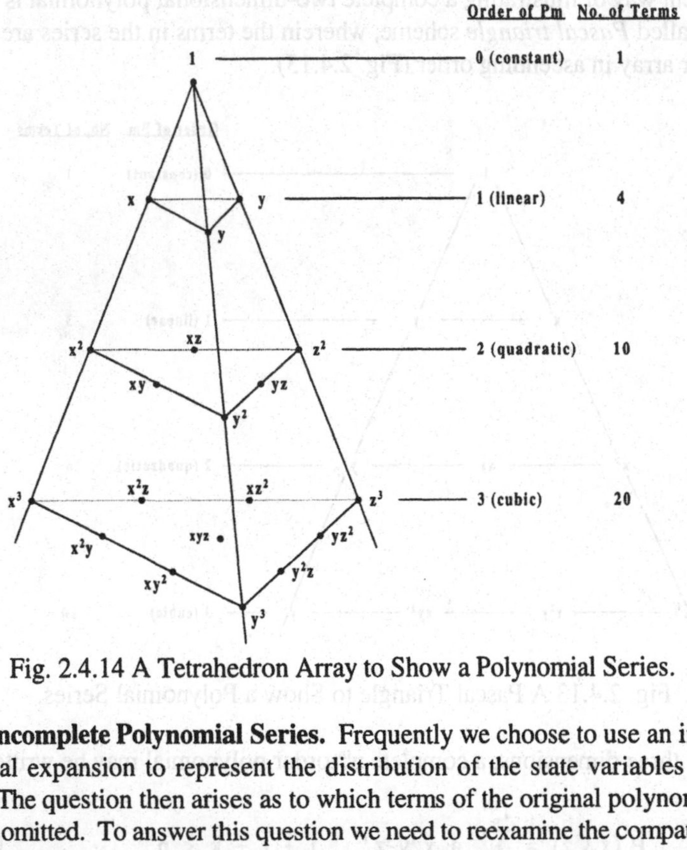

Fig. 2.4.14 A Tetrahedron Array to Show a Polynomial Series.

2.4.1.2. Incomplete Polynomial Series. Frequently we choose to use an incomplete polynomial expansion to represent the distribution of the state variables within an element. The question then arises as to which terms of the original polynomial series should be omitted. To answer this question we need to reexamine the compatibility and completeness requirements discussed earlier. Such requirements are essential to ensure continuity of the state variables and convergence to the correct solution as the mesh is refined. The completeness requirement presents no problem for elements with straight edges because it is readily met by choosing a series that includes all constant and linear terms. The compatibility requirement is met in the case of the first- or second-order operator if the unknown functions are continuous at the element interfaces (i.e, the C^0 continuity is sufficient).

In addition to satisfying the two essential requirements, the representation of the unknown variable within an element must be invariant with respect to any transformation from one Cartesian system to another. Polynomials that display such an invariance property are said to possess *geometric isotropy*. It can be shown that any complete polynomial of degree n has geometric isotropy. There is a simple guideline that permits us to choose the appropriate polynomial terms such as the resulting truncated polynomial series possesses the desired geometric isotropy (Huebner, 1975). According to this guideline, the truncated terms in the polynomial series should occur in "symmetric" pairs. For example, suppose we wish to construct a cubic polynomial for a two-dimensional element that has eight nodal variables assigned to it. In this case, a complete cubic polynomial contain 10 terms, as in Eq. (2.4.11). We need only eight terms because we have eight variables. We may drop only terms that occur in symmetric pairs. These are (a_7x^3, $a_{10}y^3$) and (a_8x^2y, a_9xy^2). Thus the resulting eight-term cubic polynomial having geometric isotropy would be

$$P_3(x,y) = a_1+a_2x+a_3y+a_4x^2+a_5xy+a_6y^2+a_8x^2+a_9xy^2 \tag{2.4.16}$$

or

$$P_3(x,y) = a_1+a_2x+a_3y+a_4x^2+a_5xy+a_6y^2+a_7x^3+a_{10}y^3. \tag{2.4.17}$$

We can readily use this idea to construct other truncated (or incomplete) polynomial expansions. This may be done conveniently via the use of the Pascal triangle and the tetrahedron array described previously.

2.4.2. *Direct Method for Deriving Base Functions*

We have shown how an unknown variable can be represented within a finite element by a polynomial series whose coefficients are generalized parameters. The number of such parameters is often chosen to be equal to the total number of degrees of freedom associated with the particular element. The evaluation of the generalized parameters in terms of the state variables at nodes and nodal coordinates is accomplished by evaluating the polynomial series of each nodal degree of freedom. The general procedure may be described in the following.

2.4.2.1. One-Dimension Case. For a linear element, consider a line segment with two end nodes located at $x = x_1$ and $x = x_2$, respectively; the corresponding functional values are f_1 and f_2, respectively (Fig. 2.4.15). Approximating the function by a linear polynomial, we obtain

$$f = a_1 + a_2x: \; f = \{P\}^T\{a\}, \quad \{P\}^T = \{1 \;\; x\}, \; \{a\} = \begin{Bmatrix} a_1 \\ a_2 \end{Bmatrix}. \tag{2.4.18}$$

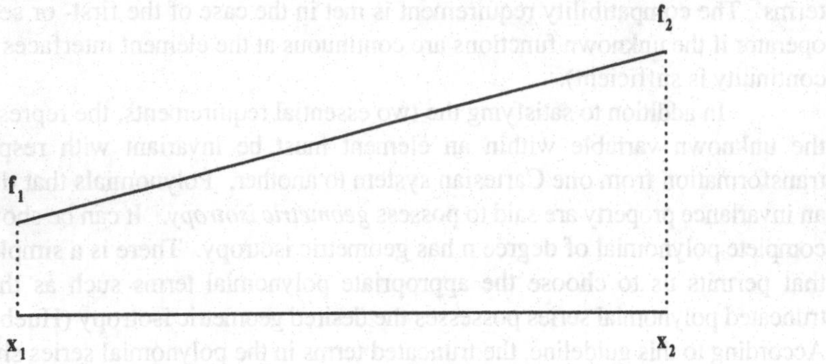

Fig. 2.4.15 A Linear Line Element.

Let the function pass through the points (x_1,f_1) and (x_2,f_2):

$$f_1 = a_1 + a_2 x_1 \qquad \{f\} = [G]\{a\}$$

$$f_2 = a_1 + a_2 x_2 \qquad [G] = \begin{bmatrix} 1 & x_1 \\ 1 & x_2 \end{bmatrix} \qquad \{a\} = \begin{Bmatrix} a_1 \\ a_2 \end{Bmatrix} \tag{2.4.19}$$

Solving for the coefficient a_1 and a_2 in terms of (x_1,f_1) and (x_2,f_2):

$$a_1 = \frac{x_2}{x_2 - x_1} f_1 - \frac{x_1}{x_2 - x_1} f_2$$

$$\{a\} = [G]^{-1}\{f\} \tag{2.4.20}$$

$$a_2 = -\frac{1}{x_2 - x_1} f_1 + \frac{1}{x_2 - x_1} f_2$$

Substituting Eq. (2.4.20) into Eq. (2.4.18), we obtain

$$f = \left(\frac{x_2 - x}{x_2 - x_1}\right) f_1 + \left(\frac{x - x_1}{x_2 - x_1}\right) f_2 \qquad f = \{P\}^T[G]^{-1}\{f\} \,. \tag{2.4.21}$$

Writing in summation, we have

$$f = \sum_{j=1}^{2} N_j f_j, \quad f = \{N\}^T\{f\}, \quad \{N\}^T = \{P\}^T[G]^{-1}, \tag{2.4.22}$$

where N_1 and N_2 are the base functions for nodes 1 and 2, respectively (Fig. 2.4.16).

The exercise above resulted in the above procedures: (1) determine the polynomial vector $\{P\}^T$, (2) set up the polynomial matrix [G], (3) obtain the inverse of the polynomial matrix $[G]^{-1}$, and (4) obtain $\{N\}^T$ by matrix multiplication. Thus, the key lies in the inverse of the matrix. We will use the following fact for the inverse of a matrix:

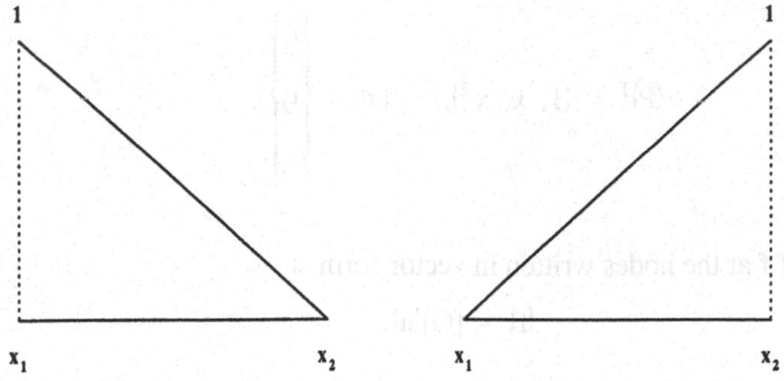

Fig. 2.4.16 Base Functions for a Linear Line Element.

$$[A]^{-1} = \frac{1}{|A|} \text{Adjoint}[A] \tag{2.4.23}$$

$$\text{Adjoint}[A] = [\text{Cofactor Matrix } A]^T \tag{2.4.24}$$

$$[\text{Cofactor Matrix } A] = [\text{Cofactor } \alpha_{ij}] \tag{2.4.25}$$

$$\text{Cofactor } \alpha_{ij} = (-1)^{i+j} \, |M_{ij}| \, . \tag{2.4.26}$$

For a quadratic element, consider a line segment with three points located at $x = x_1$, $x = x_2$, and $x = x_3$, respectively; their corresponding function values are f_1, f_2, and f_3, respectively (Fig. 2.4.17). Let f be approximated by a second-order polynomial

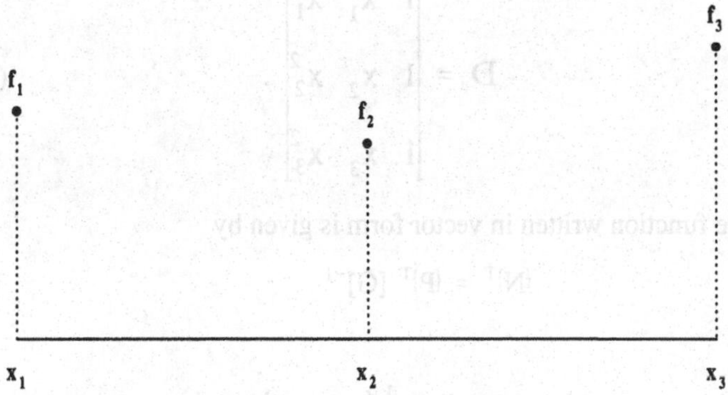

Fig. 2.4.17 A Quadratic Line Element.

$$f = \{P\}^T \{a\} , \tag{2.4.27}$$

where

$$\{P\}^T = \{1,\ x,\ x^2\}, \qquad \{a\} = \begin{Bmatrix} a_1 \\ a_2 \\ a_3 \end{Bmatrix}. \tag{2.4.28}$$

The value of f at the nodes written in vector form is

$$\{f\} = [G]\{a\}, \tag{2.4.29}$$

in which

$$[G] = \begin{bmatrix} 1 & x_1 & x_1^2 \\ 1 & x_2 & x_2^2 \\ 1 & x_3 & x_3^2 \end{bmatrix}. \tag{2.4.30}$$

The inversion of the matrix [G] is given by

$$[G]^{-1} = \frac{1}{D}\begin{bmatrix} \left(x_2x_3^2 - x_3x_2^2\right) & -\left(x_1x_3^2 - x_3x_1^2\right) & \left(x_1x_{2_2} - x_2x_1^2\right) \\ \left(x_2^2 - x_3^2\right) & \left(x_3^2 - x_1^2\right) & \left(x_1^2 - x_2^2\right) \\ \left(x_3 - x_2\right) & \left(x_1 - x_3\right) & \left(x_2 - x_1\right) \end{bmatrix}, \tag{2.4.31}$$

in which

$$D = \begin{vmatrix} 1 & x_1 & x_1^2 \\ 1 & x_2 & x_2^2 \\ 1 & x_3 & x_3^2 \end{vmatrix}. \tag{2.4.32}$$

Now the base function written in vector form is given by

$$\{N\}^T = \{P\}^T\ [G]^{-1} \tag{2.4.33}$$

or

$$\{N_1,\ N_2,\ N_3\} = \frac{1}{D}\{1,\ x,\ x^2\}$$

$$\begin{bmatrix} \left(x_2x_3^2 - x_3x_2^2\right) & \left(x_3x_1^2 - x_1x_3^2\right) & \left(x_1x_2^2 - x_2x_1^2\right) \\ \left(x_2^2 - x_3^2\right) & \left(x_3^2 - x_1^2\right) & \left(x_1^2 - x_2^2\right) \\ -\left(x_2 - x_3\right) & -\left(x_3 - x_1\right) & -\left(x_1 - x_2\right) \end{bmatrix}. \tag{2.4.34}$$

Finally, N_1, N_2, and N_3 are given as follows:

$$N_1 = \frac{1}{D}\left[\left(x_2x_3^2 - x_3x_2^2\right) + \left(x_2^2 - x_3^2\right)x + \left(x_3 - x_2\right)x^2\right] \tag{2.4.35}$$

$$N_2 = \frac{1}{D}\left[\left(x_3x_1^2 - x_1x_3^2\right) + \left(x_3^2 - x_1^2\right)x + \left(x_1 - x_3\right)x^2\right] \tag{2.4.36}$$

$$N_3 = \frac{1}{D}\left[\left(x_1x_2^2 - x_2x_1^2\right) + \left(x_1^2 - x_2^2\right)x + \left(x_2 - x_1\right)x^2\right]. \tag{2.4.37}$$

We can check if the base functions satisfy the following relationship:

$$\text{Check} \quad N_1 + N_2 + N_3 = 1, \quad \text{O.K.} \tag{2.4.38}$$

For a cubic element, consider a line segment with four points located at $x = x_1$, $x = x_2$, $x = x_3$, and $x = x_4$, respectively; their corresponding function values are f_1, f_2, f_3, and f_3, respectively (Fig. 2.4.18).

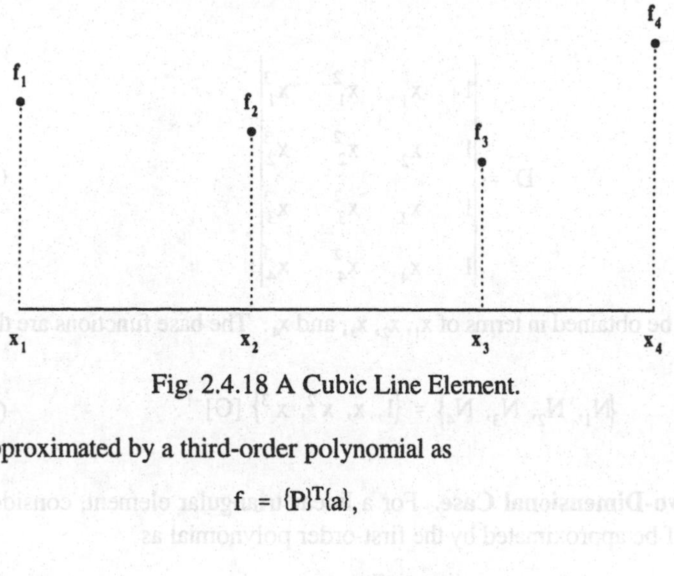

Fig. 2.4.18 A Cubic Line Element.

Let f be approximated by a third-order polynomial as

$$f = \{P\}^T\{a\}, \tag{2.4.39}$$

where

$$\{P\}^T = \{1,\ x,\ x^2,\ x^3\}, \quad \{a\} = \begin{Bmatrix} a_1 \\ a_2 \\ a_3 \\ a_4 \end{Bmatrix}. \tag{2.4.40}$$

The values of f at the nodes written in vector form is

$$\{f\} = [G]\{a\}, \tag{2.4.41}$$

in which

$$[G] = \begin{bmatrix} 1 & x_1 & x_1^2 & x_1^3 \\ 1 & x_2 & x_2^2 & x_2^3 \\ 1 & x_3 & x_3^2 & x_3^2 \\ 1 & x_4 & x_4^2 & x_4^2 \end{bmatrix}. \qquad (2.4.42)$$

The inversion of the matrix [G] is given by

$$[G]^{-1} = \frac{1}{D} \begin{bmatrix} C_{11} & C_{12} & C_{13} & C_{14} \\ C_{21} & C_{22} & C_{23} & C_{24} \\ C_{31} & C_{32} & C_{33} & C_{34} \\ C_{41} & C_{42} & C_{43} & C_{44} \end{bmatrix}, \qquad (2.4.43)$$

in which

$$D = \begin{vmatrix} 1 & x_1 & x_1^2 & x_1^3 \\ 1 & x_2 & x_2^2 & x_2^3 \\ 1 & x_3 & x_3^2 & x_3^3 \\ 1 & x_4 & x_4^2 & x_4^3 \end{vmatrix} \qquad (2.4.44)$$

and C_{ij}'s can be obtained in terms of x_1, x_2, x_3, and x_4. The base functions are then given by

$$\{N_1, N_2, N_3, N_4\} = \{1, x, x^2, x^3\} [G]^{-1}. \qquad (2.4.45)$$

2.4.2.2. Two-Dimensional Case. For a linear triangular element, consider Figure 2.4.19. Let f be approximated by the first-order polynomial as

$$f = \{P\}^T\{a\}, \qquad (2.4.46)$$

where

$$\{P\}^T = \{1, x, y\}, \qquad \{a\} = \begin{Bmatrix} a_1 \\ a_2 \\ a_3 \end{Bmatrix}. \qquad (2.4.47)$$

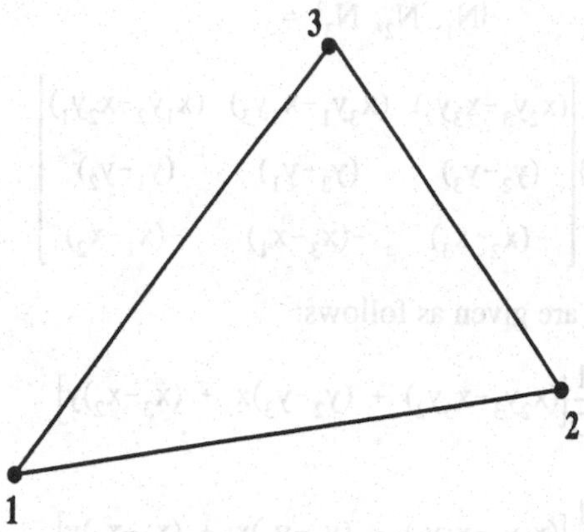

Fig. 2.4.19 A Linear Triangular Element.

The value of f at the nodes written in vector form is

$$\{f\} = [G]\{a\}, \tag{2.4.48}$$

in which

$$[G] = \begin{bmatrix} 1 & x_1 & y_1 \\ 1 & x_2 & y_2 \\ 1 & x_3 & y_3 \end{bmatrix}. \tag{2.4.49}$$

The inversion of the matrix [G] is given by

$$[G] = \frac{1}{D}\begin{bmatrix} x_2y_3-x_3y_2 & -(x_1y_3-x_3y_1) & (x_1y_2-x_2y_1) \\ (y_2-y_3) & (y_3-y_1) & (y_1-y_2) \\ (x_3-x_2) & (x_1-x_3) & (x_2-x_1) \end{bmatrix}, \tag{2.4.50}$$

in which

$$D = \begin{vmatrix} 1 & x_1 & y_1 \\ 1 & x_2 & y_2 \\ 1 & x_3 & y_3 \end{vmatrix}. \tag{2.4.51}$$

Now the base function written in vector form is given by

$$\{N\}^T = \{P\}^T\ [G]^{-1} \tag{2.4.52}$$

or

$$\{N_1, N_2, N_3\} =$$

$$\frac{1}{D}\{1, x, y\}\begin{bmatrix} (x_2y_3-x_3y_2) & (x_3y_1-x_1y_3) & (x_1y_2-x_2y_1) \\ (y_2-y_3) & (y_3-y_1) & (y_1-y_2) \\ -(x_2-x_3) & -(x_3-x_1) & -(x_1-x_2) \end{bmatrix}. \quad (2.4.53)$$

Finally, N_1, N_2, and N_3 are given as follows:

$$N_1 = \frac{1}{D}\left[(x_2y_3-x_3y_2) + (y_2-y_3)x + (x_3-x_2)y\right] \quad (2.4.54)$$

$$N_2 = \frac{1}{D}\left[(x_3y_1-x_1y_3) + (y_3-y_1)x + (x_1-x_3)y\right] \quad (2.4.55)$$

$$N_3 = \frac{1}{D}\left[(x_1y_2-x_2y_1) + (y_1-y_2)x + (x_2-x_1)y\right]. \quad (2.4.56)$$

We can check if the base functions satisfy the following relationship:

$$\text{Check} \quad N_1 + N_2 + N_3 = 1, \quad \text{O.K.} \quad (2.4.57)$$

For a quadratic triangular element consider Figure 2.4.20. First, define the polynomial vector:

$$\{P\}^T = \{1, x, y, x^2, xy, y^2\}. \quad (2.4.58)$$

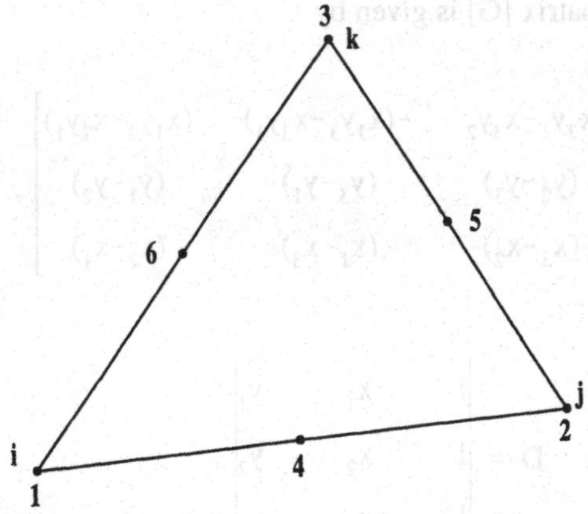

Fig. 2.4.20 A Quadratic Triangular Element.

Second, set up the matrix [G]:

$$[G] = \begin{bmatrix} 1 & x_1 & y_1 & x_1^2 & x_1y_1 & y_1^2 \\ 1 & x_2 & y_2 & x_2^2 & x_2y_2 & y_2^2 \\ 1 & x_3 & y_3 & x_3^2 & x_3y_3 & y_3^2 \\ 1 & x_4 & y_4 & x_4^2 & x_4y_4 & y_4^2 \\ 1 & x_5 & y_5 & x_5^2 & x_5y_5 & y_5^2 \\ 1 & x_6 & y_6 & x_6^2 & x_6y_6 & y_6^2 \end{bmatrix}. \tag{2.4.59}$$

Third, inverse the matrix [G]:

$$[G] = \frac{1}{D} \begin{bmatrix} C_{11} & C_{12} & C_{13} & C_{14} & C_{15} & C_{16} \\ C_{21} & C_{22} & C_{23} & C_{24} & C_{25} & C_{26} \\ C_{31} & C_{32} & C_{33} & C_{34} & C_{35} & C_{36} \\ C_{41} & C_{42} & C_{43} & C_{44} & C_{45} & C_{46} \\ C_{51} & C_{52} & c_{53} & C_{54} & C_{55} & C_{56} \\ C_{61} & C_{62} & C_{63} & C_{64} & C_{65} & C_{66} \end{bmatrix}, \tag{2.4.60}$$

where the determinant D and all the elements of the inverse matrix C_{ij}'s can be obtained in terms of $x_1, x_2, ..., x_6$ and $y_1, y_2, ...,$ and y_6. Finally, obtain the base functions

$$\{N\}^T = \{N_1, N_2, \ldots, N_6\} = \{P\}^T[G]^{-1} \tag{2.4.61}$$

For a cubic triangular element, consider Figure 2.4.21. First, define the polynomial vector:

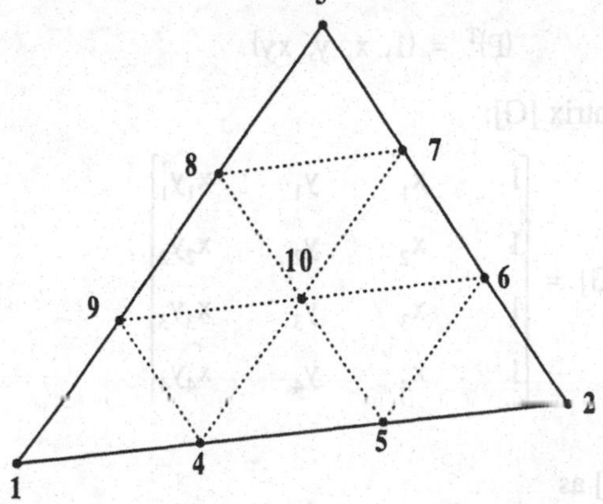

Fig. 2.4.21 A Cubic Triangular Element.

$$\{P\}^T = \{1,\ x,\ y,\ x^2,\ xy,\ y^2,\ x^3,\ x^2y,\ xy^2,\ y^3\}. \tag{2.4.62}$$

Second, set up the matrix [G]:

$$[G] = \begin{bmatrix} 1 & x_1 & y_1 & x_1^2 & x_1y_1 & y_1^2 & x_1^3 & x_1^2y_1 & x_1y_1^2 & y_1^3 \\ . & & & & & & & & & . \\ . & & & & & & & & & . \\ . & & & & & & & & & . \\ 1 & x_{10} & y_{10} & x_{10}^2 & x_{10}y_{10} & y_{10}^2 & x_{10}^3 & x_{10}^2y_{10} & x_{10}y_{10}^2 & y_{10}^3 \end{bmatrix}. \tag{2.4.63}$$

Third. inverse the matrix [G]:

$$[G]^{-1} = \frac{1}{D}\begin{bmatrix} C_{11} & C_{12} & . & . & . & . & . & . & . & C_{1,10} \\ C_{21} & C_{22} & & & & & & & & C_{2,10} \\ . & & & & & & & & & . \\ . & & & & & & & & & . \\ . & & & & & & & & & . \\ C_{91} & C_{92} & & & & & & & & C_{9,10} \\ C_{10,1} & C_{10,2} & . & . & . & . & . & . & . & C_{10,10} \end{bmatrix}, \tag{2.4.64}$$

where the determinant and all C_{ij}'s can be obtained in terms of x_1, x_2, ..., x_{10}, y_1, y_2, ..., and y_{10}. Finally, obtain the base functions:

$$\{N\}^T = \{N_1,\ N_2,\ .\ .\ .\ ,\ N_{10}\} = \{P\}^T[G]^{-1} \tag{2.4.65}$$

For a bilinear quadrilateral element, consider Figure 2.4.22. First, define the polynomial

$$\{P\}^T = \{1,\ x,\ y,\ xy\}. \tag{2.4.66}$$

Second, set up the matrix [G]:

$$[G] = \begin{bmatrix} 1 & x_1 & y_1 & x_1y_1 \\ 1 & x_2 & y_2 & x_2y_2 \\ 1 & x_3 & y_3 & x_3y_3 \\ 1 & x_4 & y_4 & x_4y_4 \end{bmatrix}. \tag{2.4.67}$$

Third, inverse the [G] as

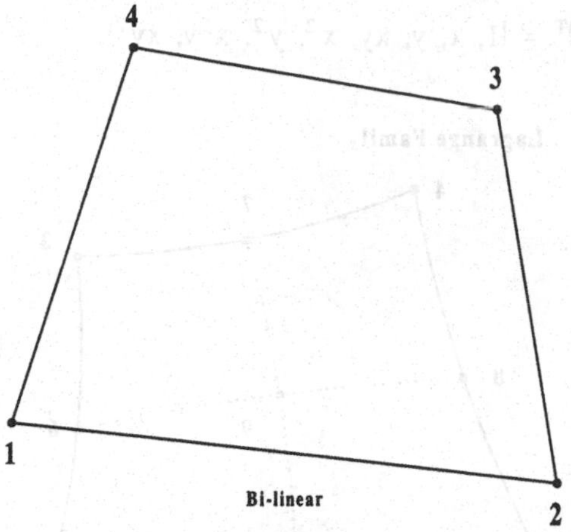

Fig. 2.4.22 A Bilinear Element.

$$[G]^{-1} = \frac{1}{D}\begin{bmatrix} C_{11} & C_{12} & C_{13} & C_{14} \\ C_{21} & C_{22} & C_{23} & C_{24} \\ C_{31} & C_{32} & C_{33} & C_{34} \\ C_{41} & C_{42} & C_{43} & C_{44} \end{bmatrix}, \tag{2.4.68}$$

in which

$$D = \begin{vmatrix} 1 & x_1 & y_1 & x_1y_1 \\ 1 & x_2 & y_2 & x_2y_2 \\ 1 & x_3 & y_3 & x_3y_3 \\ 1 & x_4 & y_4 & x_4y_4 \end{vmatrix} \tag{2.4.69}$$

and C_{ij}'s can be obtained in terms of x_1, x_2, x_3, x_4, y_1, y_2, y_3, and y_4. Finally the base functions are

$$\{N_1, N_2, N_3, N_4\} = \{1, x, y, xy\}\ [G]^{-1}. \tag{2.4.70}$$

For a biquadratic quadrilateral element, two types of elements can be considered. The first one is the Lagrange family, and the second is the serendipity family. The polynomial vector for the Lagrange element (Fig. 2.4.23) is given by the following:

$$\{P\}^T = \{1, x, y, x^2, xy, y^2, xy^2, x^2y, x^2y^2\}. \tag{2.4.71}$$

The polynomial vector for a serendipity element (Fig. 2.4.24) is given by

$$\{P\}^T = \{1, x, y, x^2, y^2, x^2y, xy^2, x^2y^2\} \tag{2.4.72}$$

or

$$\{P\}^T = \{1,\ x,\ y,\ xy,\ x^2,\ y^2,\ x^2y,\ xy^2\}. \tag{2.4.73}$$

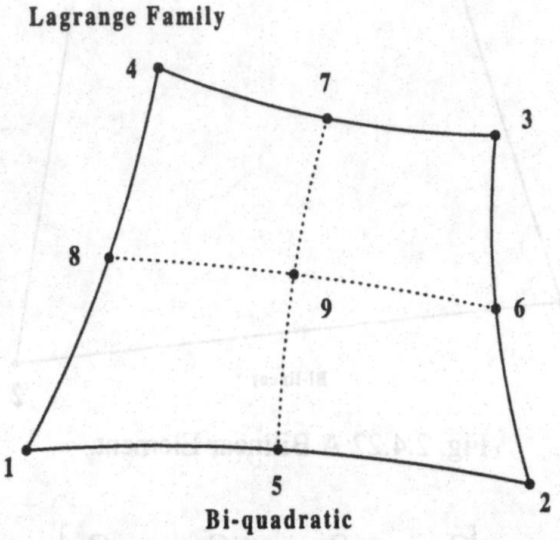

Fig. 2.4.23 A Biquadratic Lagrangian Element.

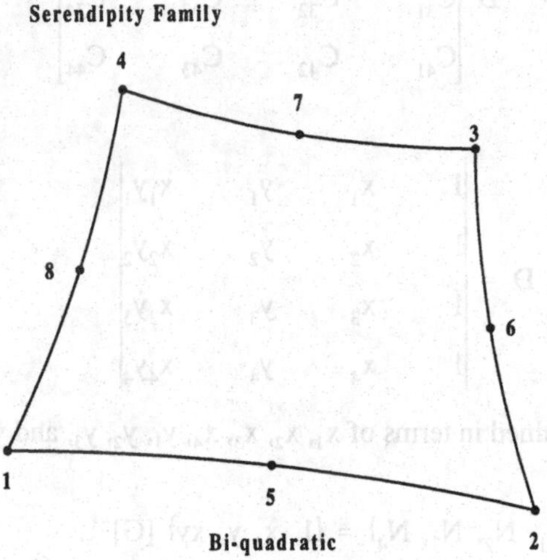

Fig. 2.4.24 A Biquadratic Serendipity Element.

For a bicubic quadrilateral element, two types of elements can be considered: one is the Lagrange family and the other is the serendipity family. The polynomial vector for the Lagrange element (Fig. 2.4.25) is given by

$$\{P\}^T = \left\{\begin{matrix} 1,\ x,\ y,\ xy,\ x^2,\ y^2,\ x^3,\ y^3,\ x^2y,\ xy^2, \\ x^3y,\ xy^3,\ x^2y^2,\ x^3y^2,\ x^2y^3,\ x^3y^3 \end{matrix}\right\}. \tag{2.4.74}$$

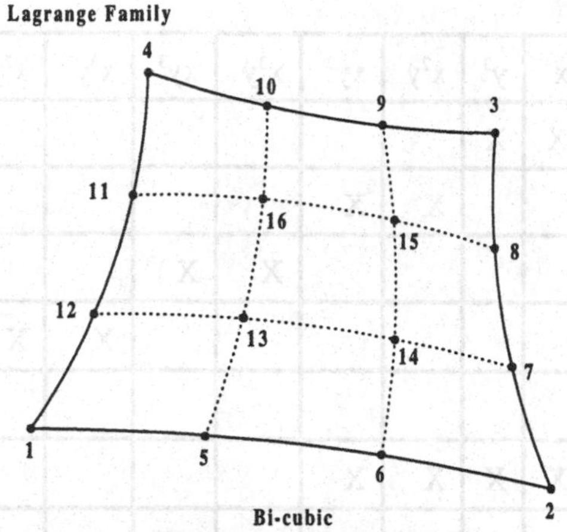

Fig. 2.4.25 A Bicubic Lagrangian Element.

The polynomial vector for the serendipity element (Fig. 2.4.26) can be obtained by deleting four terms from Eq. (2.4.73). One can choose two pairs out of the six pairs for elimination. Thus, there will be 15 possible polynomial vectors with the removed terms given in Table 2.4.1. For example, the first polynomial vector is given by

$$\{P\}^T = \{1,\ x,\ y,\ xy,\ x^2y,\ xy^2,\ x^3y,\ xy^3,\ x^2y^2,\ x^3y^2,\ x^2y^3,\ x^3y^3\}. \quad (2.4.75)$$

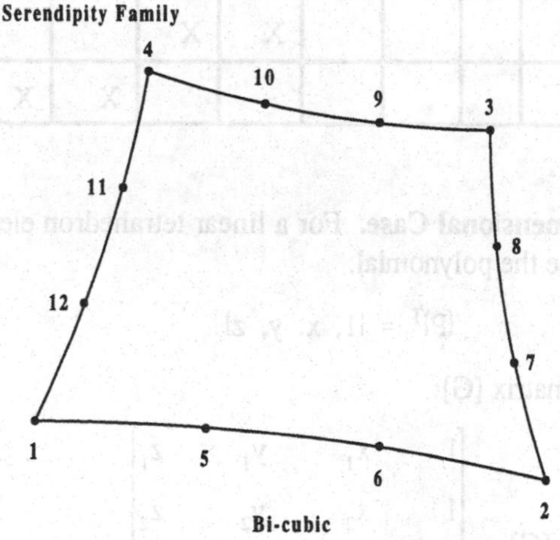

Fig. 2.4.26 A Bicubic Serendipity Element.

Table 2.4.1 Terms to Be Removed from Eq. (2.4.73).

Case	x^2	y^2	x^3	y^3	x^2y	xy^2	x^3y	xy^3	x^3y^2	x^2y^3	x^2y^2	x^3y^3
1	X	X	X	X								
2	X	X			X	X						
3	X	X					X	X				
4	X	X							X	X		
5	X	X									X	X
6			X	X	X	X						
7			X	X			X	X				
8			X	X					X	X		
9			X	X							X	X
10					X	X	X	X				
11					X	X			X	X		
12					X	X					X	X
13							X	X	X	X		
14							X	X			X	X
15									X	X	X	X

2.4.2.3. Three-Dimensional Case. For a linear tetrahedron element, consider Fig. 2.4.27. First, define the polynomial:

$$\{P\}^T = \{1,\ x,\ y,\ z\}. \tag{2.4.76}$$

Second, set up the matrix [G]:

$$[G] = \begin{bmatrix} 1 & x_1 & y_1 & z_1 \\ 1 & x_2 & y_2 & z_2 \\ 1 & x_3 & y_3 & z_3 \\ 1 & x_4 & y_4 & z_4 \end{bmatrix}. \tag{2.4.77}$$

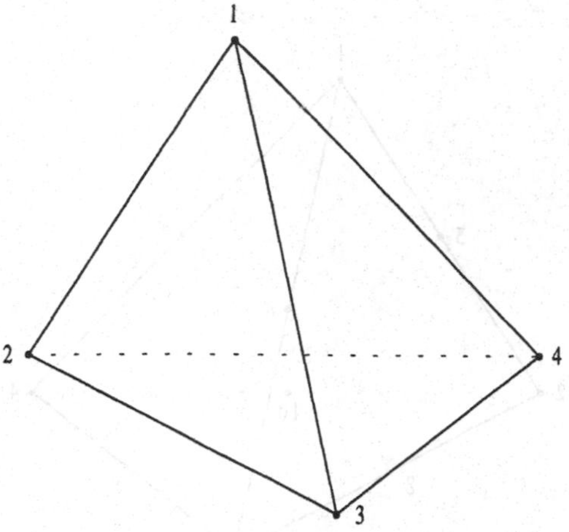

Fig. 2.4.27 A Linear Tetrahedral Element.

Third, inverse the [G] as:

$$[G]^{-1} = \frac{1}{D}\begin{bmatrix} C_{11} & C_{12} & C_{13} & C_{14} \\ C_{21} & C_{22} & C_{23} & C_{24} \\ C_{31} & C_{32} & C_{33} & C_{34} \\ C_{41} & C_{42} & C_{43} & C_{44} \end{bmatrix}, \tag{2.4.78}$$

in which

$$D = \begin{vmatrix} 1 & x_1 & y_1 & z_1 \\ 1 & x_2 & y_2 & z_2 \\ 1 & x_3 & y_3 & z_3 \\ 1 & x_4 & y_4 & z_4 \end{vmatrix} \tag{2.4.79}$$

and C_{ij}'s can be obtained in terms of x_1, x_2, x_3, x_4, y_1, y_2, y_3, y_4, z_1, z_2, z_3, and z_4. Finally the base functions are

$$\{N_1, N_2, N_3, N_4\} = \{1, x, y, z\}\, [G]^{-1}. \tag{2.4.80}$$

For a quadratic tetrahedron element (Fig. 2.4.28), the polynomial vector for the quadratic tetrahedron element is given by

$$\{P\}^T = \{1, x, y, z, x^2, xy, y^2, yz, zx, z^2\}. \tag{2.4.81}$$

For a cubic tetrahedron element (Fig. 2.4.29), the polynomial vector is given by

$$\{P\}^T = \{1, x, y, z, \ldots, z^3\}, \tag{2.4.82}$$

which has a total of 20 terms (entries).

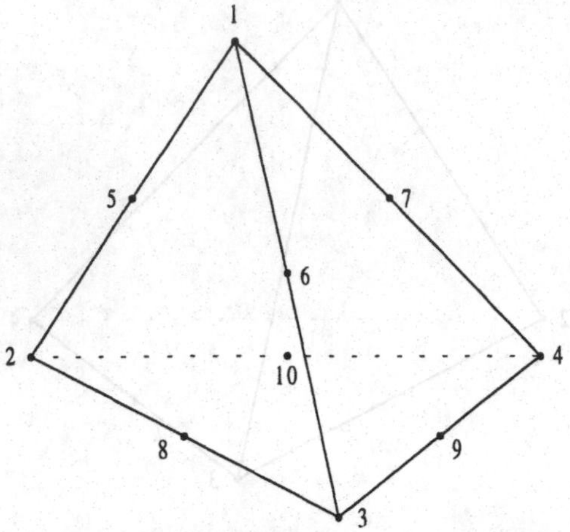

Fig. 2.4.28 A Quadratic Tetrahedral Element.

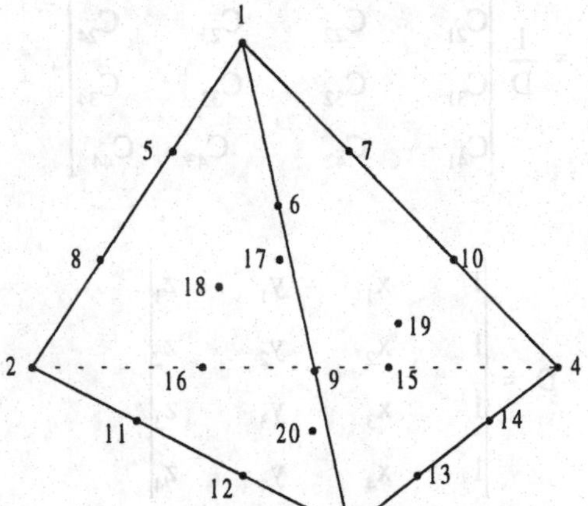

Fig. 2.4.29 A Cubic Tetrahedral Element.

For a tri-linear hexahedron element (Fig. 2.4.30), the polynomial vector is given by

$$\{P\}^T = \{1,\ x,\ y,\ z,\ xy,\ yz,\ zx,\ xyz\}. \tag{2.4.83}$$

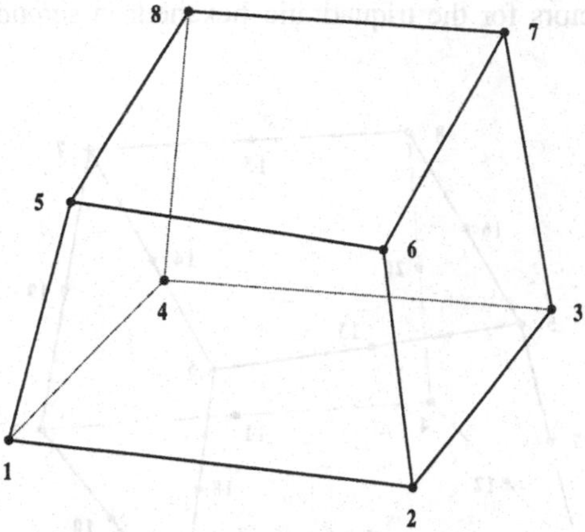

Fig. 2.4.30 A Trilinear Element.

For a triquadratic hexahedron element, two types of elements can be considered: the Lagrange and serendipity families. The polynomial vector for the triquadratic hexahedron Lagrange element (Fig. 2.4.31) is given as

$$\{P\}^T = \left\{ \begin{array}{l} 1, \ x, \ y, \ z, \ x^2, \ y^2, \ z^2, \ xy, \ yz, \ zx, \\ x^2y, \ y^2z, \ z^2x, \ xy^2, \ yz^2, \ zx^2, \ xyz, \\ x^2y^2, \ y^2z^2, \ z^2x^2, \ x^2yz, \ xy^2z, \ xyz^2, \\ x^2y^2z, \ y^2z^2x, \ z^2x^2y, \ x^2y^2z^2 \end{array} \right\}, \qquad (2.4.84)$$

which has a total of 27 entries.

Lagrandian Family

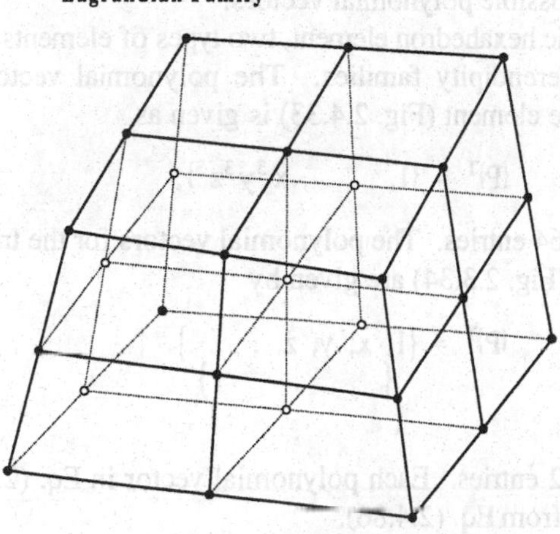

Tri-quadratic 27 Nodes (26 exterior, 1 interior)

Fig. 2.4.31. A Triquadratic Lagrangian Element.

The polynomial vectors for the triquadratic hexahedron serendipity element (Fig. 2.4.32) are given by

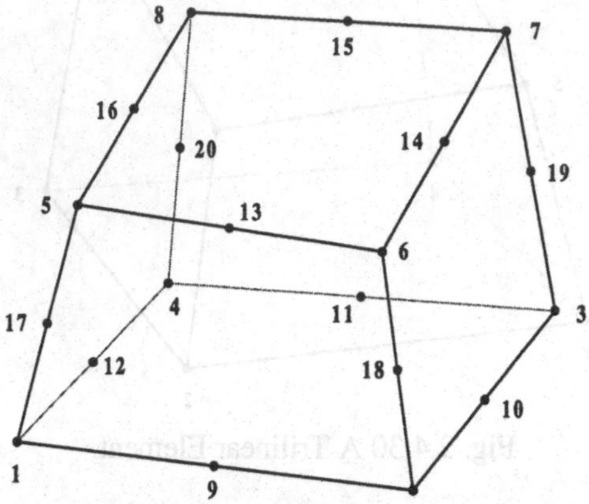

Fig. 2.4.32 A Triquadratic Hexahedron Serendipity Element.

$$\{P\}^T = \begin{matrix} \{1,\ x,\ y,\ z,\ .\ .\ .\ .\} \\ \{\quad .\quad \} \\ \{\quad .\quad \}, \end{matrix} \tag{2.4.85}$$

each of which has 20 entries. Each polynomial vector in Eq. (2.4.85) is formed by removing 7 terms from Eq. (2.4.84). Observing Eq. (2.4.85), we can remove xyz or $x^2y^2z^2$ along with a pair of triplets out of five triplets or a pair of triplets that must be removed together to preserve the geometrical isotropy. Thus, there will be a total of $2 \times ({}_5C_2 + 1) = 22$ possible polynomial vectors.

For a tricubic hexahedron element, two types of elements can be considered: the Lagrange and serendipity families. The polynomial vector for the tri-cubic hexahedron Lagrange element (Fig. 2.4.33) is given as

$$\{P\}^T = \{1,\ .\ .\ .\ ,\ x^3y^3z^3\}, \tag{2.4.86}$$

which has a total of 64 entries. The polynomial vectors for the tri-cubic hexahedron serendipity element (Fig. 2.3.34) are given by

$$\{P\}^T = \begin{matrix} \{1,\ x,\ y,\ z,\ .\ .\ .\ .\} \\ \{\quad .\quad \} \\ \{\quad .\quad \}, \end{matrix} \tag{2.4.87}$$

each of which has 32 entries. Each polynomial vector in Eq. (2.4.87) is formed by removing 32 entries from Eq. (2.4.86).

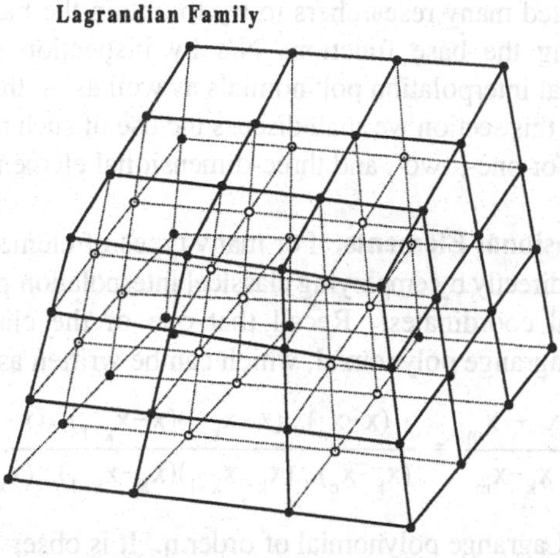

Fig. 2.4.33 A Tricubic Lagrangian Element.

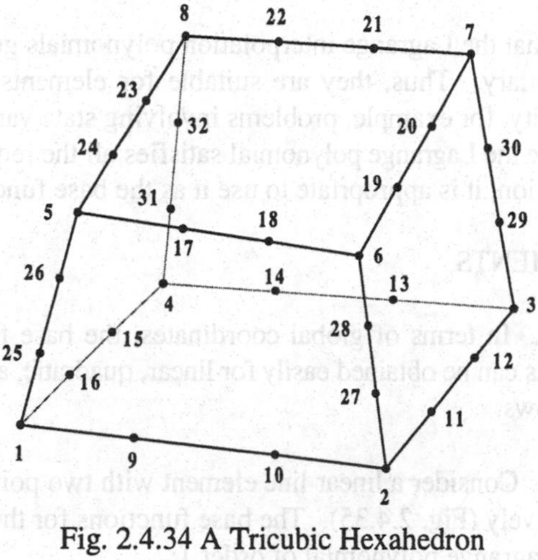

Fig. 2.4.34 A Tricubic Hexahedron Serendipity Element

2.4.3. Derivation of Base Functions by Inspection

The direct derivation of base functions is straightforward and may be carried out perfunctorily, but sometimes difficulties are encountered. In a few cases, certain element geometries can lead to a singular matrix [G]. These cases occur when the nodal arrangement is such that one degree of freedom is dependent on another or a combination of the terms in the polynomial series yields a base function that gives zero values at all nodal points. Another disadvantage of the direct method just described is the effort required to inverse the matrix [G] when the inverse does exist. These reasons have motivated many researchers to try to obtain the base functions N_i's by inspection. Deriving the base functions N_i's by inspection often relies on the adaptation of classical interpolation polynomials as well as on the use of natural and local coordinates. In this section we shall discuss the use of such procedures to obtain the shape functions for one-, two-, and three-dimensional elements.

2.4.3.1 One-Dimensional Elements. For many types of elements, it is possible to write base functions directly by employing classical interpolation polynomials, natural coordinates, or local coordinates. Recall that one of the classical interpolation polynomials is the Lagrange polynomial, which can be written as

$$L_k^{(n)}(x) = \prod_{m \neq k}^{n} \frac{x - x_m}{x_k - x_m} = \frac{(x-x_o)\ldots(x-x_{k-1})(x-x_{k+1})\ldots(x-x_n)}{(x_k-x_o)\ldots(x_k-x_{k-1})(x_k-x_{k+1})\ldots(x_k-x_n)}, \tag{2.4.88}$$

where $L_k^{(n)}(x)$ is the Lagrange polynomial of order n. It is observed that

$$L_k^{(n)}(x) = 1 \quad \text{at} \quad x = x_k \tag{2.4.89}$$

$$L_k^{(n)}(x) = 0 \quad \text{at} \quad x = x_j, \quad j \neq k. \tag{2.4.90}$$

It is further noticed that the Lagrange interpolation polynomials guarantee continuity at the element boundary. Thus, they are suitable for elements used in problems requiring C^0 continuity, for example, problems involving state variables as unknown at nodal points. Since the Lagrange polynomial satisfies all the requirements for base functions, by inspection, it is appropriate to use it as the base function.

LAGRANGE ELEMENTS

Global Coordinate. In terms of global coordinates, the base functions using the Lagrange polynomials can be obtained easily for linear, quadratic, and cubic elements, respectively, as follows.

Linear Line Element. Consider a linear line element with two points located at $x = x_1$ and $x = x_2$, respectively (Fig. 2.4.35). The base functions for the two nodes can be obtained using the Lagrange polynomial of order 1:

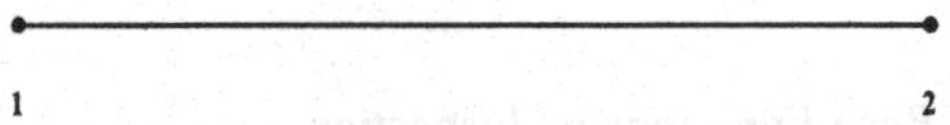

Figure 2.4.35 A Two-Node Line Element.

$$N_1(x) = L_1^{(1)}(x) = \frac{x - x_2}{x_1 - x_2} \tag{2.4.91}$$

$$N_2(x) = L_2^{(1)}(x) = \frac{x - x_1}{x_2 - x_1}. \tag{2.4.92}$$

Quadratic Line Element. For a quadratic line element with three points located at $x = x_1$, $x = x_2$, and $x = x_3$, respectively (Fig. 2.4.36), the base functions for the three nodes can be obtained using the Lagrange polynomial of order 2:

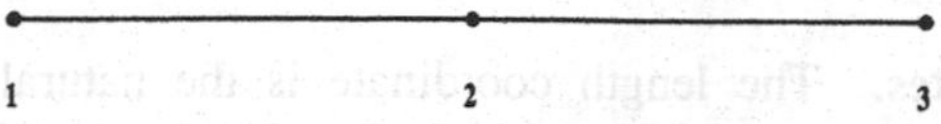

Figure 2.4.36 A Three-Node Line Element.

$$N_1(x) = L_1^{(2)}(x) = \frac{x - x_2}{x_1 - x_2}\frac{x - x_3}{x_1 - x_3} \tag{2.4.93}$$

$$N_2(x) = L_2^{(2)}(x) = \frac{x - x_1}{x_2 - x_1}\frac{x - x_3}{x_2 - x_3} \tag{2.4.94}$$

$$N_3(x) = L_3^{(2)}(x) = \frac{x - x_1}{x_3 - x_1}\frac{x - x_2}{x_3 - x_2}. \tag{2.4.95}$$

Cubic Line Element. For a cubic line element with four points located at $x = x_1$, $x = x_2$, $x = x_3$, and $x = x_4$, respectively (Fig. 2.4.37), the base functions for the four nodes can be obtained using the Lagrange polynomial of order 3:

Figure 2.4.37 A Four-Node Line Element.

$$N_1(x) = L_1^{(3)}(x) = \frac{x - x_2}{x_1 - x_2}\frac{x - x_3}{x_1 - x_3}\frac{x - x_4}{x_1 - x_4} \tag{2.4.96}$$

$$N_2(x) = L_2^{(3)}(x) = \frac{x - x_1}{x_2 - x_1}\frac{x - x_3}{x_2 - x_3}\frac{x - x_4}{x_2 - x_4} \tag{2.4.97}$$

$$N_3(x) = L_3^{(3)}(x) = \frac{x - x_1}{x_3 - x_1}\frac{x - x_2}{x_3 - x_2}\frac{x - x_4}{x_3 - x_4} \tag{2.4.98}$$

$$N_4(x) = L_4^{(3)}(x) = \frac{x - x_1}{x_4 - x_1}\frac{x - x_2}{x_4 - x_2}\frac{x - x_3}{x_4 - x_3}. \tag{2.4.99}$$

Higher-Order Elements. Using the Lagrange polynomial of order n, we can easily writte the base functions of a line element up to any order n.

Another way of deriving the base functions is by means of *natural coordinates*. The natural coordinates are termed the *length coordinates*, *area coordinates*, and *volume coordinates*, respectively for one-dimension, two-dimension, and three-dimension cases for reasons to become evident later.

Length Coordinates. The length coordinate is the natural coordinate in one dimension, which represents a local system that is related to the global system in such a way that the length coordinates will take a value of 1 or 0 at the two endpoints. Consider the line segment shown in Figure 2.4.38. Let the distances from point P to point 1 and point n + 1 be denoted by L_2 and L_1, respectively. First let us nondimensionalize L_1 and L_2 such that

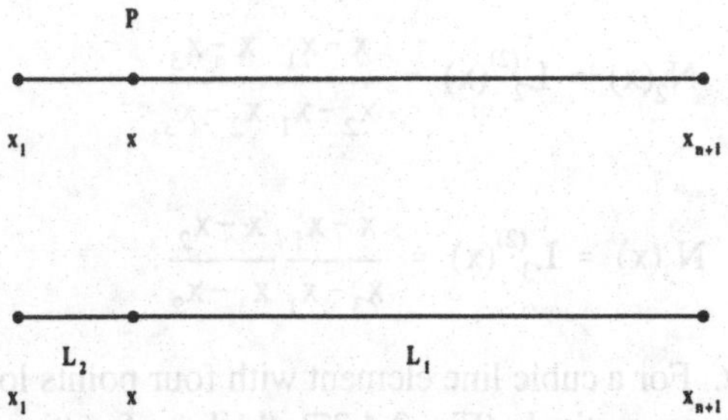

Figure 2.4.38 A Line Segment to Define Length Coordinate.

$$L_1 + L_2 = 1. \tag{2.4.100}$$

It is apparent that if we move point P to the left point x_1,

$$L_1 = 1 \quad \text{and} \quad L_2 = 0. \tag{2.4.101}$$

Similarly, if we move the point to the right point x_{n+1},

$$L_1 = 0 \quad \text{and} \quad L_2 = 1. \tag{2.4.102}$$

By inspection, we can write the function that relates L_1 and L_2 to the global coordinate x in the form

$$x = x_1 L_1 + x_{n+1} L_2. \tag{2.4.103}$$

Combining Eqs. (2.4.101) and (2.4.102) with Eq. (2.4.103), we obtain the transformation of coordinates:

$$\begin{Bmatrix} 1 \\ x \end{Bmatrix} = \begin{bmatrix} 1 & 1 \\ x_1 & x_{n+1} \end{bmatrix} \begin{Bmatrix} L_1 \\ L_2 \end{Bmatrix}. \tag{2.4.104}$$

Inversion of Eq. (2.4.104) yields the inverse transformation of coordinates:

$$\begin{Bmatrix} L_1 \\ L_2 \end{Bmatrix} = \frac{1}{x_{n+1} - x_1} \begin{bmatrix} x_{n+1} & -1 \\ -x_1 & 1 \end{bmatrix} \begin{Bmatrix} 1 \\ x \end{Bmatrix} \tag{2.4.105}$$

or

$$L_1 = \frac{x_{n+1} - x}{x_{n+1} - x_1}, \quad L_2 = \frac{x - x_1}{x_{n+1} - x_1}. \tag{2.4.106}$$

Thus, it is seen that the length coordinates are simply the ratios of lengths and hence the name of length coordinates. The variation of length coordinates is shown in Figure 2.4.39.

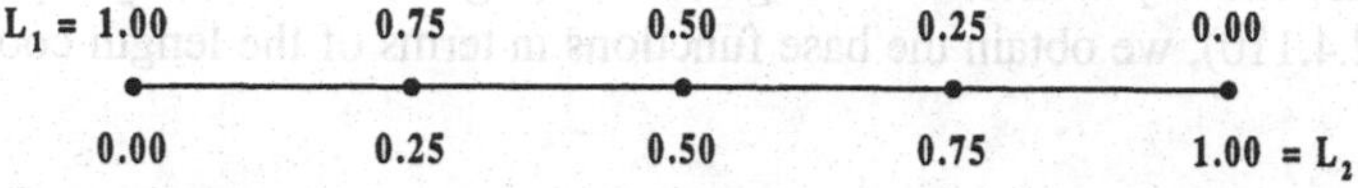

Figure 2.4.39 Length Coordinate in a Line Element.

To derive the base functions in terms of length coordinates, consider an element of order n (Fig. 2.4.40). Let us establish a particular scheme for numbering the nodal points. In this scheme, the nodes are given the two-digit label pq, where p and q are integers satisfying the relation

$$p + q = n. \tag{2.4.107}$$

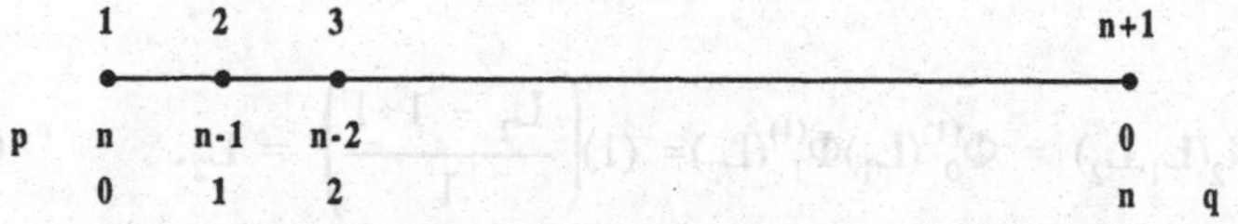

Figure 2.4.40 pq-Labeling of Nodes in a Line Element.

With the above numbering system, the base functions for node I, $N_I(L_1,L_2)$ can be written as

$$N_I(L_1,L_2) = \Phi_p^{(n)}(L_1)\Phi_q^{(n)}(L_2), \tag{2.4.108}$$

where $\Phi_p^{(n)}(L_1)$ and $\Phi_q^{(n)}(L_2)$ are functions of L_1 and L_2, respectively, defined by

$$\Phi_p^{(n)}(L_1) = \begin{cases} \prod_{k=1}^{p} \dfrac{nL_1 - k + 1}{k} & \text{for } p \geq 1 \\ 1 & \text{for } p = 0 \end{cases} \tag{2.4.109}$$

and

$$\Phi_q^{(n)}(L_2) = \begin{cases} \prod_{k=1}^{q} \dfrac{nL_2 - k + 1}{k} & \text{for } q \geq 1 \\ 1 & \text{for } q = 0 \,. \end{cases} \tag{2.4.110}$$

It is clearly seen that p denotes the number of nodes that lie to the right of node I and q denotes the number of nodes that lie to the left of node I. With the above description, the base functions for linear, quadratic, and cubic elements are now readily derived as follows.

Linear Line Element. For a linear line element with two nodes, p and q would be equal to 1 and 0 at the first (left) node, respectively. Similiarly, p and q would be equal to 0 and 1, respectively at the second (right) node (Fig. 2.4.41). Using Eqs. (2.4.108) through (2.4.110), we obtain the base functions in terms of the length coordinate as follows:

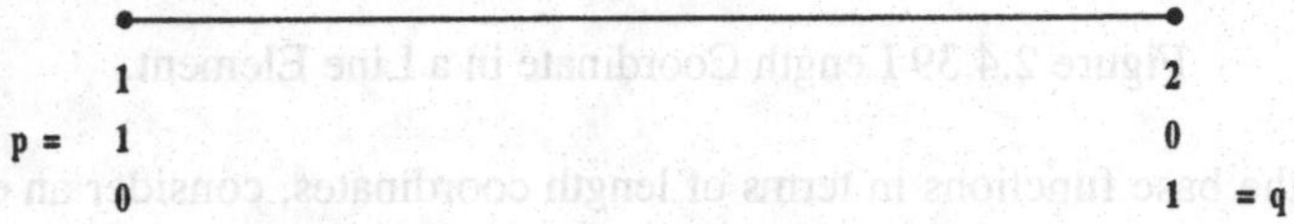

Figure 2.4.41 pq-Labeling of Nodes in a Linear Line Element.

$$N_1(L_1,L_2) = \Phi_1^{(1)}(L_1)\Phi_0^{(1)}(L_2) = \left(\frac{L_1 - 1+1}{1}\right)(1) = L_1 \tag{2.4.111}$$

$$N_2(L_1,L_2) = \Phi_0^{(1)}(L_1)\Phi_1^{(1)}(L_2) = (1)\left(\frac{L_2 - 1+1}{1}\right) = L_2 . \tag{2.4.112}$$

Quadratic Line Element. For a quadratic line element with three nodes, p and q would be equal to 2 and 0 at the first node on the left, respectively; p and q would be equal to 1 and 1, respectively in the middle node; and p and q would be equal to 0 and 2, respectively at the third node on the right (Fig. 2.4.42). Using Eqs. (2.4.108) through (2.4.110), we obtain the base functions in terms of the length coordiante as follows:

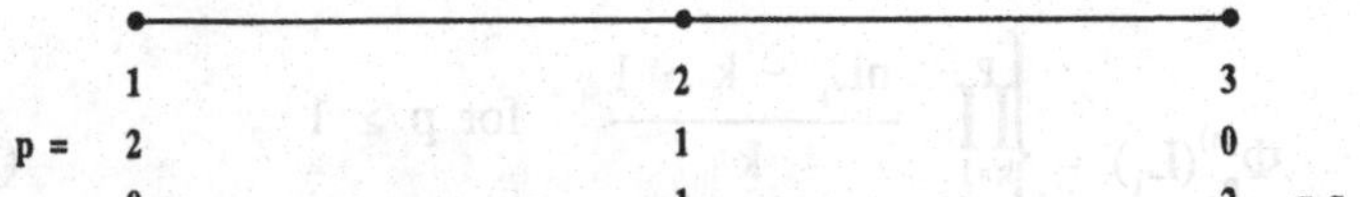

Figure 2.4.42 pq-Labeling of Nodes in a Quadratic Line Element.

$$N_1(L_1,L_2) = \Phi_2^{(2)}(L_1)\Phi_0^{(2)}(L_2)$$

$$= \left(\frac{2L_1 - 1+1}{1}\right)\left(\frac{2L_1 - 2+1}{2}\right)(1) = L_1(2L_1 - 1) \tag{2.4.113}$$

$$N_2(L_1,L_2) = \Phi_1^{(2)}(L_1)\Phi_1^{(2)}(L_2)$$

$$= \left(\frac{2L_1 - 1+1}{1}\right)\left(\frac{2L_2 - 1+1}{1}\right) = 4L_1L_2 \tag{2.4.114}$$

$$N_3(L_1,L_2) = \Phi_0^{(2)}(L_1)\Phi_2^{(2)}(L_2)$$

$$=(1)\left(\frac{2L_2 - 1+1}{1}\right)\left(\frac{2L_2 - 2+1}{2}\right) = L_2(2L_2 - 1). \tag{2.4.115}$$

Cubic Line Element. For a cubic line element with four nodes, p and q would be equal to 3 and 0, 2 and 1, 1 and 2, 0, and 3, respectively, at nodes 1, 2, 3, and 4 (Fig. 2.4.43). Using Eqs. (2.4.108) through (2.4.110), we obtain the base functions in terms of the length coordinate as follows:

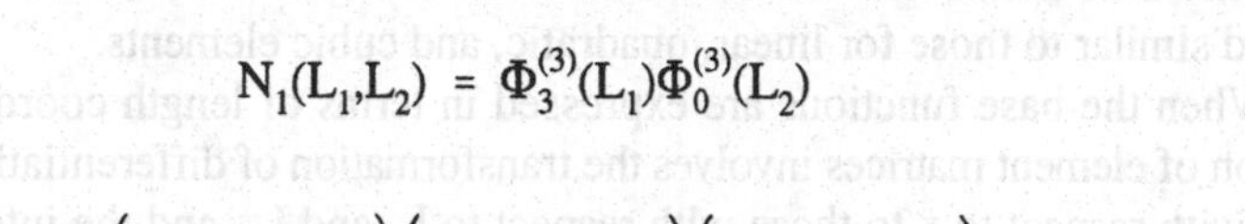

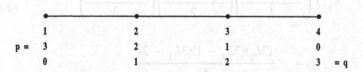

Figure 2.4.43 pq-Labeling of Nodes in a Cubic Line Element.

$$N_1(L_1,L_2) = \Phi_3^{(3)}(L_1)\Phi_0^{(3)}(L_2)$$

$$= \left(\frac{3L_1 - 1+1}{1}\right)\left(\frac{3L_1 - 2+1}{2}\right)\left(\frac{3L_1 - 3+1}{3}\right) \tag{2.4.116}$$

$$= \frac{(3L_1)(3L_1 - 1)(3L_1 -1)}{6}$$

$$N_2(L_1,L_2) = \Phi_2^{(3)}(L_1)\Phi_1^{(3)}(L_2)$$

$$= \left(\frac{3L_1 - 1+1}{1}\right)\left(\frac{3L_1 - 2+1}{2}\right)\left(\frac{3L_2 - 1+1}{1}\right) \tag{2.4.117}$$

$$= \frac{(3L_1)(3L_1 - 1)(3L_2)}{2}$$

$$N_3(L_1,L_2) = \Phi_1^{(3)}(L_1)\Phi_2^{(3)}(L_2)$$

$$= \left(\frac{3L_1 - 1+1}{1}\right)\left(\frac{3L_2 - 1+1}{1}\right)\left(\frac{3L_2 - 2+1}{2}\right) \tag{2.4.118}$$

$$= \frac{(3L_1)(3L_2)(3L_2 - 1)}{2}$$

$$N_4(L_1,L_2) = \Phi_0^{(3)}(L_1)\Phi_3^{(3)}(L_2)$$

$$=(1)\left(\frac{3L_1 - 1+1}{1}\right)\left(\frac{3L_1 - 2+1}{2}\right)\left(\frac{3L_1 - 3+1}{3}\right) \tag{2.4.119}$$

$$= \frac{(3L_2)(3L_2 - 1)(3L_2 - 2)}{6} .$$

Higher-Order Line Elements. The base functions for higher-order line elements can be obtained similar to those for linear, quadratic, and cubic elements.

When the base functions are expressed in terms of length coordinates, the computation of element matrices involves the transformation of differentiation of base functions with respect to x to those with respect to L_1 and L_2, and the integration of length coordinates over the element. The differentiation of N_i's with respect to x can be transformed to the differentiation of N_i's with respect to L_1 and L_2 using the chain rule

$$\frac{\partial N_i}{\partial x} = \frac{\partial N_i}{\partial L_1}\frac{\partial L_1}{\partial x} + \frac{\partial N_i}{\partial L_2}\frac{\partial L_2}{\partial x} = -\frac{1}{\ell}\frac{\partial N_i}{\partial L_1} + \frac{1}{\ell}\frac{\partial N_i}{\partial L_2} , \tag{2.4.120}$$

where ℓ is the length of the element. Integration of length coordinates over the element is simple with the aid of the following formula:

$$\int_e L_1^{\alpha} L_2^{\beta}\, dx = \ell \frac{\alpha!\beta!}{(\alpha+\beta+1)!} . \tag{2.4.121}$$

Local coordinates. The Lagrange polynomial has been previously written in global coordinates. Another way of writing the Lagrange polynomial is by means of local coordinates. Local coordinates represent a local system that is related to the global system in such a way that the local coordinates will take a value of +1 or -1 at the two endpoints. Consider the line segment shown in Figure 2.4.44. Let us define a local coordinate by normalizing the x between two end points as depicted in Figure 2.4.44.

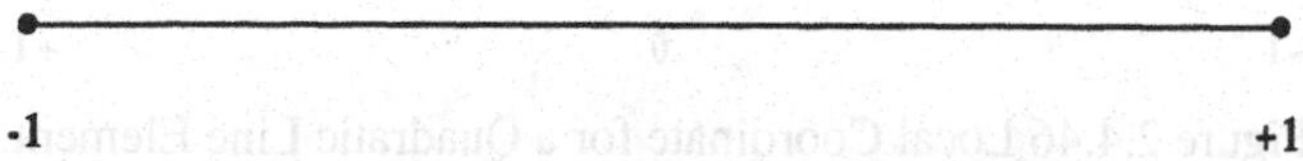

Figure 2.4.44 Local Coordinate for a Line Element.

Now we are ready to derive base functions in terms of the local coordinate using the Lagrange polynomial again.

Linear Line Element. Consider a linear line element with the first point located at ξ = -1 and the second point located at ξ = 1, respectively (Fig. 2.4.45); the base functions for the two points can be obtained using the Lagrange polynomial of order 1:

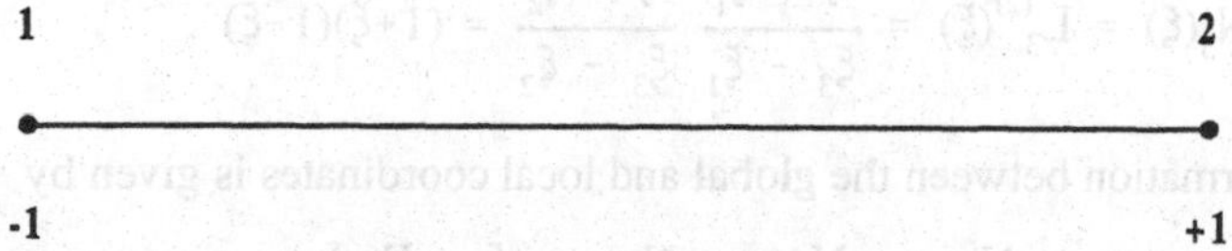

Figure 2.4.45 Local Coordinate for a Linear Line Element.

$$N_1(\xi) = L_1^{(1)}(\xi) = \frac{\xi - \xi_2}{\xi_1 - \xi_2} = \frac{1}{2}(1 - \xi) \tag{2.4.122}$$

$$N_2(\xi) = L_2^{(1)}(\xi) = \frac{\xi - \xi_1}{\xi_2 - \xi_1} = \frac{1}{2}(1 + \xi) . \tag{2.4.123}$$

The transformation between global coordinate and local coordinate is given by

$$x = x_1 N_1 + x_2 N_2, \quad \xi = [2x - (x_1 + x_2)]/L_e \tag{2.4.124}$$

$$dx = \frac{L_e}{2} d\xi \tag{2.4.125}$$

$$\frac{dN_i}{dx} = \frac{dN_i}{d\xi}\frac{d\xi}{dx} = \frac{2}{L_e}\frac{dN_i}{d\xi}, \qquad (2.4.126)$$

where L_e is the length of the element.

Quadratic Line Element. For a quadratic line element with the first, second, and third points located at $\xi = -1$, $\xi = 1$, and $\xi = 0$, respectively (Fig. 2.4.46), the base functions for the three points are obtained using the Lagrange polynomial of order 2:

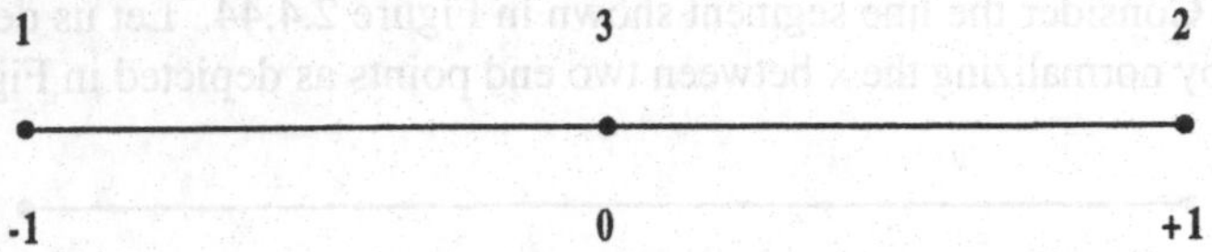

Figure 2.4.46 Local Coordinate for a Quadratic Line Element.

$$N_1(\xi) = L_1^{(2)}(\xi) = \frac{\xi - \xi_2}{\xi_1 - \xi_2}\frac{\xi - \xi_3}{\xi_1 - \xi_3} = -\frac{\xi(1 - \xi)}{2} \qquad (2.4.127)$$

$$N_2(\xi) = L_2^{(2)}(\xi) = \frac{\xi - \xi_1}{\xi_2 - \xi_1}\frac{\xi - \xi_3}{\xi_2 - \xi_3} = \frac{\xi(1 + \xi)}{2} \qquad (2.4.128)$$

$$N_3(\xi) = L_3^{(2)}(\xi) = \frac{\xi - \xi_1}{\xi_3 - \xi_1}\frac{\xi - \xi_2}{\xi_3 - \xi_2} = (1+\xi)(1-\xi) . \qquad (2.4.129)$$

The transformation between the global and local coordinates is given by

$$x = N_1x_1 + N_2x_2 + N_3x_3, \quad \xi = F(x) \qquad (2.4.130)$$

Cubic Line Element. For a cubic line element with the first, second, third, and fourth points located at $\xi = -1$, $\xi = 1$, $\xi = -1/3$, and $\xi = 1/3$, respectively (Fig. 2.4.47); the four base functions are obtained using the Lagrange polynomial of order 3:

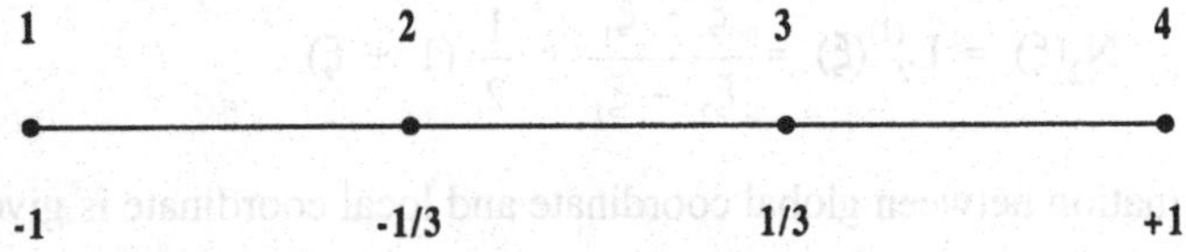

Figure 2.4.47 Local Coordinate for a Cubic Line Element.

$$N_1(\xi) = L_1^{(3)}(\xi) = \frac{\xi - \xi_2}{\xi_1 - \xi_2}\frac{\xi - \xi_3}{\xi_1 - \xi_3}\frac{\xi - \xi_4}{\xi_1 - \xi_4} = \tag{2.4.131}$$

$$-\frac{1}{16}(1+3\xi)(1-3\xi)(1-\xi)$$

$$N_2(\xi) = L_2^{(3)}(\xi) = \frac{\xi - \xi_1}{\xi_2 - \xi_1}\frac{\xi - \xi_3}{\xi_2 - \xi_3}\frac{\xi - \xi_4}{\xi_2 - \xi_4} = \tag{2.4.132}$$

$$-\frac{1}{16}(1+3\xi)(1-3\xi)(1+\xi)$$

$$N_3(\xi) = L_3^{(3)}(\xi) = \frac{\xi - \xi_1}{\xi_3 - \xi_1}\frac{\xi - \xi_2}{\xi_3 - \xi_2}\frac{\xi - \xi_4}{\xi_3 - \xi_4} = \tag{2.4.133}$$

$$\frac{9}{16}(1-3\xi)(1+\xi)(1-\xi)$$

$$N_4(\xi) = L_4^{(3)}(\xi) = \frac{\xi - \xi_1}{\xi_4 - \xi_1}\frac{\xi - \xi_2}{\xi_4 - \xi_2}\frac{\xi - \xi_3}{\xi_4 - \xi_3} = \tag{2.4.134}$$

$$\frac{9}{16}(1+3\xi)(1+\xi)(1-\xi) \ .$$

The transformation between the global and local coordinates is given by

$$x = N_1 x_1 + N_2 x_2 + N_3 x_3 + N_4 x_4, \quad \xi = F(x) \ . \tag{2.4.135}$$

Higher-Order Line Elements. Using the Lagrange polynomial of order n, we can easily write the base functions of a line element up to any order n in terms of local coordinate.

HERMITE ELEMENTS

Although problems in subsurface media generally involve the solution of a state variable f, there may be occasions when it is desirable to introduce the first derivative of f as an additional unknown at the nodes. If f and its first derivatives are to be specified at the nodes, the base functions must then be chosen such that the first derivatives of f are continuous at the element interfaces, that is, C^1 continuity. We demonstrate here how the desired base functions can be constructed using Hermite polynomials. In general, an n-th order Hermite polynomial in the local coordinate ξ

is a polynomial of degree (2n + 1) and may be denoted as $H^n(\xi)$,

$$H^n(\xi) = P_{2n+1}(\xi) . \tag{2.4.136}$$

Hermite polynomials are useful as interpolation functions because their values and the values of their derivatives up to order n are either unity or zero at the end points of the closed interval [-1,1]. This property can be represented symbolically if we assign two subscripts and write $H^n_{mi}(\xi)$, where m denotes the order of derivatives and i refers to either node of the end nodes. The symbolic representation is thus given by

$$\frac{d^k H^n_{mi}(\xi_j)}{d\xi^k} = \begin{cases} \delta_{ij} & \text{for } k = m, \quad m = 0, 1, \ldots, n \\ 0 & \text{for } k \neq m . \end{cases} \tag{2.4.137}$$

Consider the element given in Figure 2.4.48, we seek to construct the approximate function that takes the form

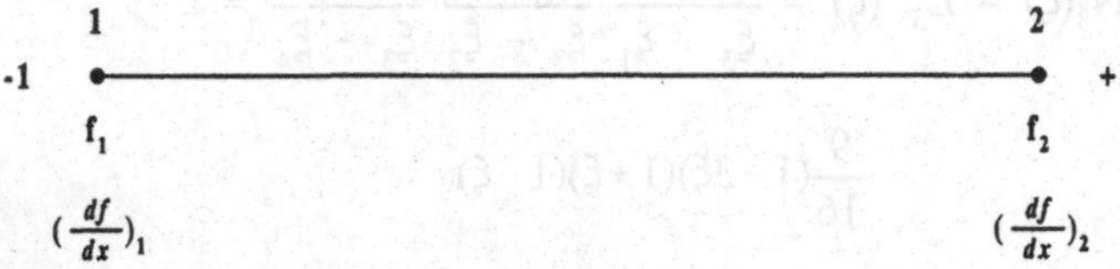

Figure 2.4.48 Local Coordinate of a Hermite Line Element.

$$f = f_1 H^1_{01} + \left(\frac{df}{dx}\right)_1 H^1_{11} + f_2 H^1_{02} + \left(\frac{df}{df}\right)_2 H^1_{12} \tag{2.4.138}$$

$$f = \sum_{j=1}^{2} \left[f_j N_{0j} + \left(\frac{df}{dx}\right)_j N_{1j} \right] . \tag{2.4.139}$$

To derive the expression for the function N_{01}, we first write it in the form of a third-order polynomial:

$$N_{01} = a_1 + a_2\xi + a_3\xi^2 + a_4\xi^3 . \tag{2.4.140}$$

To determine the coefficients a_1, a_2, a_3, and a_4, we use the four conditions

$$N_{01} = 1 \text{ at } \xi = -1, \qquad 1 = a_1 - a_2 + a_3 - a_4 \tag{2.4.141}$$

$$N_{01} = 0 \text{ at } \xi = 1, \qquad 0 = a_1 + a_2 + a_3 + a_4 \tag{2.4.142}$$

$$\frac{dN_{01}}{dx} = 0 \text{ at } \xi = -1, \qquad 0 = (2a_2 - 4a_3 + 6a_4)/L_e \tag{2.4.143}$$

$$\frac{dN_{01}}{dx} = 0 \text{ at } \xi = 1, \qquad 0 = (2a_2 + 4a_3 + 6a_4)/L_e \,. \tag{2.4.144}$$

The solution of Eq. (2.4.141) through (2.4.144) yields

$$a_1 = \frac{1}{2}, \; a_2 = -\frac{3}{4}, \; a_3 = 0, \; a_4 = \frac{1}{4} \,. \tag{2.4.145}$$

Thus,

$$N_{01} = \frac{1}{4}\left(2 - 3\xi + \xi^3\right) = \frac{1}{4}(\xi - 1)^2(\xi + 2) \,. \tag{2.4.146}$$

Similarly,

$$N_{02} = \frac{1}{4}\left(2 + 3\xi - \xi^3\right) = -\frac{1}{4}(\xi + 1)^2(\xi - 2) \,. \tag{2.4.147}$$

To derive the expression for the function N_{11}, we first write it in the form of a third-order polynomial

$$N_{11} = a_1 + a_2\xi + a_3\xi^2 + a_4\xi^3 \,. \tag{2.4.148}$$

To determine the coefficients a_1, a_2, a_3, and a_4, we use the four conditions:

$$N_{11} = 0 \text{ at } \xi = -1, \qquad 0 = a_1 - a_2 + a_3 - a_4 \tag{2.4.149}$$

$$N_{11} = 0 \text{ at } \xi = 1, \qquad 0 = a_1 + a_2 + a_3 + a_4 \tag{2.4.150}$$

$$\frac{dN_{11}}{dx} = 1 \text{ at } \xi = -1, \qquad 1 = (2a_2 - 4a_3 + 6a_4)/L_e \tag{2.4.151}$$

$$\frac{dN_{11}}{dx} = 0 \text{ at } \xi = 1, \qquad 0 = (2a_2 + 4a_3 + 6a_4)/L_e \,. \tag{2.4.152}$$

The solution of Eqs. (2.4.149) through (2.4.152) yields

$$a_1 = \frac{L_e}{8}, \quad a_2 = \frac{-L_e}{8}, \quad a_3 = \frac{-L_e}{8}, \quad a_4 = \frac{-L_e}{8} \,. \tag{2.4.153}$$

Thus,

$$N_{11} = \frac{L_e}{8}\left(1 - \xi - \xi^2 + \xi^3\right) = \frac{L_e}{8}(\xi + 1)(\xi - 1)^2 \,. \tag{2.4.154}$$

Similarly,

$$N_{12} = \frac{L_e}{8}\left(-1 - \xi + x^2 + \xi^3\right) = \frac{L_e}{8}(\xi + 1)^2(\xi - 1) \; . \qquad (2.4.155)$$

2.4.3.2. Two-Dimensional Elements. In this subsection, we present families of triangular and quadrilateral elements that can be used to deal with problems requiring C^0 continuity of the function f. For such problems we usually choose the nodal values of the unknown function to be the degrees of freedom of the elements. To ensure interelement continuity, we require that the number of nodes along the side of the element be just sufficient to determine uniquely the variation of f along that side. For example, if f is assigned to have a quadratic representation, then three nodes must be specified along each element side. To derive the base functions for triangular elements, the natural coordinates (in this case the area coordinates) will be used (Fig. 2.4.49). For quadrilateral elements, the local coordinates will be used:

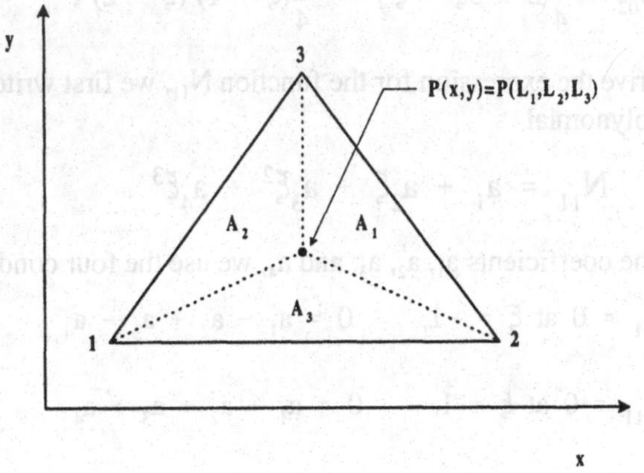

Fig. 2.4.49 Area Coordinate of a Triangular Element.

$$L_1 = \frac{A_1}{A}, \quad L_2 = \frac{A_2}{A}, \quad \text{and } L_2 = \frac{A_3}{A} \; , \qquad (2.4.156)$$

where A is the total area of the triangle and A_i (i = 1, 2, 3) denotes the areas of the sub-triangle i. It is clearly seen that

$$L_1 = 1, \; L_2 = 0, \; \text{ and } \; L_3 = 0 \; \text{ at node 1} \qquad (2.4.157)$$

$$L_2 = 1, \; L_3 = 0, \; \text{ and } \; L_1 = 0 \; \text{ at node 2} \qquad (2.4.158)$$

$$L_3 = 1, \; L_1 = 0, \; \text{ and } \; L_2 = 0 \; \text{ at node 3} \; . \qquad (2.4.159)$$

The transformation from global coordinates (x, y) to area coordinates (L_1, L_2, L_3) is clearly given by

$$L_1 + L_2 + L_3 = 1 \tag{2.4.160}$$

$$x_1L_1 + x_2L_2 + x_3L_3 = x \tag{2.4.161}$$

$$y_1L_1 + y_2L_2 + y_3L_3 = y \tag{2.4.162}$$

or

$$\begin{Bmatrix} 1 \\ x \\ y \end{Bmatrix} = \begin{bmatrix} 1 & 1 & 1 \\ x_1 & x_2 & x_3 \\ y_1 & y_2 & y_3 \end{bmatrix} \begin{Bmatrix} L_1 \\ L_2 \\ L_3 \end{Bmatrix}. \tag{2.4.163}$$

The inversion of Eq. (2.4.163) yields the transformation from area coordinates (L_1, L_2, L_3) to global coordinate (x, y);

$$L_i = \frac{1}{2A}(a_i + b_ix + c_iy),\ i = 1, 2, 3\ , \tag{2.4.164}$$

where

$$a_1 = x_2y_3 - x_3y_2,\ b_1 = y_2 - y_3,\ c_1 = c_3 - c_2 \tag{2.4.165}$$

$$a_2 = x_3y_1 - x_1y_3,\ b_2 = y_3 - y_1,\ c_2 = c_1 - c_3 \tag{2.4.166}$$

$$a_3 = x_1y_2 - x_2y_1,\ b_3 = y_1 - y_2,\ c_3 = c_2 - c_1\ . \tag{2.4.167}$$

Having the transformation between the global coordinates and the area coordinates, we can now derive the base functions in terms of area coordinates. As in the one-dimensional case, to construct the base functions, we establish a particular scheme for numbering the nodal points of triangular elements (Fig. 2.4.50). Via this scheme, every node I is given a three-digit label pqr where p, q, and r are integers satisfying the relation

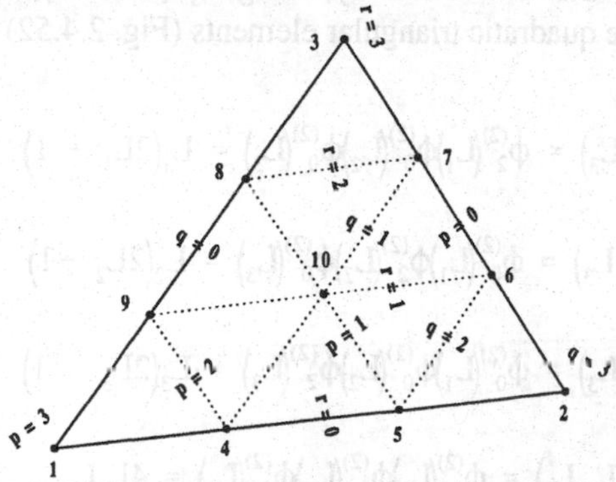

Fig. 2.4.50 pqr-Labeling of a Triangular Element.

$$p + q + r = n \; . \tag{2.4.168}$$

With the above convention, the base function for any node I is given by

$$N_I(L_1,L_2,L_3) = \phi_p^{(n)}(L_1)\phi_q^{(n)}(L_2)\phi_r^{(n)}(L_3) \; . \tag{2.4.169}$$

Linear Triangular Element. Using Eq. (2.4.169), we may derive the three base functions for the linear triangular elements (Fig. 2.4.51) as

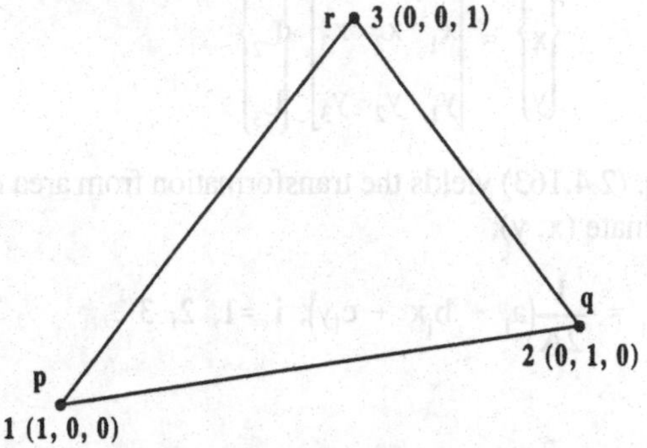

Fig. 2.4.51 pgq-Labeling of Nodes in a Linear Triangular Element.

$$N_1(L_1,L_2,L_3) = \phi_1^{(1)}(L_1)\phi_0^{(1)}(L_2)\phi_0^{(1)}(L_3) = L_1 \tag{2.4.170}$$

$$N_2(L_1,L_2,L_3) = \phi_0^{(1)}(L_1)\phi_1^{(1)}(L_2)\phi_0^{(1)}(L_3) = L_2 \tag{2.4.171}$$

$$N_3(L_1,L_2,L_3) = \phi_0^{(1)}(L_1)\phi_0^{(1)}(L_2)\phi_1^{(1)}(L_3) = L_3 \; . \tag{2.4.172}$$

Quadratic Triangular Element. Similarly, using Eq. (2.4.169), we can derive the six base functions for the quadratic triangular elements (Fig. 2.4.52) as

$$N_1(L_1,L_2,L_3) = \phi_2^{(2)}(L_1)\phi_0^{(2)}(L_2)\phi_0^{(2)}(L_3) = L_1(2L_1 - 1) \tag{2.4.173}$$

$$N_2(L_1,L_2,L_3) = \phi_0^{(2)}(L_1)\phi_2^{(2)}(L_2)\phi_0^{(2)}(L_3) = L_2(2L_2 - 1) \tag{2.4.174}$$

$$N_3(L_1,L_2,L_3) = \phi_0^{(2)}(L_1)\phi_0^{(2)}(L_2)\phi_2^{(2)}(L_3) = L_3(2L_3 - 1) \tag{2.4.175}$$

$$N_4(L_1,L_2,L_3) = \phi_1^{(2)}(L_1)\phi_1^{(2)}(L_2)\phi_0^{(2)}(L_3) = 4L_1L_2 \tag{2.4.176}$$

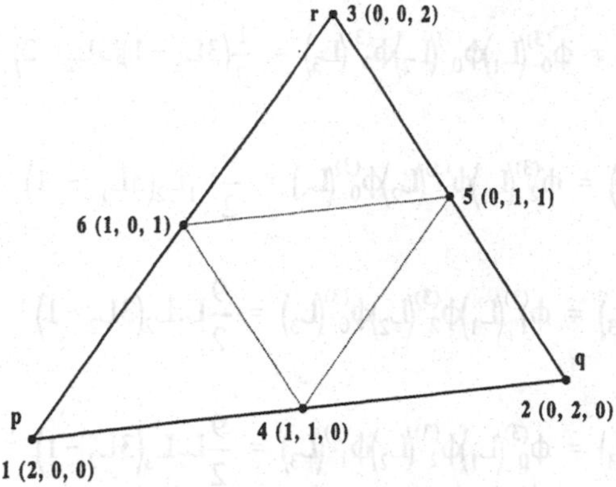

Fig. 2.4.52 pqr-Labeling of Nodes in a Quadratic Triangular Element.

$$N_5(L_1,L_2,L_3) = \phi_0^{(2)}(L_1)\phi_1^{(2)}(L_2)\phi_1^{(2)}(L_3) = 4L_2L_3 \tag{2.4.177}$$

$$N_6(L_1,L_2,L_3) = \phi_1^{(2)}(L_1)\phi_0^{(2)}(L_2)\phi_1^{(2)}(L_3) = 4L_1L_3. \tag{2.4.178}$$

Cubic Triangular Element. Finally, using Eq. (2.4.169), we can derive the 10 base functions for cubic triangular elements (Fig. 2.4.53) as

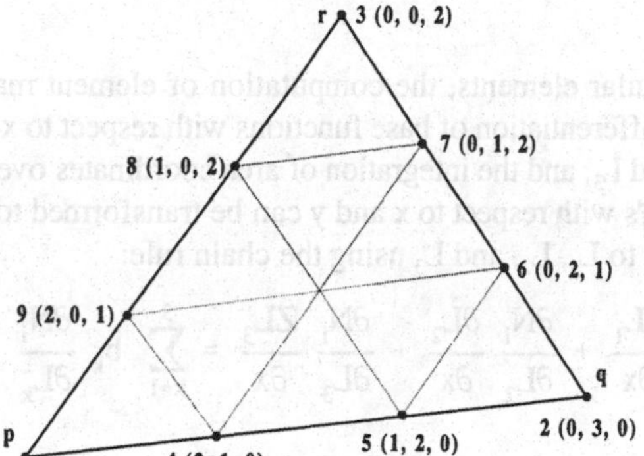

Fig. 2.4.53 pqr-Labeling of Nodes in a Cubic Triangular Element.

$$N_1(L_1,L_2,L_3) = \phi_3^{(3)}(L_1)\phi_0^{(3)}(L_2)\phi_0^{(3)}(L_3) = \frac{1}{2}L_1(3L_1 -)(3L_1 - 2) \tag{2.4.179}$$

$$N_2(L_1,L_2,L_3) = \phi_0^{(3)}(L_1)\phi_3^{(3)}(L_2)\phi_0^{(3)}(L_3) = \frac{1}{2}L_2(3L_2 - 1)(3L_2 - 2) \tag{2.4.180}$$

$$N_3(L_1,L_2,L_3) = \phi_0^{(3)}(L_1)\phi_0^{(3)}(L_2)\phi_3^{(3)}(L_3) = \frac{1}{2}(3L_3 - 1)(3L_2 - 2) \tag{2.4.181}$$

$$N_4(L_1,L_2,L_3) = \phi_2^{(3)}(L_1)\phi_1^{(3)}(L_2)\phi_0^{(3)}(L_3) = \frac{9}{2}L_1L_2(3L_1 - 1) \tag{2.4.182}$$

$$N_5(L_1,L_2,L_3) = \phi_1^{(3)}(L_1)\phi_2^{(3)}(L_2)\phi_0^{(3)}(L_3) = \frac{9}{2}L_1L_2(3L_2 - 1) \tag{2.4.183}$$

$$N_6(L_1,L_2,L_3) = \phi_0^{(3)}(L_1)\phi_2^{(3)}(L_2)\phi_1^{(3)}(L_3) = \frac{9}{2}L_2L_3(3L_2 - 1) \tag{2.4.184}$$

$$N_7(L_1,L_2,L_3) = \phi_0^{(3)}(L_1)\phi_1^{(3)}(L_2)\phi_2^{(3)}(L_3) = \frac{9}{2}L_2L_3(3L_3 - 1) \tag{2.4.185}$$

$$N_8(L_1,L_2,L_3) = \phi_1^{(3)}(L_1)\phi_0^{(3)}(L_2)\phi_2^{(3)}(L_3) = \frac{9}{2}L_1L_3(3L_3 - 1) \tag{2.4.186}$$

$$N_9(L_1,L_2,L_3) = \phi_2^{(3)}(L_1)\phi_0^{(3)}(L_2)\phi_1^{(3)}(L_3) = \frac{9}{2}L_1L_3(3L_1 - 1) \tag{2.4.187}$$

$$N_{10}(L_1,L_2,L_3) = \phi_1^{(3)}(L_1)\phi_1^{(3)}(L_2)\phi_1^{(3)}(L_3) = 27L_1L_2L_3 \, . \tag{2.4.188}$$

For triangular elements, the computation of element matrices involves the transformation of differentiation of base functions with respect to x and y to those with respect to L_1, L_2, and L_3, and the integration of area coordinates over the element. The differentiation of N_i's with respect to x and y can be transformed to the differentiation of N_i's with respect to L_1, L_2, and L_3 using the chain rule:

$$\frac{\partial N_i}{\partial x} = \frac{\partial N_i}{\partial L_1}\frac{\partial L_1}{\partial x} + \frac{\partial N_i}{\partial L_2}\frac{\partial L_2}{\partial x} + \frac{\partial N_i}{\partial L_3}\frac{ZL_3}{\partial x} = \sum_{k=1}^{3} b_k \frac{\partial N_i}{\partial L_k} \tag{2.4.189}$$

$$\frac{\partial N_i}{\partial y} = \frac{\partial N_i}{\partial L_1}\frac{\partial L_1}{\partial y} + \frac{\partial N_i}{\partial L_2}\frac{\partial L_2}{\partial y} + \frac{\partial N_i}{\partial L_3}\frac{ZL_3}{\partial y} = \sum_{k=1}^{3} c_k \frac{\partial N_i}{\partial L_k} \, . \tag{2.4.190}$$

Integration of area coordinates over the element is simple with the aid of the following formula:

$$\int_e L_1^{\alpha}L_2^{\beta}L_3^{\gamma}dA = 2A\,\frac{\alpha!\beta!\gamma!}{(\alpha+\beta+\gamma+2)!} \, . \tag{2.4.191}$$

QUADRILATERAL ELEMENTS

The construction of base functions for quadrilateral elements is best accomplished using the local coordinates (ξ,η). In the local coordinates, the original quadrilateral element is mapped into a square whose corners are located at $\xi = \pm 1$ and $\eta = \pm 1$,as shown in Figure 2.4.54. For bilinear quadrilateral elements, the four base functions can be obtained by taking the tensor product of the two base functions for the linear line elements as

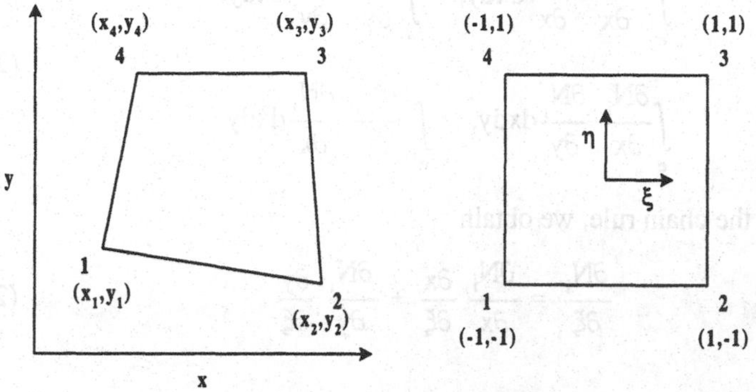

Fig. 2.4.54 Transformation from Global to Local Coordinates for Linear Quadrilateral Elements.

$$N_1(\xi,\eta) = L_1^{(1)}(\xi)\, L_1^{(1)}(\eta) = \frac{1}{4}(1-\xi)(1-\eta) \tag{2.4.192}$$

$$N_2(\xi,\eta) = L_2^{(1)}(\xi)\, L_1^{(1)}(\eta) = \frac{1}{4}(1+\xi)(1-\eta) \tag{2.4.193}$$

$$N_3(\xi,\eta) = L_2^{(1)}(\xi)\, L_2^{(1)}(\eta) = \frac{1}{4}(1+\xi)(1+\eta) \tag{2.4.194}$$

$$N_4(\xi,\eta) = L_1^{(1)}(\xi)\, L_2^{(1)}(\eta) = \frac{1}{4}(1-\xi)(1+\eta)\ . \tag{2.4.195}$$

The required transformation from global coordinate to local coordinate is obtained via the base functions:

$$x = \sum_{j=1}^{4} x_j N_j(\xi,\eta) \quad \textit{and} \quad y = \sum_{j=1}^{4} y_j N_j(\xi,\eta)\ . \tag{2.4.196}$$

Since the coordinate transformation uses the base functions, the element is termed the *isoparametric* element. In computing the element matrices for quadrilateral elements, the following types of integrals are often encountered in subsurface hydrology problems:

$$\int_e dxdy \quad \text{or} \quad \int_e N_i N_j dxdy \tag{2.4.197}$$

$$\int_e N_i \frac{\partial N_j}{dx} dxdy \quad \text{or} \quad \int_e N_i \frac{\partial N_j}{\partial y} dxdy \tag{2.4.198}$$

$$\begin{aligned} &\int_e \frac{\partial N_i}{\partial x}\frac{\partial N_j}{\partial x} dxdy, \quad \int_e \frac{\partial N_i}{\partial y}\frac{\partial N_j}{\partial y} dxdy, \\ &\int_e \frac{\partial N_i}{\partial x}\frac{\partial N_j}{\partial y} dxdy, \quad \int_e \frac{\partial N_i}{\partial y}\frac{\partial N_j}{\partial x} dxdy\ . \end{aligned} \tag{2.4.199}$$

Now using the chain rule, we obtain

$$\frac{\partial N_i}{\partial \xi} = \frac{\partial N_i}{\partial x}\frac{\partial x}{\partial \xi} + \frac{\partial N_i}{\partial y}\frac{\partial y}{\partial \xi} \tag{2.4.200}$$

and

$$\frac{\partial N_i}{\partial \eta} = \frac{\partial N_i}{\partial x}\frac{\partial x}{\partial \eta} + \frac{\partial N_i}{\partial y}\frac{\partial y}{\partial \eta}\ . \tag{2.4.201}$$

Written in matrix notation, Eqs. (2.4.200 and (2.4.201)are

$$\begin{Bmatrix} \dfrac{\partial N_i}{\partial \xi} \\ \dfrac{\partial N_i}{\partial \eta} \end{Bmatrix} = \begin{bmatrix} \dfrac{\partial x}{\partial \xi} & \dfrac{\partial y}{\partial \xi} \\ \dfrac{\partial x}{\partial \eta} & \dfrac{\partial y}{\partial \eta} \end{bmatrix} \begin{Bmatrix} \dfrac{\partial N_i}{\partial x} \\ \dfrac{\partial N_i}{\partial y} \end{Bmatrix} = [J] \begin{Bmatrix} \dfrac{\partial N_i}{\partial x} \\ \dfrac{\partial N_i}{\partial y} \end{Bmatrix}. \tag{2.4.202}$$

Inversion of Eq. (2.4.202) yields

$$\begin{Bmatrix} \dfrac{\partial N_i}{\partial x} \\ \dfrac{\partial N_i}{\partial y} \end{Bmatrix} = [J]^{-1} \begin{Bmatrix} \dfrac{\partial N_i}{\partial \xi} \\ \dfrac{\partial N_i}{\partial \eta} \end{Bmatrix}, \quad [J] = \begin{bmatrix} \dfrac{\partial x}{\partial \xi} & \dfrac{\partial y}{\partial \xi} \\ \dfrac{\partial x}{\partial \eta} & \dfrac{\partial y}{\partial \eta} \end{bmatrix}, \quad [J]^{-1} = \frac{1}{|J|} \begin{bmatrix} \dfrac{\partial y}{\partial \eta} & -\dfrac{\partial y}{\partial \xi} \\ -\dfrac{\partial x}{\partial \eta} & \dfrac{\partial x}{\partial \xi} \end{bmatrix}. \tag{2.4.203}$$

Finally, the integration of a differential area can be written as

$$\int_e dxdy = \int_{-1}^{1}\int_{-1}^{1} |J|\, d\xi d\eta\ . \tag{2.4.204}$$

For quadrilateral elements higher than bilinear, two families of elements will be dealt with: one is the Lagrange family and the other is the serendipity family.

Lagrange Family. The Lagrange family is so called because the base functions for the elements in this family can be derived simply by taking the tensor product of one-dimensional Lagrange polynomials. The base functions and the transformation between the local coordinates and global coordinates for a biquadratic and bicubic Lagrange family will be presented in the following section.

Biquadratic Lagrange Quadrilateral Elements. Using the tensor product, we can obtain the nine base functions for the biquadratic Lagrange quadrilateral elements (Fig. 2.4.55) as

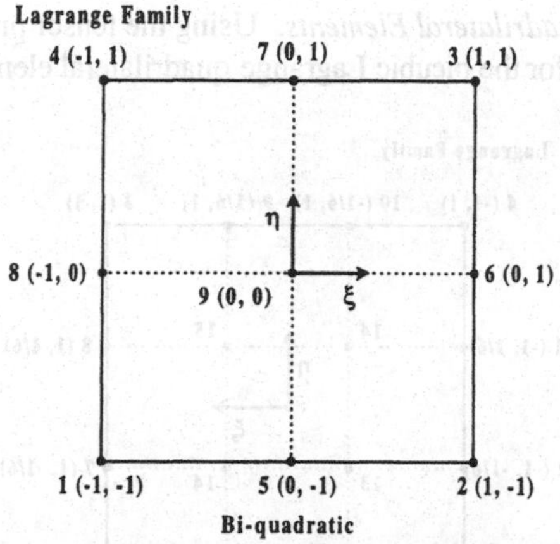

Fig. 2.4.55 Local Coordinate of Nodes in a Biquadratic Element.

$$N_1(\xi,\eta) = L_1^{(2)}(\xi)\, L_1^{(2)}(\eta) = \xi(1-\xi)\eta(1-\eta)/4 \tag{2.4.205}$$

$$N_2(\xi,\eta) = L_2^{(2)}(\xi)\, L_1^{(2)}(\eta) = -\xi(1+\xi)\eta(1-\eta)/4 \tag{2.4.206}$$

$$N_3(\xi,\eta) = L_2^{(2)}(\xi)\, L_2^{(2)}(\eta) = \xi(1+\xi)\eta(1+\eta)/4 \tag{2.4.207}$$

$$N_4(\xi,\eta) = L_1^{(2)}(\xi)\, L_2^{(2)}(\eta) = -\xi(1-\xi)\eta(1+\eta)/4 \tag{2.4.208}$$

$$N_5(\xi,\eta) = L_3^{(2)}(\xi)\, L_1^{(2)}(\eta) = -(1+\xi)(1-\xi)\eta(1-\eta)/2 \tag{2.4.209}$$

$$N_6(\xi,\eta) = L_2^{(2)}(\xi)\, L_3^{(2)}(\eta) = \xi(1+\xi)(1+\eta)(1-\eta)/2 \tag{2.4.210}$$

$$N_7(\xi,\eta) = L_3^{(2)}(\xi)\, L_2^{(2)}(\eta) = (1+\xi)(1-\xi)\eta(1+\eta)/2 \qquad (2.4.211)$$

$$N_8(\xi,\eta) = L_1^{(2)}(\xi)\, L_3^{(2)}(\eta) = -\xi(1-\xi)(1+\eta)(1+\eta)/2 \qquad (2.4.212)$$

$$N_9(\xi,\eta) = L_3^{(2)}(\xi)\, L_3^{(2)}(\eta) = (1+\xi)(1-\xi)(1+\eta)(1-\eta)\,. \qquad (2.4.213)$$

The required transformation from global coordinate to local coordinate is given as

$$x = \sum_{j=1}^{9} x_j N_j(\xi,\eta) \quad \textit{and} \quad y = \sum_{j=1}^{9} y_j N_j(\xi,\eta)\,. \qquad (2.4.214)$$

Bicubic Lagrange Quadrilateral Elements. Using the tensor product, we can obtain the 16 base functions for the bicubic Lagrange quadrilateral elements (Fig. 2.4.56) as

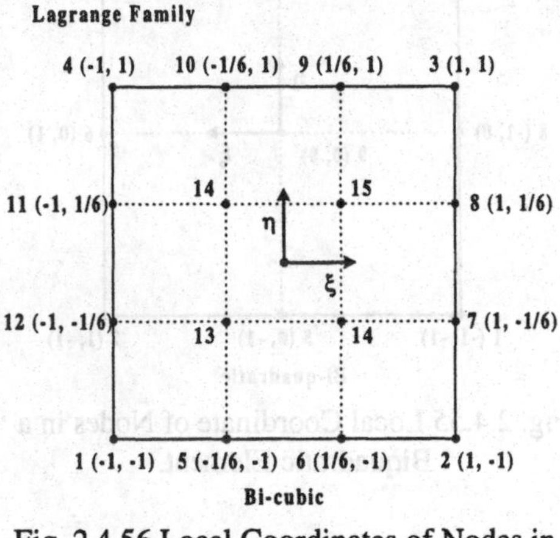

Fig. 2.4.56 Local Coordinates of Nodes in a Bicubic Element.

$$N_1(\xi,\eta) = L_1^{(3)}(\xi)\, L_1^{(3)}(\eta), \quad N_2(\xi,\eta) = L_2^{(3)}(\xi)\, L_1^{(3)}(\eta) \qquad (2.4.215)$$

$$N_3(\xi,\eta) = L_2^{(3)}(\xi)\, L_2^{(3)}(\eta), \quad N_4(\xi,\eta) = L_1^{(3)}(\xi)\, L_2^{(3)}(\eta) \qquad (2.4.216)$$

$$N_5(\xi,\eta) = L_3^{(3)}(\xi)\, L_1^{(3)}(\eta), \quad N_6(\xi,\eta) = L_4^{(3)}(\xi)\, L_1^{(3)}(\eta) \qquad (2.4.217)$$

$$N_7(\xi,\eta) = L_2^{(3)}(\xi)\, L_3^{(3)}(\eta), \quad N_8(\xi,\eta) = L_2^{(3)}(\xi)\, L_4^{(3)}(\eta) \qquad (2.4.218)$$

$$N_9(\xi,\eta) = L_4^{(3)}(\xi)\, L_2^{(3)}(\eta), \quad N_{10}(\xi,\eta) = L_3^{(3)}(\xi)\, L_2^{(3)}(\eta) \qquad (2.4.219)$$

$$N_{11}(\xi,\eta) = L_1^{(3)}(\xi)\, L_4^{(3)}(\eta), \quad N_{12}(\xi,\eta) = L_1^{(3)}(\xi)\, L_3^{(3)}(\eta) \qquad (2.4.220)$$

$$N_{13}(\xi,\eta) = L_3^{(3)}(\xi)\, L_3^{(3)}(\eta), \quad N_{14}(\xi,\eta) = L_4^{(3)}(\xi)\, L_3^{(3)}(\eta) \qquad (2.4.221)$$

$$N_{15}(\xi,\eta) = L_4^{(3)}(\xi)\, L_4^{(3)}(\eta), \quad N_{16}(\xi,\eta) = L_3^{(3)}(\xi)\, L_4^{(3)}(\eta)\ . \qquad (2.4.222)$$

The required transformation from global coordinate to local coordinate is given as

$$x = \sum_{j=1}^{16} x_j N_j(\xi,\eta) \quad \textit{and} \quad y = \sum_{j=1}^{16} y_j N_j(\xi,\eta)\ . \qquad (2.4.223)$$

Serendipity Family. The elements in this family contain only exterior nodes (Fig. 2.4.57). Base functions in this family can be obtained by the principal of linear blending for corner codes as

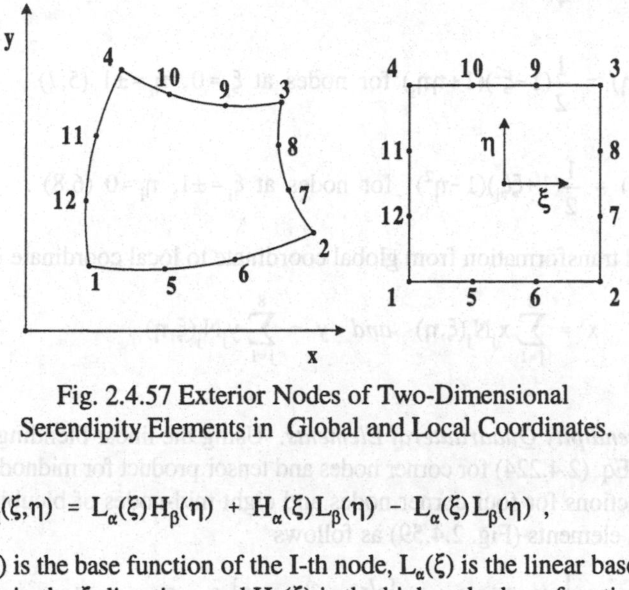

Fig. 2.4.57 Exterior Nodes of Two-Dimensional Serendipity Elements in Global and Local Coordinates.

$$N_I(\xi,\eta) = L_\alpha(\xi) H_\beta(\eta) + H_\alpha(\xi) L_\beta(\eta) - L_\alpha(\xi) L_\beta(\eta)\ , \qquad (2.4.224)$$

where $N_I(\xi,\eta)$ is the base function of the I-th node, $L_\alpha(\xi)$ is the linear base function of the α-th node in the ξ-direction, and $H_\alpha(\xi)$ is the high-order base function of the α-th node in the ξ-direction. For midnodes, the base functions are obtained by the tensor product of one-dimensional base functions. The following sections demonstrate the use of the linear blending and the tensor product to derive base functions for the serendipity family.

Bi-quadratic Serendipity Quadrilateral Elements. Using the linear blending approach as specified by Eq. (2.4.224) for corner nodes and tensor product for mid-nodes, we obtain the base functions for four corner nodes and four midnodes of biquadratic serendipity quadrilateral elements (Fig. 2.4.58) as follows:

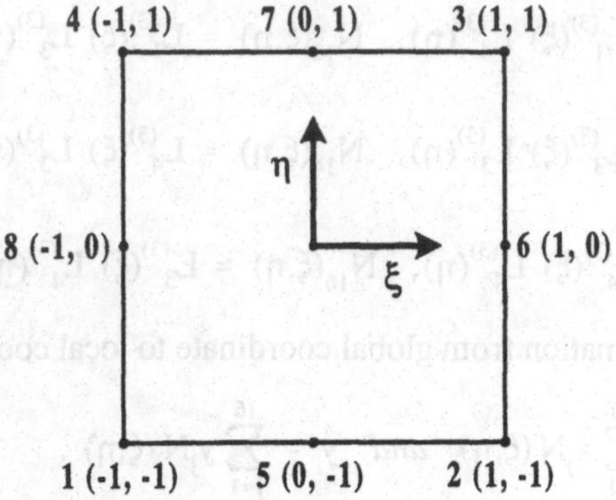

Fig. 2.4.58 Local Coordinate of Nodes in a Biquadratic Serendipity Element.

$$N_i(\xi,\eta) = \frac{1}{4}(1+\xi\xi_i)(1+\eta\eta_i)(\xi\xi_i+\eta\eta_i - 1) \text{ for Corner Nodes} \quad (2.4.225)$$

$$N_i(\xi,\eta) = \frac{1}{2}(1-\xi^2)(1+\eta\eta_i) \text{ for nodes at } \xi_i=0,\ \eta_i=\pm 1\ (5,7) \quad (2.4.226)$$

$$N_1(\xi,\eta) = \frac{1}{2}(1+\xi\xi_i)(1-\eta^2) \text{ for nodes at } \xi_i=\pm 1,\ \eta_i=0\ (6,8)\ . \quad (2.4.227)$$

The required transformation from global coordinate to local coordinate is given as

$$x = \sum_{j=1}^{8} x_j N_j(\xi,\eta) \quad \textit{and} \quad y = \sum_{j=1}^{8} y_j N_j(\xi,\eta)\ . \quad (2.4.228)$$

Bi-cubic Serendipity Quadrilateral Elements. Using the linear blending approach as specified by Eq. (2.4.224) for corner nodes and tensor product for midnodes, we obtain the base functions for four corner nodes and eight mid-nodes of bicubic serendipity quadrilateral elements (Fig. 2.4.59) as follows

$$N_i(\xi,\eta) = \frac{1}{32}(1+\xi\xi_i)(1+\eta\eta_i)\left[9(\xi^2+\eta^2) - 10\right] \text{ for Corner Nodes} \quad (2.4.229)$$

$$N_i(\xi,\eta) = \frac{9}{32}(1+\xi\xi_i)(1+9\eta\eta_i)(1-\eta^2) \text{ for nodes at } \xi_i=\pm 1,\ \eta_i=\pm\frac{1}{3} \quad (2.4.230)$$

$$N_i(\xi,\eta) = \frac{9}{32}(1+9\xi\xi_i)(1+\eta\eta_i)(1-\xi^2) \text{ for nodes at } \xi_i=\pm\frac{1}{3},\ \eta_i=\pm 1\ . \quad (2.4.231)$$

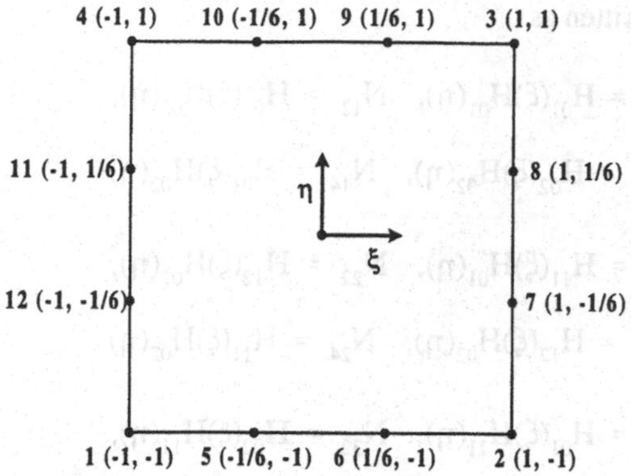

Fig. 2.4.59 Local Coordinate of Nodes in a Bicubic Serendipity Element.

The required transformation from global coordinate to local coordinate is given as

$$x = \sum_{j=1}^{12} x_j N_j(\xi,\eta) \quad and \quad y = \sum_{j=1}^{12} y_j N_j(\xi,\eta) \; . \qquad (2.4.232)$$

HERMITE ELEMENTS

The base functions for Hermite quadrilateral elements can be constructed by taking the tensor product of one-dimensional Hermite polynomials. The trial function for a typical element in Figure 2.4.60 is written as

$$f = \sum_{j=1}^{4} \left[f_j N_{1j} + \left(\frac{\partial f}{\partial \xi}\right)_j N_{2j} + \left(\frac{\partial f}{\partial \eta}\right)_j N_{3j} + \left(\frac{\partial^2 f}{\partial \xi \partial \eta}\right)_j N_{4j} \right] , \qquad (2.4.233)$$

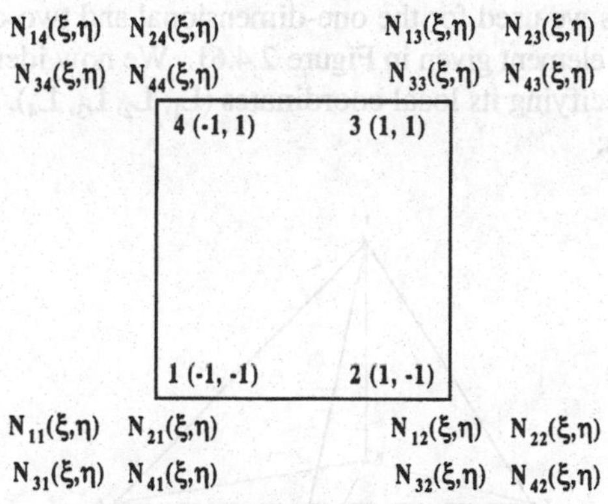

Fig. 2.4.60 Local Coordinate of Nodes in a Hermite Element.

where the mixed second derivative is required as a nodal parameter because of the combination of the Hermite polynomials that arises in the product. The 16 base functions may be written as

$$N_{11} = H_{01}(\xi)H_{01}(\eta),\quad N_{12} = H_{02}(\xi)H_{01}(\eta),$$
$$N_{13} = H_{02}(\xi)H_{02}(\eta),\quad N_{14} = H_{01}(\xi)H_{02}(\eta) \tag{2.4.234}$$

$$N_{21} = H_{11}(\xi)H_{01}(\eta),\quad N_{22} = H_{12}(\xi)H_{01}(\eta),$$
$$N_{23} = H_{12}(\xi)H_{02}(\eta),\quad N_{24} = H_{11}(\xi)H_{02}(\eta) \tag{2.4.235}$$

$$N_{31} = H_{01}(\xi)H_{11}(\eta),\quad N_{32} = H_{02}(\xi)H_{11}(\eta),$$
$$N_{33} = H_{02}(\xi)H_{12}(\eta),\quad N_{34} = H_{01}(\xi)H_{12}(\eta) \tag{2.4.236}$$

$$N_{41} = H_{11}(\xi)H_{11}(\eta),\quad N_{42} = H_{12}(\xi)H_{11}(\eta),$$
$$N_{43} = H_{12}(\xi)H_{12}(\eta),\quad N_{44} = H_{11}(\xi)H_{12}(\eta)\ . \tag{2.4.237}$$

2.4.3.3. Three-Dimensional Elements. The formulation of base functions for three-dimensional elements is a direct extension of the two-dimensional elements. The simplest three-dimensional element is the four-node tetrahedron. To derive the base functions for tetrahedral elements, the natural coordinates (in this case the volume coordinates) will be used. For hexahedral elements, the local coordinates will be used.

TETRAHEDRAL ELEMENTS

The base functions for tetrahedral elements can be best described by using the natural coordinates. The development of natural coordinates for tetrahedral elements follows the same procedure as we used for the one-dimensional and two-dimensional cases. Consider a tetrahedral element given in Figure 2.4.61. We now identify a point within the tetrahedron by specifying its local coordinates (L_1, L_2, L_3, L_4). These coordinates are defined as follows:

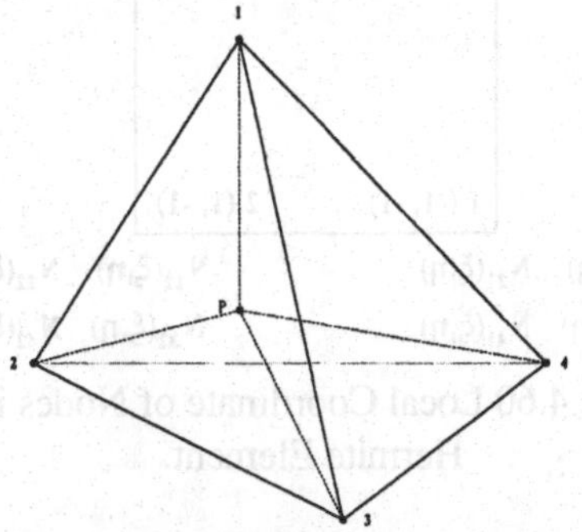

Fig. 2.4.61 Volume Coordinate of Tetrahedral Elements.

$$L_i = \frac{V_i}{V}, \quad i = 1, 2, 3, 4 \ , \tag{2.4.238}$$

where V is the total area of the tetrahedron and V_i denotes the volume defined by the point P and the face opposite to the vertex i. The transformation from global coordinates (x, y, z) to area coordinates (L_1, L_2, L_3, L_4) is clearly given by

$$\begin{Bmatrix} 1 \\ x \\ y \\ z \end{Bmatrix} = \begin{bmatrix} 1 & 1 & 1 & 1 \\ x_1 & x_2 & x_3 & x_4 \\ y_1 & y_2 & y_3 & y_4 \\ z_1 & z_2 & z_3 & z_4 \end{bmatrix} \begin{Bmatrix} L_1 \\ L_2 \\ L_3 \\ L_4 \end{Bmatrix} . \tag{2.4.239}$$

The inversion of Eq. (2.4.239) yields the transformation from volume coordinates (L_1, L_2, L_3, L_4) to global coordinate (x, y, z):

$$L_i = \frac{1}{6V}(a_i + b_i x + c_i y + d_i z) \ , \tag{2.4.240}$$

where

$$a_1 = \begin{vmatrix} x_2 & y_2 & z_2 \\ x_3 & y_3 & z_3 \\ x_4 & y_4 & z_4 \end{vmatrix}, \quad b_1 = -\begin{vmatrix} 1 & y_2 & z_2 \\ 1 & y_3 & z_3 \\ 1 & y_4 & z_4 \end{vmatrix} \tag{2.4.241}$$

$$c_1 = \begin{vmatrix} x_2 & 1 & z_2 \\ x_3 & 1 & z_3 \\ x_4 & 1 & z_4 \end{vmatrix}, \quad d_1 = \begin{vmatrix} x_2 & y_2 & 1 \\ x_3 & y_3 & 1 \\ x_4 & y_4 & 1 \end{vmatrix} \tag{2.4.242}$$

$$V = \begin{vmatrix} 1 & x_1 & y_1 & z_1 \\ 1 & x_2 & y_2 & z_2 \\ 1 & x_3 & y_3 & z_3 \\ 1 & x_4 & y_4 & z_4 \end{vmatrix}, \tag{2.4.243}$$

and the remaining coefficients are obtained by a cyclic permutation of the subscripts.

Having the transformation between the global coordinates and the volume coordinates, we can now derive the base functions in terms of volume coordinates. As in the one-dimensional case, to construct the base functions, we establish a particular scheme for numbering the nodal points of the tetrahedral elements (Fig. 2.4.62). Via this scheme, every node I is given a four-digit label pqrs where p, q, r, and s are integers satisfying the relation

$$p + q + r + s = n \; . \tag{2.4.244}$$

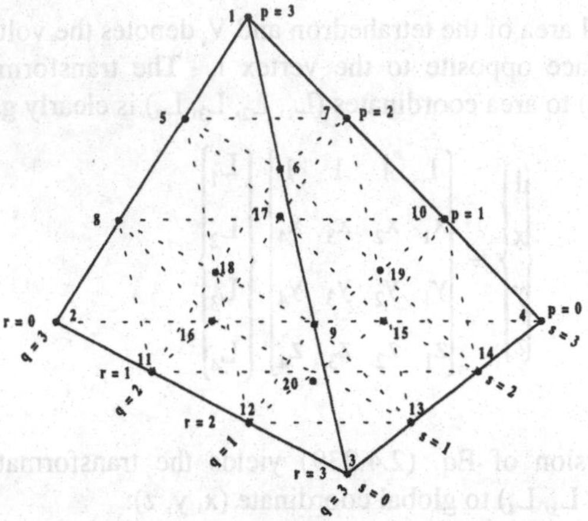

Fig. 2.4.62 pqrs-Labeling of a Tetrahedral Elements.

With the above convention, the base function for any node I is given by

$$N_I(L_1,L_2,L_3,L_4) \; = \; \phi_p^{(n)}(L_1)\phi_q^{(n)}(L_2)\phi_r^{(n)}(L_3)\phi_s^{(n)}(L_4) \; . \tag{2.4.245}$$

Linear Tetrahedral Elements. For a linear tetrahedral element (Fig. 2.4.63), we can use Eq. (2.4.245) to obtain the base functions as follows. For node 1, we have p = 1, q = 0, r = 0, and s = 0. For node 2, we have p = 0, p = 1, q = 0, r = 0, and s = 0. For node 2, we have p = 0, q = 0, r = 1, and s = 0. For node 4, we have p = 0, q = 0, r = 0, and s = 1. Thus, the four base functions are given by

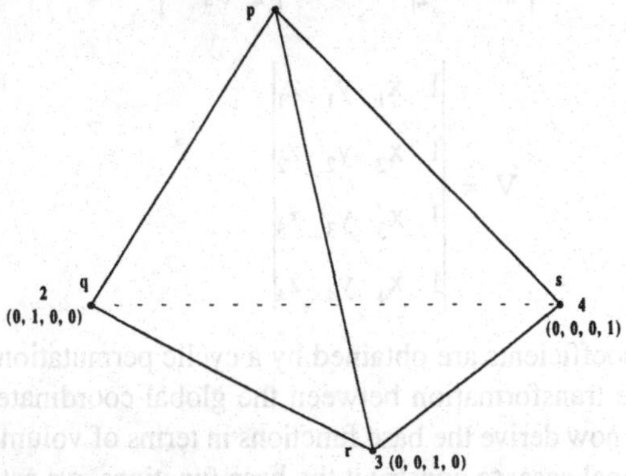

Fig. 2.4.63 pqrs-Labeling of Nodes in a Linear Tetrahedral Element.

$$N_i = L_i, \quad i = 1, 2, 3, 4 \,. \tag{2.4.246}$$

Quadratic Tetrahedral Elements. For a quadratic element (Fig. 2.4.64), the four digit integers for the vertex nodes are obvious. For a midside node, for example, node 5, we have p = 1, q = 1, r = 0, and s = 0. Thus, the base functions are obtained as follows:

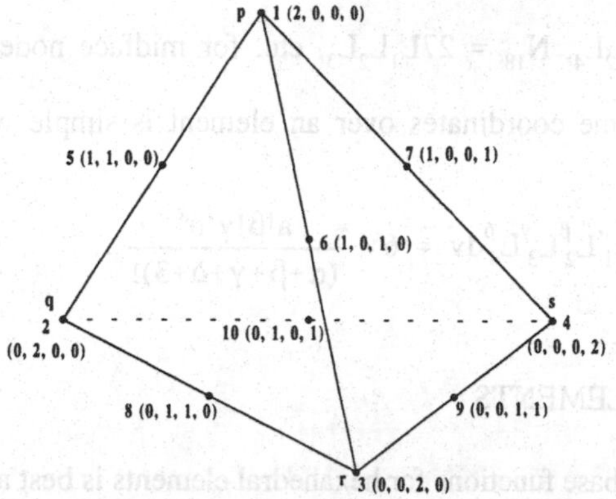

Fig. 2.4.64 pqrs-Labeling of Nodes in a Quadratic Element.

$$N_1 = (2L_1 - 1)L_1 \quad \text{for vertex nodes} \tag{2.4.247}$$

$$N_5 = 4L_1L_2, \quad N_6 = 4L_1L_3, \text{ etc. for midside nodes .} \tag{2.4.248}$$

Cubic Tetrahedral Element. Similarly, we can make up a four-digit integer for every node of a cubic tetrahedral element (Fig. 2.4.65) and thus obtain the base functions as follows:

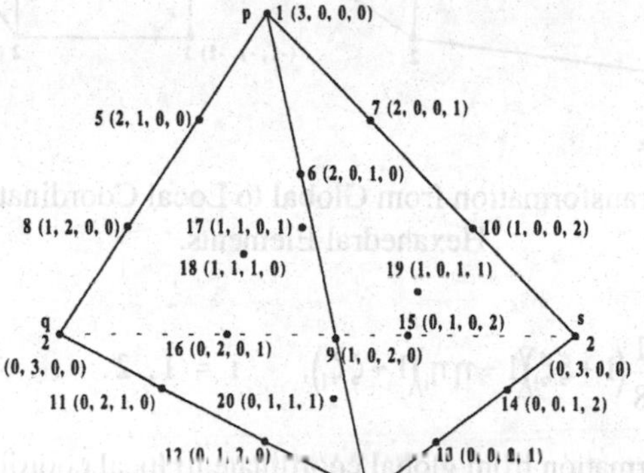

Fig. 2.4.65 pqrs-Labeling of Nodes in a Cubic Tetrahedral Element.

$$N_i = \frac{1}{2}(3L_i-1)(3L_i-2)L_i \quad \text{for vortex nodes } (i = 1, 2, 3, 4) \tag{2.4.249}$$

$$N_5 = \frac{9}{2}L_1L_2(3L_1-2), \quad \text{etc. for side nodes} \tag{2.4.250}$$

$$N_{17} = 27L_1L_2L_4,\ N_{18} = 27L_1L_2L_3, \quad \text{etc. for midface nodes .} \tag{2.4.251}$$

Integration of volume coordinates over an element is simple with the aid of the following formula:

$$\int_E L_1^{\alpha}L_2^{\beta}L_3^{\gamma}L_4^{\delta}dv = 6V\,\frac{a!\beta!\gamma!\delta!}{(\alpha+\beta+\gamma+\delta+3)!} \,. \tag{2.4.252}$$

HEXAHEDRAL ELEMENTS

The construction of base functions for hexahedral elements is best accomplished using the local coordinates (ξ,η,ζ). In the local coordinates, the original hexahedral element is mapped into a cubic element whose corners are located at $\xi = \pm1$, $\eta = \pm1$, and $\zeta = \pm1$, as shown in Figure 2.4.66. For trilinear hexahedral elements, the eight base functions can be obtained by taking the tensor product of the two base functions for the linear line elements as

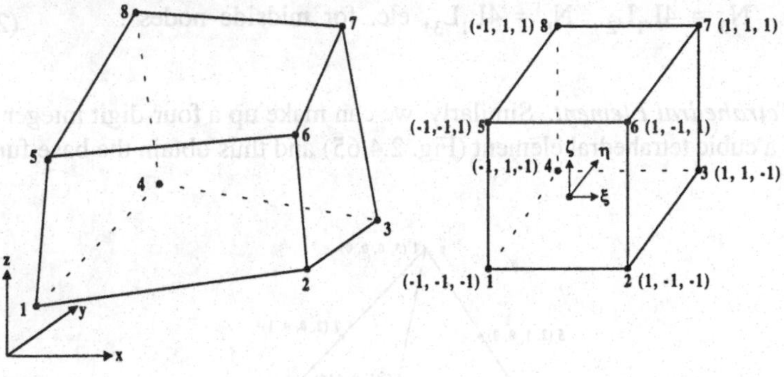

Fig. 2.4.66 Transformation from Global to Local Coordinate for Linear Hexahedral Elements.

$$N_i(\xi,\eta,\zeta) = \frac{1}{8}(1+\xi\xi_i)(1+\eta\eta_i)(1+\zeta\zeta_i), \qquad i = 1, 2, \ldots, 8. \tag{2.4.253}$$

The required transformation from global coordinate to local coordinate is obtained by via the base functions as

$$x = \sum_{j=1}^{8} x_j N_j(\xi,\eta,\zeta), \quad y = \sum_{j=1}^{8} y_j N_j(\xi,\eta,\zeta), \quad z = \sum_{j=1}^{8} z_j N_j(\xi,\eta,\zeta) \ . \tag{2.4.254}$$

Since the coordinate transformation uses the base functions, the element is termed the *isoparametric* element. In computing the element matrices for hexahedral elements, the following types of integrals are often encountered in subsurface hydrology problems:

$$\int_e dxdydz, \quad \int_e N_i N_j dxdydz \tag{2.4.255}$$

$$\int_e N_i \frac{\partial N_j}{dx} dxdydz, \quad \int_e N_i \frac{\partial N_j}{\partial y} dxdydz \ \text{ or } \ \int_e N_i \frac{\partial N_j}{\partial z} dxdydz \tag{2.4.256}$$

and

$$\int_e \frac{\partial N_i}{\partial x}\frac{\partial N_j}{\partial x} dxdydz, \quad \int_e \frac{\partial N_i}{\partial y}\frac{\partial N_j}{\partial y} dxdyz, \quad \int_e \frac{\partial N_i}{\partial z}\frac{\partial N_j}{\partial z} dxdydz$$

$$\int_e \frac{\partial N_i}{\partial x}\frac{\partial N_j}{\partial y} dxdydz, \quad \int_e \frac{\partial N_i}{\partial y}\frac{\partial N_j}{\partial z} dxdydx, \quad \int_e \frac{\partial N_i}{\partial z}\frac{\partial N_j}{\partial x} dxdydz \ . \tag{2.4.257}$$

Now using the chain rule, we obtain

$$\frac{\partial N_i}{\partial \xi} = \frac{\partial N_i}{\partial x}\frac{\partial x}{\partial \xi} + \frac{\partial N_i}{\partial y}\frac{\partial y}{\partial \xi} + \frac{\partial N_i}{\partial z}\frac{\partial z}{\partial \xi} \tag{2.4.258}$$

$$\frac{\partial N_i}{\partial \eta} = \frac{\partial N_i}{\partial x}\frac{\partial x}{\partial \eta} + \frac{\partial N_i}{\partial y}\frac{\partial y}{\partial \eta} + \frac{\partial N_i}{\partial z}\frac{\partial z}{\partial \eta} \tag{2.4.259}$$

$$\frac{\partial N_i}{\partial \zeta} = \frac{\partial N_i}{\partial x}\frac{\partial x}{\partial \zeta} + \frac{\partial N_i}{\partial y}\frac{\partial y}{\partial \zeta} + \frac{\partial N_i}{\partial z}\frac{\partial z}{\partial \zeta} \ . \tag{2.4.260}$$

Written in matrix notation, Eqs. (2.4.258) through (2.4.260) are

$$\begin{Bmatrix} \dfrac{\partial N_i}{\partial \xi} \\[2ex] \dfrac{\partial N_i}{\partial \eta} \\[2ex] \dfrac{\partial N_i}{\partial \zeta} \end{Bmatrix} = \begin{bmatrix} \dfrac{\partial x}{\partial \xi} & \dfrac{\partial y}{\partial \xi} & \dfrac{\partial z}{\partial \xi} \\[2ex] \dfrac{\partial x}{\partial \eta} & \dfrac{\partial y}{\partial \eta} & \dfrac{\partial y}{\partial \eta} \\[2ex] \dfrac{\partial x}{\partial \zeta} & \dfrac{\partial y}{\partial \zeta} & \dfrac{\partial z}{\partial \zeta} \end{bmatrix} \begin{Bmatrix} \dfrac{\partial N_i}{\partial x} \\[2ex] \dfrac{\partial N_i}{\partial y} \\[2ex] \dfrac{\partial N_i}{\partial z} \end{Bmatrix} = [J] \begin{Bmatrix} \dfrac{\partial N_i}{\partial x} \\[2ex] \dfrac{\partial N_i}{\partial y} \\[2ex] \dfrac{\partial N_i}{\partial z} \end{Bmatrix} . \tag{2.4.261}$$

Inversion of Eq. (2.4.261) yields

$$\begin{Bmatrix} \dfrac{\partial N_i}{\partial x} \\[2ex] \dfrac{\partial N_i}{\partial y} \\[2ex] \dfrac{\partial N_i}{\partial z} \end{Bmatrix} = [J]^{-1} \begin{Bmatrix} \dfrac{\partial N_i}{\partial \xi} \\[2ex] \dfrac{\partial N_i}{\partial \eta} \\[2ex] \dfrac{\partial N_i}{\partial \zeta} \end{Bmatrix} \quad \textit{where} \quad [J] = \begin{bmatrix} \dfrac{\partial x}{\partial \xi} & \dfrac{\partial y}{\partial \xi} & \dfrac{\partial z}{\partial \xi} \\[2ex] \dfrac{\partial x}{\partial \eta} & \dfrac{\partial y}{\partial \eta} & \dfrac{\partial z}{\partial \eta} \\[2ex] \dfrac{\partial x}{\partial \zeta} & \dfrac{\partial y}{\partial \zeta} & \dfrac{\partial z}{\partial \zeta} \end{bmatrix} . \tag{2.4.262}$$

Finally, the integration of a differential volume can be written as

$$\int_e dxdydz = \int_{-1}^{1}\int_{-1}^{1}\int_{-1}^{1} |J|\, d\xi d\eta d\zeta \ . \tag{2.4.263}$$

For hexahedral elements higher than trilinear, the Lagrange family involves too many interior nodes; they are seldom used. Thus, for three-dimensional hexahedral elements, only the serendipity family needs to be considered. Base functions in this serendipity family (Fig. 2.4.67) can be obtained by the principle of linear blending for corner nodes, as follows:

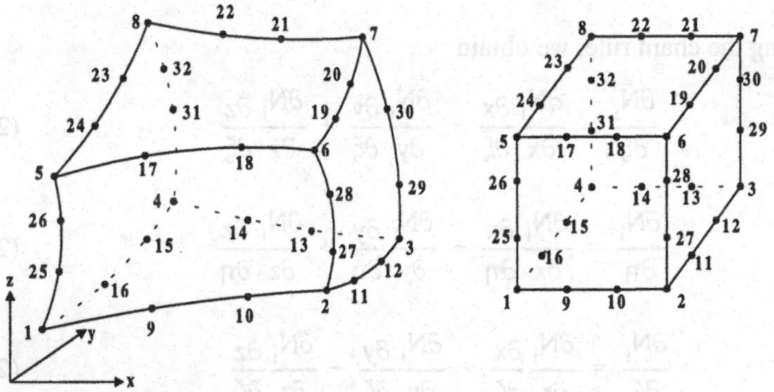

Fig. 2.4.67 Exterior Nodes of Three-Dimensional Serendipity Elements in Global and Local Coordinates.

$$\begin{aligned} N_I(\xi,\eta,\zeta) = {} & H_\alpha(\xi)L_\beta(\eta)L_\gamma(\eta) + L_\alpha(\xi)H_\beta(\eta)L_\gamma(\eta) + \\ & L_\alpha(\xi)L_\beta(\eta)H_\gamma(\zeta) - 2L_\alpha(\xi)L_\beta(\eta)L_\gamma(\eta) \ , \end{aligned} \tag{2.4.264}$$

where $N_\alpha(\xi,\eta,\zeta)$ is the base function of the I-th node, $L_\alpha(\xi)$ is the linear base function of the α-th node in the ξ-th direction, $H_\beta(\eta)$ is the high-order base functions of the β-th node in the η-th direction, and $H_\gamma(\zeta)$ is the high-order base functions of the γ-th node in the ζ-direction. For midside nodes, the base functions are obtained by the tensor product of the one-dimensional base functions. We shall demonstrate the use of the linear blending and tensor product to derive the base functions for the serendipity family as follows.

Triquadratic Hexahedral Elements. Using the method of linear blending for corner nodes and the tensor product for midnodes, respectively, we obtain the base functions for eight corner nodes and 12 midnodes, respectively, for the triquadratic hexahedral elements (Fig. 2.4.68) as follows:

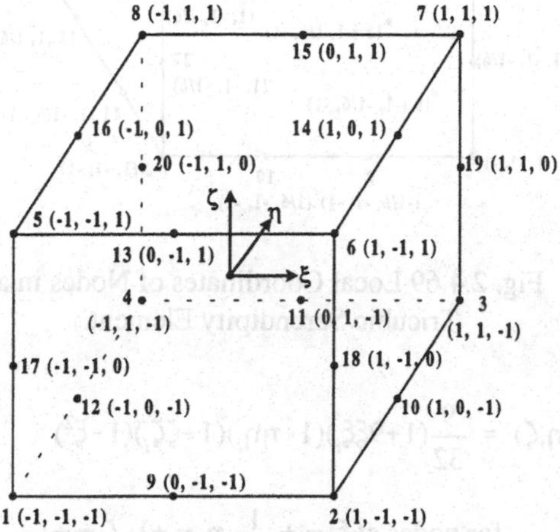

Fig. 2.4.68 Local Coordinates of Nodes in a Triquadratic Serendipity Element.

$$N_i(\xi,\eta,\zeta) = \frac{1}{8}(1+\xi\xi_i)(1+\eta\eta_i)(1+\zeta\zeta_i)(\xi\xi_i+\eta\eta_i+\zeta\zeta_i-2) \qquad (2.4.265)$$

for Corner nodes (1, 2, 3, 4, 5, 6, 7, 8)

$$N_i(\xi,\eta,\zeta) = \frac{1}{4}(1-\xi^2)(1+\eta\eta_i)(1+\zeta\zeta_i) \ . \qquad (2.4.266)$$

for nodes at $\xi_i = 0$, $\eta_i = \pm 1$, $\zeta = \pm 1$

Tri-cubic Element. Using the method of linear blending for corner nodes and tensor product for midnode, respectively, we obtain the base functions for eight corner and 24 midnodes, respectively, for the tricubic hexahedral elements (Fig. 2.4.69) as follows:

$$N_i(\xi,\eta,\zeta) = \frac{1}{64}(1+\xi\xi_i)(1+\eta\eta_i)(1+\zeta\zeta_i)\left[9(\xi^2+\eta^2+\zeta^2) - 19\right] \qquad (2.4.267)$$

for Corner nodes (1, 2, 3, 4, 5, 6, 7, 8)

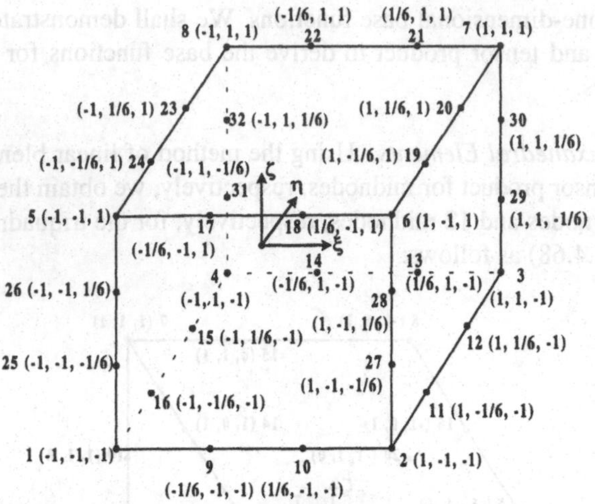

Fig. 2.4.69 Local Coordinates of Nodes in a Tricubic Serendipity Element.

$$N_i(\xi,\eta,\zeta) = \frac{9}{32}(1+9\xi\xi_i)(1+\eta\eta_i)(1+\zeta\zeta_i)(1-\xi^2) \ . \qquad (2.4.268)$$

for nodes at $\xi_i = \pm\ \frac{1}{3},\ \eta_i = \pm 1,\ \zeta = \pm 1$

HERMITE ELEMENTS

Hermite elements in three-dimensional cases introduce too many degrees of freedom. They are seldom used in practice.

2.4.4. Numerical Integration in Local Coordinates: Gaussian Quadratures

In forming the element matrix equation, we often need to evaluate integral over each element subdomain. Often the integrals contain lengthy complicated expressions and can be very tedious to evaluate. When the natural coordinates are used, exact integrations can be obtained with Eqs. (2.4.120), (2.4.191), and (2.4.252), respectively, for line, triangular, and tetrahedral elements. However, when the local coordinates are used, exact integration is often not easy to obtain because of the involvement of the Jacobian. Under such circumstances, numerical integration is normally used. Several types of numerical integration formulas are available. We shall discuss only the *Gaussian* (or *Gauss-Legendre*) quadrature for lines, squares, and cubes.

In the Gaussian quadrature technique, a definite integral of the function f is approximated by the weighted sum of values of f at selected points. Consider that the following integral formula is sought in the form of

$$\int_{-1}^{1} f(x)dx = w_1 f(x_1) + w_2 f(x_2) + \dots + w_n f(x_n) , \quad (2.4.269)$$

where the weight w_i's and the arguments x_i's are to be determined so as to obtain the best possible accuracy. A formula is said to be of best possible accuracy if it is exact for all polynomials of as high a degree as possible. Since there are altogether 2n arbitrary parameters, we can hope to obtain a formula that is exact for all polynomials up to and including the degree of (2n - 1).

Let us consider first the case n = 2; we are to determine w_1, w_2, x_1, and x_2 such that the formula

$$\int_{-1}^{1} f(x)dx = w_1 f(x_1) + w_2 f(x_2) \quad (2.4.270)$$

is exact for all polynomials of degree less than or equal to 3. Because integration is a linear operation, it is actually sufficient to take for f(x) the special polynomials x^k (k = 0, 1, 2, 3). Thus, letting $f(x) = x^k$ (x = 0, 1, 2, 3) in Eq. (2.4.269) leads to the following four equations for determining w_1, w_2, x_1, and x_2:

$$k = 0; \quad 2 = w_1 + w_2 \quad (2.4.271)$$

$$k = 1; \quad 2 = w_1 x_1 + w_2 x_2 \quad (2.4.272)$$

$$k = 2; \quad \frac{2}{3} = w_1 x_1^2 + w_2 x_2^2 \quad (2.4.273)$$

$$k = 3; \quad 0 = w_1 x_1^3 + w_2 x_2^3 \quad (2.4.274)$$

The solution of Eq. (2.4.271) through (2.4.274) is

$$w_1 = 1; \quad w_2 = 1; \quad x_1^2 = \frac{1}{3}; \quad x_2^2 = \frac{1}{3} \quad (2.4.275)$$

The same procedure can be used to derive the general formula Eq. (2.4.269). We take for f(x) the special polynomials x^k (k = 0, 1, 2, 3, ..., 2n-1). Observing that

$$\int_{-1}^{1} x^k dx = 0 \quad \text{if } k \text{ is odd} \quad (2.4.276)$$

$$\int_{-1}^{1} x^k dx = \frac{2}{k+1} \quad \text{if } k \text{ is even} , \quad (2.4.277)$$

we obtain the following 2n equations:

$$w_1 + w_2 + \ldots + w_n = 2$$

$$w_1x_1 + w_2x_2 + \ldots + w_nx_n = 0$$

$$w_1x_1^2 + w_2x_2^2 + \ldots + w_nx_n^2 = \frac{2}{3} \tag{2.4.278}$$

$$\ldots\ldots\ldots\ldots$$

$$w_1x_1^{2n-1} + w_2x_2^{2n-1} + \ldots + w_nx_n^{2n-1} = 0$$

We have, therefore, a system of 2n nonlinear equations in 2n unknowns w_k's (k = 1, 2, . . . , n) and x_k's (k = 1, 2, . . . , n). The solution of this system by the method used above for the case n > 2 is now too complicated. It can, however, be shown that these equations do possess a unique solution. Using the theory of orthogonal polynomials, we can show that the x_k's are the zeros of a set of polynomials known as the Legendre polynomials. The Gaussian weights w_k's are then determined by solving the system of Eq. (2.4.278).

Since the Legendre polynomials play such an important role in Gaussian quadrature, we shall define these polynomials and state some of the simplest properties. The Legendre polynomials $P_m(x)$ are defined by the following recursion formula:

$$P_0(x) = 1 \quad P_1(x) = x$$

$$P_{m+1}(x) = \frac{1}{m+1}\left[(2m+1)xP_m(x) - mP_{m-1}(x)\right] \tag{2.4.279}$$

The useful properties of these polynomials for our purposes areas follows:

- *Property 1.* $P_m(x)$ is a polynomial of degree m in x. This follows directly from Eq. (2.4.279).

- *Property 2.* The zeros of $P_m(x)$ are all real and distinct and lie on the interval [-1,1]. It is easily verified that the zeros of $P_2(x)$ are $x_1 = -\sqrt{3}/3$, $x_2 = \sqrt{3}/3$ and that the zeros of $P_3(x)$ are $x_1 = -\sqrt{3}/\sqrt{5}$, $x_2 = 0$, $x_3 = \sqrt{3}/\sqrt{5}$.

- *Property 3.* The zeros of $P_m(x)$ are symmetrically placed with respect to the origin. If m is odd, one zero of $P_m(x)$ is always at x = 0.

Using the zeros of the Legendre polynomials and solving the system of Eq. (2.4.278), we list in Table 2.4.2 the locations and weighting coefficients of the Gaussian points for n = 2, 3, and 4.

The two-dimensional analogy of the above procedure requires the evaluation of an integral

$$I = \int_{-1}^{1}\int_{-1}^{1} f(x,y)dxdy \ . \qquad (2.4.280)$$

The integration proceeds stepwise by first considering the inner integral holding y constant, that is,

Table 2.4.2 One-Dimensional Gaussian-Legendre Quadrature Points and Coefficients.

$\pm x_i$	n	w_i	
0.577350269189629	2	1.000000000000000	Linear Element
0.774596669241483	3	0.555555555555556	Quadratic Element
0.000000000000000		0.888888888888889	
0.861136311594053	4	0.347854845137454	Cubic Element
0.339981043584856		0.652145154862546	

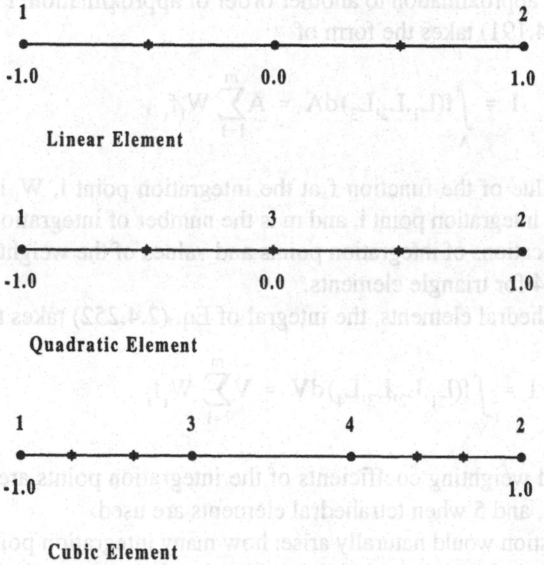

$$\int_{-1}^{1} f(x,y)dx = \sum_{j=1}^{n} w_j f(x_j,y) = F(y) \qquad (2.4.281)$$

The outer integral can now be evaluated:

$$\int_{-1}^{1} F(y)dy = \sum_{i=1}^{n} w_i F(y_j) = \sum_{i=1}^{n}\sum_{j=1}^{n} w_i w_j f(x_y, y_i) \ . \qquad (2.4.282)$$

Thus

$$\int_{-1}^{1}\int_{-1}^{1} f(x,y)dxdy = \sum_{i=1}^{n}\sum_{j=1}^{n} w_i \ w_j f(x_j, y_i) \ , \qquad (2.4.283)$$

where for n = 3, a polynomial up to the fifth degree in both x and y can be integrated exactly. In a similar manner, we can extend the procedure to three-dimensions; thus,

$$\int_{-1}^{1}\int_{-1}^{1}\int_{-1}^{1} f(x,y,z)dxdydz = \sum_{i=1}^{n}\sum_{j=1}^{n}\sum_{k=1}^{n} w_i \ w_j w_k f(x_i, y_j, z_k) \ . \qquad (2.4.284)$$

2.4.5. *Numerical Integration in Natural Coordinate*

Although analytical integrations can be obtained for two-dimensional and three-dimensional natural coordinates, sometimes it is advantageous to evaluate the integration numerically; for example, numerical integration provides easy modification from one-order of approximation to another order of approximation. For triangles, the integral of Eq.(2.4.191) takes the form of

$$I = \int_A f(L_1, L_2, L_3)dA = A\sum_{i=1}^{m} W_i f_i \ , \qquad (2.4.285)$$

where f_i is the value of the function f at the integration point i, W_i is the weighting coefficient for the integration point i, and m is the number of integration points. Table 2.4.3 shows the locations of integration points and values of the weighting coefficients for m = 1, 3, and 4 for triangle elements.

For tetrahedral elements, the integral of Eq. (2.4.252) takes the form of

$$I = \int_V f(L_1, L_2, L_3, L_4)dV = V\sum_{i=1}^{m} W_i f_i \ . \qquad (2.4.286)$$

The locations and weighting coefficients of the integration points are given in Table 2.4.4 for m = 1, 4, and 5 when tetrahedral elements are used.

The question would naturally arise: how many integration points are required for a given finite element approximation with triangle or tetrahedral elements? To answer this question, we should know that the finite element approximation is exact to the order 2(p-d), where p is the order complete polynomial present and d is the order of differentials occurring in the appropriate expression. For C^0 problems, d = 1; thus finite element approximation is exact to the order 0 with linear element (p=1), exact to

Table 2.4.3 Gaussian-Legendre Quadrature Points and Coefficients for Triangular Elements.

m	Order	Error	Points	Coordinate	Weight
1	Linear	$R=O(h^2)$	a	1/3, 1/3, 1/3	1
	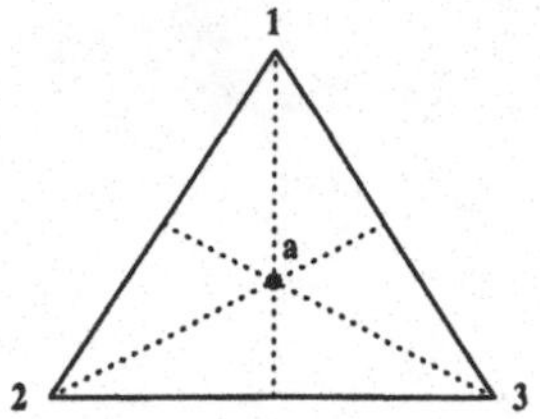				
3	Quadratic	$R=O(h^3)$	a	1/2, 1/2, 0	1/3
	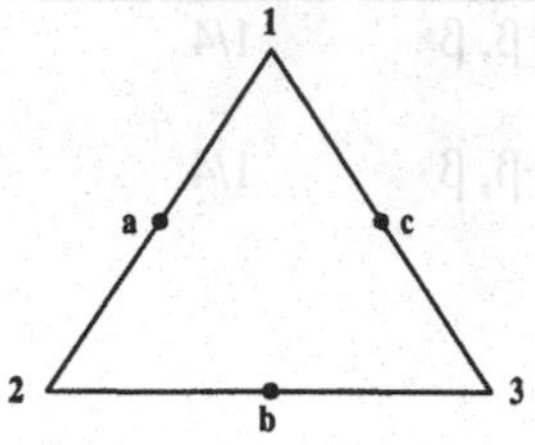		b	0, 1/2, 1/2	1/3
			c	1/2, 0, 1/2	1/3
3	Quadratic	$R=O(h^3)$	a	2/3, 1/6, 1/6	1/3
	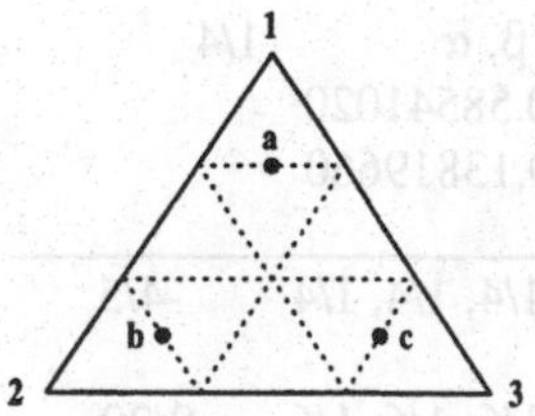		b	1/6, 2/3, 1/6	1/3
			c	1/6, 1/6, 2/3	1/3
4	Cubic	$R=O(h^4)$	a	1/3, 1/3, 1/3	-27/48
	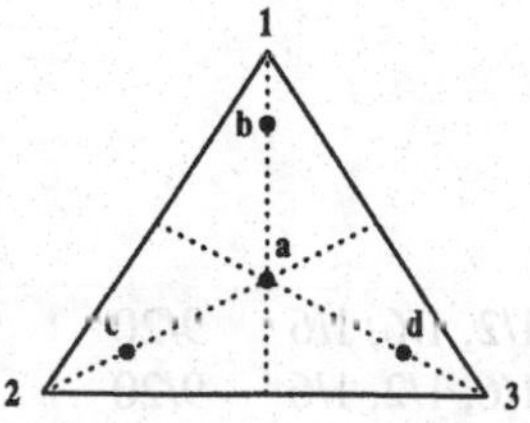		b	3/5, 1/5, 1/5	25/48
			c	1/5, 3/5, 1/5	25/48
			d	1/5, 1/5, 3/5	25/48

Table 2.4.4 Gaussian-Legendre Quadrature Points and Coefficients for Tetrahedral Elements.

m	Order	Error	Points	Coordinate	Weight
1	Linear	$R=O(h^2)$	a	1/4, 1/4, 1/4, 1/4	1

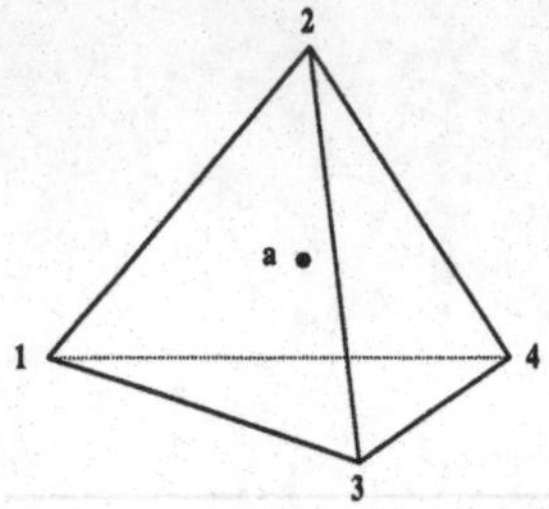

m	Order	Error	Points	Coordinate	Weight
4	Quadratic	$R=O(h^3)$	a	$\alpha, \beta, \beta, \beta$	1/4
			b	$\beta, \alpha, \beta, \beta$	1/4
			c	$\beta, \beta, \alpha, \beta$	1/4
			d	$\beta, \beta, \beta, \alpha$	1/4
				$\alpha = 0.58541020$	
				$\beta = 0.13819660$	

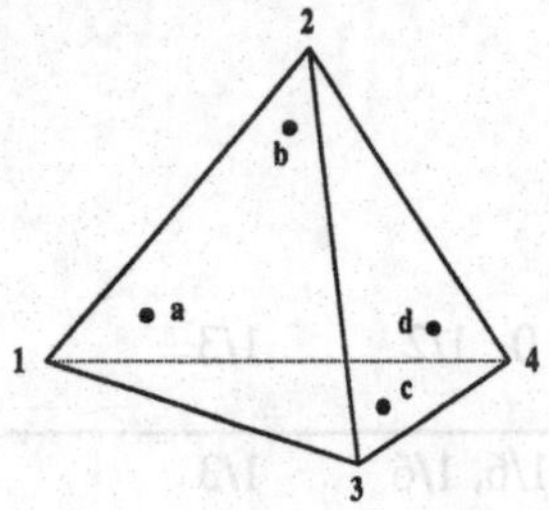

m	Order	Error	Points	Coordinate	Weight
5	Cubic	$R=O(h^4)$	a	1/4, 1/4, 1/4, 1/4	-4/5
			b	1/2, 1/6, 1/6, 1/6	9/20
			c	1/6, 1/2, 1/6, 1/6	9/20
			d	1/6, 1/6, 1/2, 1/6	9/20
			e	1/6, 1/6, 1/6, 1/2	9/20

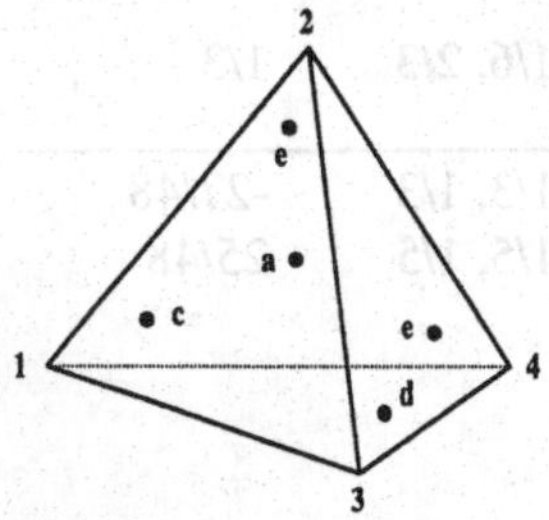

the order 2 with quadratic element (p=2), and exact to the order 4 with cubic elements (p=3). Therefore, if we choose m such that the integration is exact to the order 2(p-d), we will not have loss of convergence due to numerical integration. We can use this fact to choose appropriate m.

2.5. Boundary-Element Methods

The boundary-element method (BEM) came into wide use during the 1970s. Although it is considered by some to be a subset of the finite-element method (FEM), there appears to be sufficient differences that it should be classified as a separate technique. Although the BEM can be formulated to solve a variety of problems, its advantageous use is not as wide as the FEM. Its advantage is not that it is a powerful technique capable of solving the most complex problem but that it is extremely efficient. The efficiency of the BEM stems from the fact that it is a boundary method and thus does not require discretization of the entire domain as in the FEM or finite-difference methods (FDM). The dimension of the computational domain is reduced by one; a two-dimensional problem is solved by a line integration and a three-dimensional problem is solved by an area integration. This efficiency is contingent on one's ability to obtain fundamental solutions to the original two- and three-dimensional problems though. Failure to obtain fundamental solutions would make the reduction of dimensionality impossible. Therefore, application of BEM to real-world problems would remain limited.

2.5.1. The Basic Method

The BEM stems from a simple application of basic calculus, the well-known divergence theorem

$$\int_R \nabla\cdot\mathbf{F}dR = \int_A \mathbf{F}\cdot\mathbf{n}dA\ , \tag{2.5.1}$$

in which **F** is a vector in region R with boundary surface A. The vector **n** is the outward unit normal to R. Using first $\mathbf{F} = U\nabla V$ and second $\mathbf{F} = V\nabla U$ in Eq. (2.5.1), we may produce

$$\int_R \nabla\cdot(U\nabla V)dR = \int_A U\nabla V\cdot\mathbf{n}dA \tag{2.5.2}$$

$$\int_R \nabla\cdot(V\nabla U)dR = \int_A V\nabla U\cdot\mathbf{n}dA\ . \tag{2.5.3}$$

Subtracting Eq. (2.5.2) from Eq. (2.5.3) produces Green's second identity:

$$\int_R \left(U\nabla^2 V - V\nabla^2 U\right)dR = \int_A \left(U\frac{\partial V}{\partial n} - V\frac{\partial U}{\partial n}\right)dA\ , \tag{2.5.4}$$

where $\partial V/\partial n = \nabla V \cdot \mathbf{n}$. If U and V are chosen so that they are harmonic, the domain integral vanishes, leaving

$$\int_A \left(U\frac{\partial V}{\partial n} - V\frac{\partial U}{\partial n} \right) dA = 0 \; . \tag{2.5.5}$$

For potential problems it is convenient to associate U with the potential Φ and take V as a singular solution. Usually V = ln(r) in two-dimensions and V = 1/r in three dimensions where r is the distance from an arbitrary point P to the boundary element dA. In order to include both two- and three-dimensional cases, let

$$G = \begin{cases} \ln(r) & \text{in two-dimension} \\ \dfrac{1}{r} & \text{in three-dimension} \; . \end{cases} \tag{2.5.6}$$

Then

$$\int_A \left(\phi\frac{\partial G}{\partial n} - G\frac{\partial \phi}{\partial n} \right) dA = 0 \; . \tag{2.5.7}$$

In order to find the solution at some point P, the point is excluded from R by a circle (2-D) or sphere (3-D), which constitutes part of the boundary A, A = a + S as shown in Figure 2.5.1. Taking the radius of the circle or sphere to zero with P remaining in the center produces

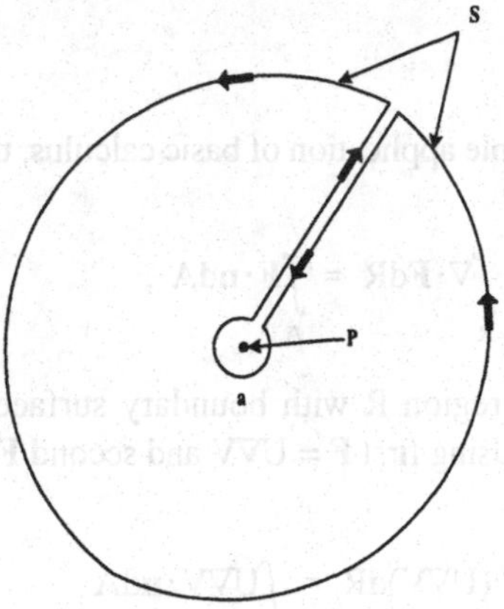

Fig. 2.5.1 Integration Path
Around a Singular Point.

$$-\alpha\phi_P = \int_S \left(\phi\frac{\partial G}{\partial n} - G\frac{\partial \phi}{\partial n} \right) dS \; , \tag{2.5.8}$$

in which $\alpha = 2\pi$ in two dimensions or $\alpha = 4\pi$ in three dimensions.

Equation (2.5.8) indicates that the solution, Φ_P, at any point can be formed from a simple line integration around the boundary provided that Φ and $\partial\Phi/\partial n$ are known everywhere on the boundary. Unfortunately, a well-posed potential flow problem has as boundary conditions either Φ or $\partial\Phi/\partial n$ (or a relationship between them) at each boundary point but not both. The "missing" condition can be found directly from Eq. (2.5.7) or Eq. (2.5.8). Using Eq. (2.5.8) and taking P to the boundary produces

$$-\beta\phi_b = \int_S \left(\phi\frac{\partial G}{\partial n} - G\frac{\partial \phi}{\partial n} \right) dS \ , \qquad (2.5.9)$$

where ß is the angle of the boundary formed inside the domain R. For a straight line (in 2D) or plane (in 3D), ß is equal to π or 2π.

2.6. Solution of Simultaneous Linear Algebraic Equations

Each numerical approximation leads to a system of algebraic equations for the discretized domain. The system of algebraic equations written in matrix form is

$$[A]\{x\} = \{b\} \ , \qquad (2.6.1)$$

where [A] is a matrix containing coefficients related to grid discretization and media properties, such as transmisivity, hydraulic conductivity, storativity, porosity, dispersivity, etc.; {x} is a column vector containing the dependent variables such as head, concentration, temperature, etc.; and {b} is a column vector containing all known information, for example, specified pumpage and boundary and initial condition information. Although solving the matrix equation is a mathematical problem, we must be aware of its important aspects, since generally the solution of matrix equations is the most expensive part of the computer codes. In general, Eq. (2.6.1) can be solved numerically by one of two basic ways: (1) direct and (2) iteration. In general iterative techniques are more efficient than direct solution techniques for large problems, but they may not yield convergent solutions. The relative merits of direct and iterative methods are given in Table 2.6.1. It should be pointed out that for some problems, the matrix [A] does not have to be decomposed each time step. For such cases, the efficiency of direct methods is improved considerably.

2.6.1. Direct Methods to Solve Matrix Equations

In direct methods, a sequence of operations is performed only once, providing a solution that is exact, except for round-off error. Direct methods can further be divided into (1) solution by determinants (Cramer rules), (2) solution by matrix inversion, and (3) solution by successive elimination of the unknowns. The first two categories of methods are not practical. Thus, we will focus our attention here on the class of solution algorithms based on the efficient elimination methods of C. F. Gauss.

2.6.1.1. Gaussian Elimination Method. Many direct methods for solving linear algebraic equations are variants of the Gaussian elimination. Although they are algebraically identical, they differ in how matrices are stored, the elimination sequence, or the approach taken to minimize large round-off errors. The following example illustrates the Gaussian elimination:

Table 2.6.1 Advantages and Disadvantages of Direct and Iterative Methods.

Methods	Advantages	Disadvantages
Direct	Sequence of operations only performed once No initial estimates required No iteration parameters required No tolerance required	May be inefficient in terms of CPU storage and time for large problems Can have round-off errors
Iterative	Efficient in terms of CPU storage and time for large problems	Required initial estimates Required iteration parameters Required tolerance Matrix must be well conditioned

$$4x_1 + 4x_2 + 4x_3 = 16 \tag{2.6.2}$$

$$3x_1 + 4x_2 + x_3 = 8 \tag{2.6.3}$$

$$6x_1 + 9x_2 + 3x_3 = 21 \tag{2.6.4}$$

Use Eq. (2.6.2) to eliminate x_1 from Eqs. (2.6.2) and (2.6.3):

$$(2.6.3) - (2.6.2)\times\frac{3}{4}: \quad x_2 - 2x_3 = -4 \tag{2.6.5}$$

$$(2.6.4) - (2.6.2)\times\frac{3}{2}: \quad 3x_2 - 3x_3 = -8\ . \tag{2.6.6}$$

Use Eq. (2.6.5) to eliminate x_2 from Eq. (2.6.6):

$$(2.6.6) - (2.6.5)\times 3: \quad 3x_3 = 9 \tag{2.6.7}$$

Now we use Eqs. (2.6.7), (2.6.5), and (2.6.2) solving backward to get $x_3 = 3$, $x_2 = 2$, $x_1 = -1$.

The above elementary method has a much more elegant description in terms of matrices. The coefficient matrix,

$$[A] = \begin{bmatrix} 4 & 4 & 4 \\ 3 & 4 & 1 \\ 6 & 9 & 3 \end{bmatrix} \qquad (2.6.8)$$

is decomposed as [A] = [L][U], where

$$[L] = \begin{bmatrix} 1 & 0 & 0 \\ \ell_{21} & 1 & 0 \\ \ell_{31} & \ell_{32} & 1 \end{bmatrix} \qquad (2.6.9)$$

$$[U] = \begin{bmatrix} u_{11} & u_{12} & u_{13} \\ 0 & u_{22} & u_{23} \\ 0 & 0 & u_{33} \end{bmatrix}. \qquad (2.6.10)$$

[L] is the lower triangular with unit diagonal, and [U] is upper triangular. The generation to nxn matrices is obvious. In our specific example,

$$[L] = \begin{bmatrix} 1 & 0 & 0 \\ \frac{3}{4} & 1 & 0 \\ \frac{3}{2} & 3 & 1 \end{bmatrix}, \quad [U] = \begin{bmatrix} 4 & 4 & 4 \\ 0 & 1 & -2 \\ 0 & 0 & 3 \end{bmatrix} \qquad (2.6.11)$$

[U] is the matrix of the coefficients of Eqs. (2.6.2), (2.6.5), and (2.6.7), and [L] is the matrix of the multiples of the original equations used to get Eqs. (2.6.2), (2.6.5), and (2.6.7).

The problem [A]{x} = {b} is easier to solve as [L]([U]{x}) = {b} by writing the system as

$$[L]\{y\} = \{b\}, \quad [U]\{x\} = \{y\} \qquad (2.6.12)$$

each of which is trivial to solve by forward substitution for the first equation of Eq. (2.6.12) and backward substitution for the second equation of Eq. (2.6.12). Consider the first part, [L]{y} = {b} as above:

$$\begin{aligned} y_1 \qquad\qquad &= 16 \\ \frac{3}{4}y_1 + y_2 \qquad &= 8 \\ \frac{2}{3}y_1 + 3y_2 + y_3 &= 21 \end{aligned}, \qquad (2.6.13)$$

which yields

$$y_1 = 16, \quad y_2 = -4, \quad y_3 = 9 , \tag{2.6.14}$$

which are simply the right sides of Eqs. (2.6.2), (2.6.5), and (2.6.7). The second part of the system [U]{x} = {y} is the same as Eqs. (2.6.2), (2.6.5), and (2.6.7).

2.6.1.2. LU Decomposition of a General Matrix. The basis for Gaussian elimination is that any nonsingular matrix [A] can be decomposed as

$$[A] = \left[a_{ij}\right] = [L][U] , \tag{2.6.15}$$

where

$$[L] = \left[\ell_{ik}\right], \quad \ell_{ii} = 1, \quad \ell_{ik} = 0 \text{ if } i < k \tag{2.6.16}$$

and

$$[U] = \left[u_{kj}\right], \quad u_{kj} = 0 \text{ if } k > j . \tag{2.6.17}$$

In fact, we can write

$$a_{ij} = \sum_{k=1}^{n} \ell_{ik} u_{kj} = \sum_{k=1}^{\min(i,j)} \ell_{ik} u_{kj} \tag{2.6.18}$$

because of the triangular condition on ℓ_{ij} and u_{kj}.

We want to determine ℓ_{ik} and u_{kj}. Table 2.6.2 shows the order of computation using the known a_{ij}'s. Each line computes one unknown (underlined); the remaining terms were previously computed. The total operation count for all rows is

$$\text{Total Operation Count} = \sum_{i=1}^{n} (i-1)(n - \frac{i}{2} + 1) = \frac{1}{3} n(n^2 - 1) . \tag{2.6.19}$$

The operation count refers to the total number of multiplications and divisions required to solve for one unknown in that row. The row operational count is the total number of operations required for that whole row in Table 2.6.2. The total operation count, the crucial figure for determining computer running time, is the actual number of operations necessary to compute the LU decomposition.

But we have not found the solution yet. We must solve [L]{y} = {b}, which expands as

$$y_i + \sum_{j=1}^{i-1} \ell_{ij} y_j = b_i, \quad i = 1, 2, \ldots, n . \tag{2.6.20}$$

Writing Eq. (2.6.20) in tabular form, we obtain Table 2.6.3. And then we solve for [U]{x} = {y} or

$$\sum_{j=1}^{n} u_{ij} x_j = y_i . \tag{2.6.21}$$

Table 2.6.2 Computation of ℓ_{ik} and u_{kj} and Operational Counts.

Row	i	j	Equation	OC	Row Op Count
1	1	any	$a_{1j}=u_{1j}$	0	0
2	2	1	$a_{21}=\ell_{21}u_{11}=\underline{\ell}_{21}a_{11}$	1	1
2	2	≥ 2	$a_{2j}=\ell_{21}u_{1j}+\underline{u}_{2j}$	1	n-1
2	TOTAL				n
3	3	1	$a_{31}=\ell_{31}u_{11}=\underline{\ell}_{31}a_{11}$	1	1
3	3	2	$a_{32}=\ell_{31}u_{12}+\underline{\ell}_{32}u_{22}$	2	2
3	3	≥ 3	$a_{3j}=\ell_{31}u_{1j}+\ell_{32}u_{22}+\underline{u}_{3j}$	2	2(n-2)
3	TOTAL				2(n-1)
i	i	1	$a_{i1}=\ell_{i1}u_{11}=\underline{\ell}_{i1}a_{11}$	1	1
i	i	2	$a_{i2}=\ell_{i1}u_{12}+\underline{\ell}_{i2}u_{22}$	2	2
			.		
			.		
i	i	i-1	$a_{ij}=\ell_{i1}u_{1j}+\ell_{i2}u_{2j}+\,.\,\ell_{i,i-1}\underline{u}_{i-1,j}$	i-1	i-1
i	i	≥ i	$a_{ij}=\ell_{i1}u_{1j}+\ell_{i2}u_{2j}+\ldots+\underline{u}_{ij}$	i-1	(i-1)(n-i+1)
i	TOTAL				(i-1)(n-i/2+1)

Table 2.6.3 Operation Count for Forward Substitution.

Row	Equation	Op Count
1	$y_1 = b_1$	0
2	$y_2 = b_2 - \ell_{21}y_1$	1
3	$y_3 = b_3 - \ell_{31}y_1 - \ell_{32}y_2$	2
i		i-1
TOTAL		$\frac{1}{2}n(n-1)$

Writing Eq. (2.6.21) as a table, we obtain Table 2.6.4. This is a larger total than previously because of the n extra divisions required by u_{ii} not being unity. Thus once we have found [L] and [U], the solution for a specific {b} requires n^2 further operations. If we solve the same system repetitively with different b's,

$$[A]\{x^{(k)}\} = \{b^{(k)}\}, \qquad k=1,2,\ldots, m. \tag{2.6.22}$$

then we need only one LU decomposition requiring $n(n^2-1)/3$ operations and m individual solutions requiring n^2 steps each. The total effort is $[n(n^2-1) + mn^2]$ operations. Most of the time use comes from the $n^3/3$ in the LU decomposition, unless m is large.

Table 2.6.4 Operation Count for Backward Substitution.

Row	Equation	Op Count
n	$u_{nn} x_n = y_n$	1
n-1	$u_{n-1,n-1}x_{n-1} + u_{n-1,n}x_n = y_{n-1}$	2
i		n+1-i
TOTAL		$\frac{1}{2}n(n+1)$

2.6.1.3. Computer Implementation. The procedure we have been studying is formally rigorous, but what happens when it is implemented on a computer? There will be trouble for many occasions! Consider the following example worked on a machine that rounds to ten decimal digits:

$$3\cdot 10^{-11} x_1 + x_2 = 7 \tag{2.6.23}$$

$$x_1 + x_2 = 9\,. \tag{2.6.24}$$

Using the standard elimination, we take

$$\begin{aligned} &(2.6.24) - 3.333333333\cdot 10^{10} \times (2.6.23): \\ &(1 - 3.333333333\cdot 10^{10})x_2 = 9 - 2.333333333\cdot 10^{11}\,, \end{aligned} \tag{2.6.25}$$

which leads

$$x_2 = 7 \tag{2.6.26}$$

to 10 decimal places because 1 and 9 are lost in the roundoff. Using the back substitution into Eq. (2.6.23), we obtain $x_1 = 0$. But this is wrong! It is not consistent with (2.6.24); in fact, by inspection, we see that x_1 is approximately equal to 2.

The wrong answer is caused by using a small "pivot" (the number here on the diagonal, which is used to eliminate the column). This invariably leads to trouble. To remedy this problem we have two options to avoid small pivots:

- *Full Pivoting*. Always use the largest remaining coefficient in the matrix for the next elimination. For example, if the matrix for the next elimination is as follows:

$$\begin{bmatrix} 5 & 0 & 9 \\ 4 & 7 & 1 \\ 6 & 0 & 4 \end{bmatrix}, \tag{2.6.27}$$

  then x_2 is already eliminated, so we choose the largest coefficient not in the second row or column. Thus, we use the 9 to eliminate the 4. In Eqs. (2.6.23) and (2.6.24), we may use x_2 instead of x_1 as the pivot, yielding $x_1 = 2$ and $x_2 = 7$ to ten places. We could also pivot around x_1 in Eq.(2.6.24) with the same result. Note that full pivoting eliminates the unknowns in random order.

- *Partial Pivoting*. Eliminate $x_1, ..., x_n$ in order, but pick the largest element from column k to eliminate x_k. Partial pivoting would use x_1 in Eq. (2.6.24) as the pivot. This would result in the correct solution of $x_1 = 2$ and $x_2 = 7$.

Full pivoting works for any matrix the computer can solve. Partial pivoting may still fail. Suppose instead of Eqs. (2.6.23) and (2.6.24), we had the equivalent equation

$$3{\cdot}10^{-11}x_1 + x_2 = 7 \tag{2.6.28}$$

$$3{\cdot}10^{-11}x_1 + 3{\cdot}10^{-11}x_2 = 2.7{\cdot}10^{-10}\ . \tag{2.6.29}$$

Using x_1 in Eq. (2.6.28) as the pivot we get to the ten places

$$-x_2 = -7 \tag{2.6.30}$$

from which

$$x_2 = 7 \quad \text{and} \quad x_1 = 0\ . \tag{2.6.31}$$

Wrong again. To avoid this problem, we treat the system prior to starting the elimination, as follows:

- *Row Equilibration*. Multiply each row by a factor so that its largest coefficient is unity. This ensures that no line like (2.6.29) has all small coefficients.

- *Column Equilibration*. Define new variables

$$\bar{x}_k = c_k x_{ck} , \tag{2.6.32}$$

where c_k is chosen so that the largest coefficient in each column of the equation for $\bar{x}$ is unity, and solve for $\bar{x}$.

Column equilibration will define

$$\overline{x_1} = 3\cdot 10^{-11} x_1, \quad \text{and} \quad \overline{x_2} = x_2 \tag{2.6.33}$$

so that Eqs. (2.6.28) and (2.6.29) become

$$\overline{x_1} + \overline{x_2} = 7 \tag{2.6.34}$$

$$\overline{x_1} + 3\cdot 10^{-11}\overline{x_2} = 2.7\cdot 10^{-10} \tag{2.6.35}$$

After elimination, we obtain solutions of Eqs. (2.6.34) and (2.6.35) of $\bar{x}_1 = 0$ and $\bar{x}_2 = 7$ (which is correct to ten places!) but wrong when we convert to x_1 and x_2. However, if we always use partial pivoting with row and/or column equilibration, it works nearly as well as full pivoting. For example, the row equilibration will restore Eqs. (2.6.28) and (2.6.29) to Eqs. (2.6.23) and (2.6.24), which would yield a correct solution when a partial pivoting is used.

2.6.1.4. The Crout Method. Gaussian elimination has been shown to be one form of the LU decomposition. An important variant of this fundamental approach is a compact Crout method. The principal advantage of the Crout method is that the intermediate reduced matrices need not to be recorded. As a result the elements ℓ_{ij} and u_{ij} can be calculated in a single machine operation by a continuous accumulation of products. To see how the Crout method works, it is convenient to consider the coefficient matrix to be divided into three parts; part one being the region that is fully reduced, part two the region that is currently being reduced (called active zone), and part three the region that contains the original unreduced coefficients. These regions are shown in Figure 2.6.1. The algorithm for Crout method is as follows:

$$u_{11} = a_{11}, \qquad \ell_{11} = 1 . \tag{2.6.36}$$

For each active zone j from 2 to n

$$\ell_{j1} = \frac{a_{j1}}{u_{11}}, \qquad U_{1j} = a_{1j} . \tag{2.6.37}$$

Then

$$\ell_{ji} = \left(a_{ji} - \sum_{m=1}^{i-1} \ell_{jm} u_{mi} \right) / u_{ii} , \quad u_{ij} = \left(a_{ij} - \sum_{m=1}^{i-1} \ell_{im} u_{mj} \right) \tag{2.6.38}$$

$$i = 1, 2, \ldots, j-1$$

and finally

$$\ell_{jj} = 1, \qquad u_{jj} = \left(a_{jj} - \sum_{m=1}^{j-1} \ell_{jm} U_{mj} \right) . \tag{2.6.39}$$

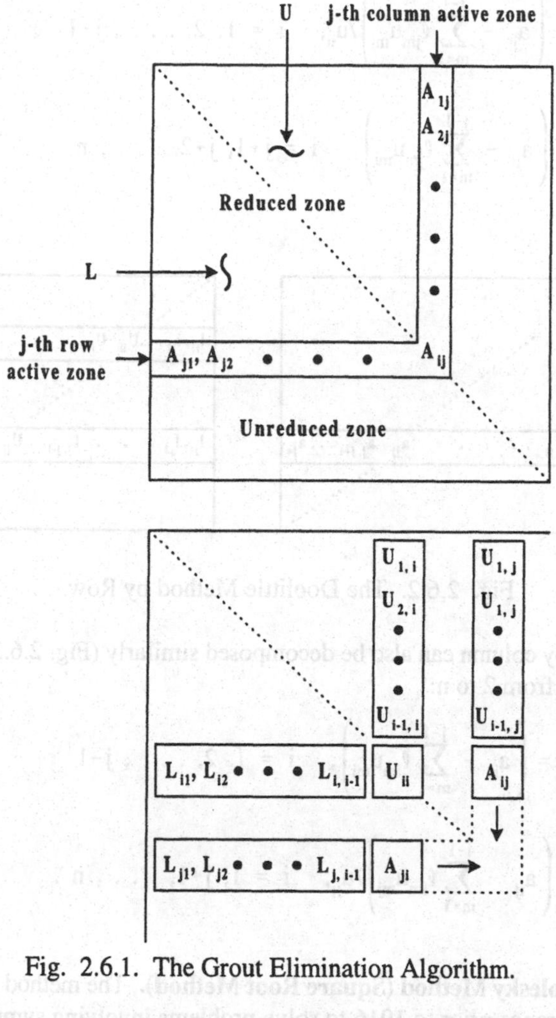

Fig. 2.6.1. The Grout Elimination Algorithm.

The ordering of the reduction process and the terms used are shown in Figure 2.6.1. Equations (2.6.36) through (2.6.39) give the computation procedure. Graphically from Figure 2.6.1, the computation of any L_{ji} along the j-th row is A_{ji} minus the summation of the product of each of the terms up to the immediately left term to A_{ji} on the same row as A_{ji} starting from the left-most term with the corresponding term on the same column starting from the top-most term, then divide the result by the diagonal term on the i-th column U_{ii}. The computation of any U_{ij} on the j-th column can be done similarly with the word column replaced by row and row replaced by column, but the final division by L_{ii} is not necessary because all L_{ii}'s are 1.

2.6.1.5. The Doolittle Method. The Doolittle method is a minor modification of the Crout algorithm designed to facilitate the forward elimination step when the coefficient matrix [A] is stored by rows. In this scheme only the j-th rows of [L] and [U] are generated at the j-th stage. The computation proceeds from left to right along each row (Fig. 2.6.2). For the j-th row, the formula is for j from 2 to n:

$$\ell_{ji} = \left(a_{ji} - \sum_{m=1}^{i-1} \ell_{jm} u_{mi} \right) / u_{ii}, \quad i = 1, 2, \ldots, j-1 \tag{2.6.40}$$

$$u_{ji} = \left(a_{ji} - \sum_{m=1}^{j-1} \ell_{jm} u_{mi} \right), \quad i = j+1, j+2, \ldots, n . \tag{2.6.41}$$

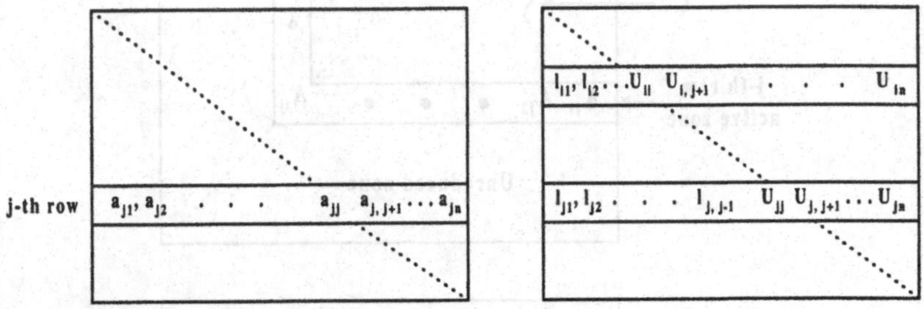

Fig. 2.6.2. The Doolittle Method by Row.

A matrix stored by column can also be decomposed similarly (Fig. 2.6.3). For the j-th column, j ranges from 2 to n:

$$u_{ij} = \left(a_{ij} - \sum_{m=1}^{j-1} \ell_{im} u_{mj} \right), \quad i = 1, 2, \ldots, j-1 \tag{2.6.42}$$

$$\ell_{ij} = \left(a_{ij} - \sum_{m=1}^{i-1} \ell_{im} u_{mj} \right) / u_{jj}, \quad i = 1, j+1, \ldots, n . \tag{2.6.43}$$

2.6.1.6. The Cholesky Method (Square Root Method). The method was used by A. L. Cholesky in France prior to 1916 to solve problems involving symmetric matrices (Salvadori and Baron, 1961). It is sometimes called the *Banachiewicz method* after T. Banachiewicz, who derived the method in matrix form in Poland in 1938. The method is applicable to symmetric positive definite matrices, and for such matrices is the best method for triangular decomposition (Wilkinson, 1965). Positive definite implies the nonsingularity of the minors necessary for the LU decomposition. The symmetric positive definite matrices can be decomposed in the form

$$[A] = [L][L]^T . \tag{2.6.44}$$

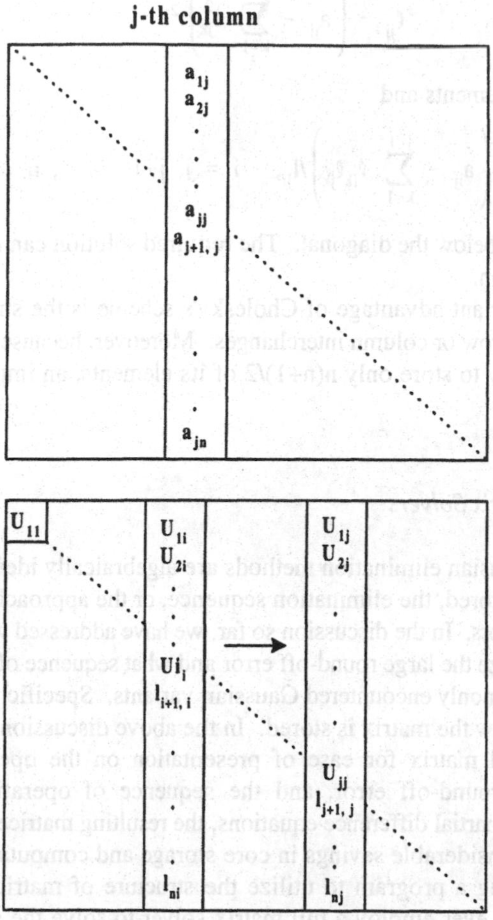

Fig. 2.6.3 The Doolittle Method by Column.

where [L] is a lower triangular matrix with positive diagonal elements. Expansion of Eq. (2.6.44) in terms of its individual elements yields for the main diagonal

$$a_{jj} = \ell_{j1}^2 + \ell_{j2}^2 + \cdots + \ell_{jj}^2 \tag{2.6.45}$$

and for a_{ij} below the main diagonal

$$a_{ij} = \ell_{i1}\ell_{j1} + \ell_{i2}\ell_{j2} + \cdots + \ell_{ij}\ell_{jj}, \quad (i > j) \ . \tag{2.6.46}$$

We can solve Eqs. (2.6.45) and (2.6.46) for the element of [L] provided all elements for the k-th column are obtained, beginning with the diagonal, before proceeding to the (k+1)th. The appropriate formula for j = 1 to n-th column is

$$\ell_{jj} = \left(a_{jj} - \sum_{k=1}^{j-1} \ell_{jk}^2 \right)^{\frac{1}{2}} \tag{2.6.47}$$

for the diagonal elements and

$$\ell_{ij} = \left(a_{ij} - \sum_{k=1}^{j-1} \ell_{ik}\ell_{jk} \right) / l_{jj}, \quad i = j, j+1, \ldots, n \tag{2.6.48}$$

for those elements below the diagonal. The required solution can now be obtained by forward substitution.

An important advantage of Cholesky's scheme is the small round-off error generated without row or column interchanges. Moreover, because of the symmetry of [A], it is necessary to store only n(n+1)/2 of its elements, an important reduction in computer storage.

2.6.2. Direct Matrix Solvers

All variants of Gausian elimination methods are algebraically identical; they differ in how matrices are stored, the elimination sequence, or the approach taken to minimize large round-off errors. In the discussion so far, we have addressed what approaches can be taken to minimize the large round-off error and what sequence of elimination is done with the most commonly encountered Gaussian variants. Specific references have not been made as to how the matrix is stored. In the above discussions, we have used the structure of a full matrix for ease of presentation on the operational count, the minimization of round-off error, and the sequence of operation. In numerical approximations to partial difference equations, the resulting matrices are almost always not full. Thus, considerable savings in core storage and computational effort can be achieved by coding a program to utilize the structure of matrices. Of course, no practitioner would ever employ a full matrix solver to solve the system of algebraic equations resulting from finite-element or finite difference-approximations to partial differential equations. Full matrix solvers remain in textbooks used as teaching material (Wang and Anderson, 1982) for students to understand the procedures. Practical matrix solving codes generally fall into one the following categories: (1) fixed-band-matrix solvers, (2) variable-band-matrix solvers, (3) sparse-matrix solvers, (4) frontal methods, and (5) out-of-core methods. The advantages and disadvantages of these solveers have been discussed elsewhere (Yeh, 1985). In the following, we will briefly discuss how the matrix is stored for each solution method.

2.6.2.1. Fixed Band-Matrix Solvers. Matrices generated using finite-element or finite difference-methods are generally sparse and banded. Because of simplicity in assembling the coefficient matrix in fix banded form and easy access to band-matrix solver software, this type of matrix solver has been used extensively; many industrial codes have employed fixed band-matrix solvers. Only the terms within a nonzero fixed band are stored in this scheme. A typical storage method of fixed band-matrix solvers is given in Figure 2.6.4. In this figure, all nonzero diagonal entries are marked with a

D, all nonzero off-diagonal entries are marked with an X, all zero entries that may become nonzero after elimination are marked with a ?, and all zero entries that would still remain zero after elimination are marked with 0. All other entries outside the band are initially zero and remain zero throughout the solution process.

Storage can be reduced from n^2 words for the full matrix to $n(2m+1)$ (where m is the maximum difference between node number for any element and $2m + 1$ is the band width) words for the banded matrix. For example, the storage in Figure 2.6.4 is reduced from 400 to 260. Decomposition can be reduced from $n^3/3$ operations for the full matrix to nm^2 operations for the banded matrix.

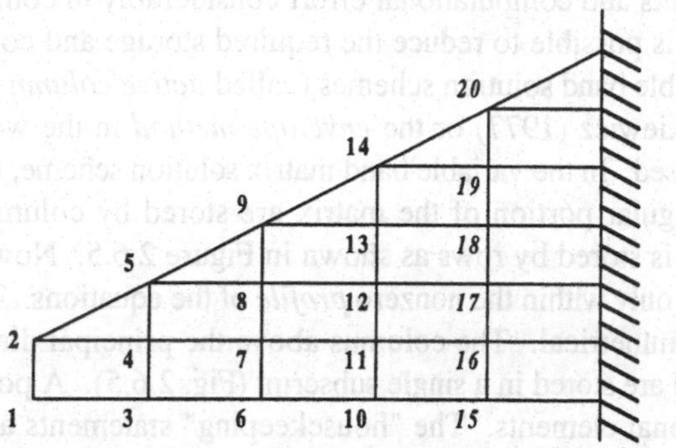

Fig. 2.6.4. Example for a Fixed Band-Matrix Storage Scheme.

```
   1 2 3 4 5 6 7 8 9 0 1 2 3 4 5 6 7 8 0 0
1  D X X X 0 0 0
2  X D X X X 0 0 0
3  X X D X ? X X 0 0
4  X X X D X X X X 0 0
5  0 X ? X D ? X X X 0 0
6  0 0 X X ? D X ? ? X X 0
7  0 0 X X X X D X ? X X X 0
8    0 0 X X ? X D X ? X X X 0
9      0 0 X ? ? X D ? ? X X X 0
0        0 0 X X ? ? D X ? ? ? X X
1          0 X X X ? X D X ? ? X X X
2            0 X X X ? X D X ? ? X X X
3              0 X X ? ? X D X ? ? X X X
4                0 X ? ? ? X D ? ? ? X X X
5                  0 X X ? ? ? D X ? ? ? ?
6                    X X X ? ? X D X ? ? ?
7                      X X X ? ? X D X ? ?
8                        X X X ? ? X D X ?
9                          X X ? ? ? X D X
0                            X ? ? ? ? X D
```

The mapping from full matrix to banded matrix is accomplished by the following simple formula:

$$a_{ij} \dashrightarrow b_{i,j-i+m}, \qquad (2.6.49)$$

which indicates that the complication in the assembling routine is minor. The use of banded storage could, nevertheless, complicate the routine that solves the matrix equation, but it is well within the capacity of an average engineering programmer.

2.6.2.2. Variable Band-Matrix Solvers. Fixed band-matrix solvers have reduced storage requirements and computational effort considerably in comparison to the full matrix solvers. It is possible to reduce the required storage and computational effort still further if variable band solution schemes (called *active column profile solution* in the work by Zienkiewicz (1977) or the *envelope method* in the work by Pinder and Gray (1977)) are used. In the variable band matrix solution scheme, the necessary parts of the upper triangular portion of the matrix are stored by columns, and the lower triangular portion is stored by rows as shown in Figure 2.6.5. Now it is necessary to store and compute only within the nonzero *profile* of the equations. The profiles of the matrix must be symmetrical. The columns above the principal diagonal or the rows below the diagonal are stored in a single subscript (Fig. 2.6.5). A pointer array is used to locate the diagonal elements. The "housekeeping" statements associated with the variable storage scheme are much more complicated than those required with a fixed banded-storage scheme. The matrix equation solution routine would also become much more complicated than that using a fixed banded-storage scheme, but one can easily pull out a solution routine from the book by Zienkiewicz (1977).

This method of storage has definite advantage over fixed banded storage. First, it always requires less storage (unless the matrix is diagonal!); second, the storage requirements are not severely affected by a few very long columns; and last, it is very easy to use a vector dot product routine to effect the triangular decomposition and forward reduction. This last fact is extremely important to modern machines, which are vector oriented. The only disadvantage of this method compared to the fixed band-matrix solution scheme is the increase in programming complexity (Livesley, 1983), but it still should be within the capacity of an average programmer.

2.6.2.3. Sparse Matrix Solvers. Sparse matrix solvers can be used to greatly reduce initial storage requirements. In this type of solution scheme, only the nonzero elements of the matrix need to be stored. The easiest way to store such elements is by means of a row and column index, so that each entry, in a one-dimensional array containing all nonzero elements of the matrix, is associated with a column number and a row number. However, during the decomposition processes, some of the zero entries will be filled in with nonzero entries (Duff, 1977). Means of minimizing the "fill-ins" have been explored at the expense of considerable program complexity and substantial "bookkeeping" (Tewarson, 1973).

k	C_k
1	a_{11}

2	a_{12}
3	a_{22}

4	a_{13}
5	a_{23}
6	a_{33}

7	a_{14}
.	
.	
16	a_{66}

17	a_{67}
18	a_{77}

19	a_{18}
20	a_{28}
21	a_{38}
22	a_{48}
23	a_{58}
24	a_{68}
25	a_{78}
26	a_{88}

j	JD_j
1	1
2	3
3	6
4	10
5	13
6	16
7	18
8	26

$$
\begin{matrix}
a_{11} & a_{12} & a_{13} & a_{14} & & & & a_{18} \\
a_{21} & a_{22} & a_{23} & a_{24} & & & & a_{28} \\
a_{31} & a_{32} & a_{33} & a_{34} & a_{35} & & & a_{38} \\
a_{41} & a_{42} & a_{43} & a_{44} & a_{45} & a_{46} & & a_{48} \\
 & & & a_{54} & a_{55} & a_{56} & & a_{58} \\
 & & & a_{64} & a_{65} & a_{66} & a_{67} & a_{68} \\
 & & & & & a_{76} & a_{77} & a_{78} \\
a_{81} & a_{82} & a_{83} & a_{84} & a_{85} & a_{86} & a_{87} & a_{88}
\end{matrix}
$$

i	1	2	3	4	5	6	7	8
ID_i	1	3	6	10	13	16	18	26

k	1	2 3	4 5 6	7 ... 16	17 18	19 20 .. 26
R_k	a_{11}	a_{21} a_{22}	a_{31} a_{32} a_{33}	a_{41} ... a_{66}	a_{76} a_{78}	a_{81} a_{82} .. a_{88}

Fig. 2.6.5 Variable Band-Matrix Storage Scheme.

2.6.2.4. Frontal Methods. The frontal methods have been successfully employed in finite-element solution packages for structural engineering problems (Melosh and Bamford, 1969; Mondkar and Powell, 1974; Irons, 1970; Hood, 1976; Hinton and Owen, 1977). It appears that the application of frontal methods to subsurface hydrological problems are lacking.

The frontal method is essentially a version of Gaussian elimination in which the coefficients in the equations are assembled and processed in a sequence determined by the advance of a wave-front through the finite-element mesh. In contrast to the methods described previously, the sequence followed by the frontal method is determined by the order in which the elements are specified, rather than by the order in which the nodes are numbered. The advantage of the method lies in the fact that at no time during the calculation is the complete coefficient matrix present in the computer store. For the example shown in Figure 2.6.6, the procedure is as follows.

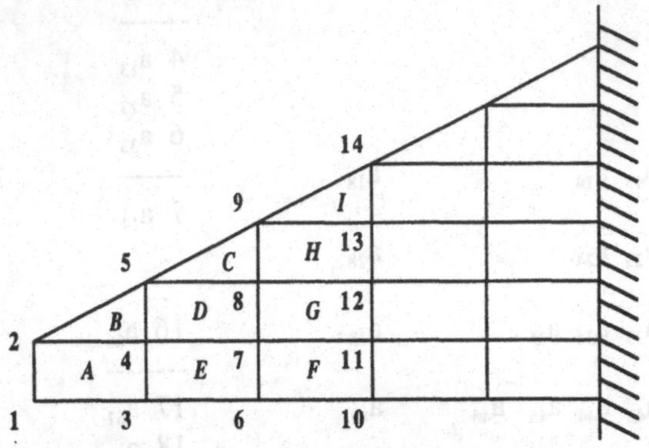

Fig. 2.6.6. An Example for the Frontal Method.

The processing of the elements is assumed to take place in the order A, B, C, D .... The first step is the addition of the coefficients from element A to the space allocated to the nodal equations, in accordance with the finite element assembly procedure. Since node 1 is not connected to nodes in other elements the equation for node 1 is complete, and since the equations are symmetric, the coefficients in the equations for node 1 give the row multipliers required for the elimination of the unknowns associated with node 1. Thus, the correct multiples of the equations for node 1 may be added to equations 2, 3, and 4, even though the latter equations are still incomplete. Once the additions have taken place, the equations for node 1 play no further part in the elimination and may be transferred to subsidiary storage ready for the final back-substitution phase of the solution process. At this stage, the wave-front follows the boundary 2-4-3 as shown in Figure 2.6.7 (a). Nodes on the wave-front are termed frontal nodes.

The addition of element B completes the equations for node 2, so that the displacements associated with this node can be eliminated and the equations transferred to subsidiary storage. Node 5 is added to the wave-front as shown in Figure 2.6.7 (b).

The addition of element C causes coefficients to be inserted in the equations for nodes 8 and 9 but does not complete the equations for any of the active nodes. No elimination takes place, but nodes 8 and 9 are added to the wave-front, as shown in Figure 2.6.7 (c). The addition of element D adds node 7 to the wave-front and allows the elimination of the unknowns associated with node 5, and so on.

When all the elements have been processed, the back-substitution phase of the solution scans through the nodes in an order that is reverse of the order in which they were eliminated.

	Element Introduced	Nodes Eliminated	Frontal Nodes
(a)	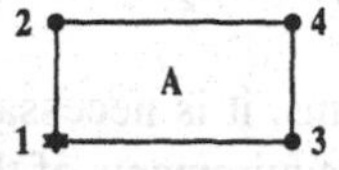	1	2-4-3
(b)	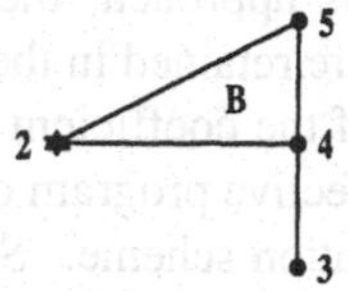	2	5-4-3
(c)	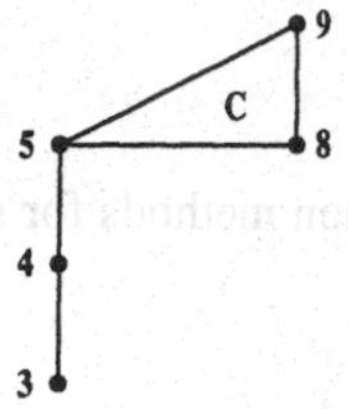	None	9-8-5-4-3
(d)	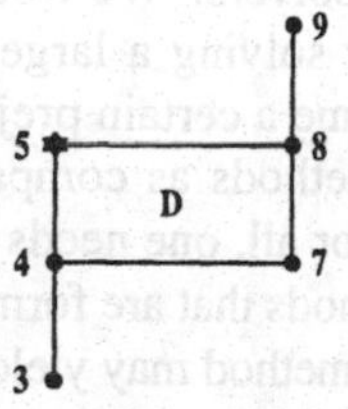	5	9-8-7-4-3
(e)	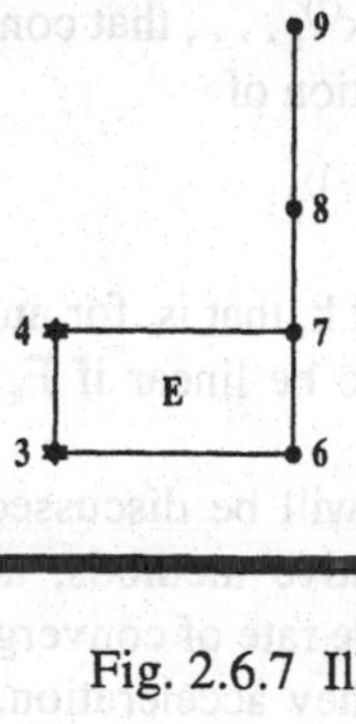	3,4	9-8-7-6

Fig. 2.6.7 Illustration of Frontal Solution Procedure.

From this example, it is seen that frontal methods are closely connected to finite-element methods. Frontal solution packages involve simultaneously assembling and solving the matrix. The programming of the frontal methods involves a considerable amount of bookkeeping in comparison to band methods and is a task an inexperienced programmer should not undertake (Livesley, 1983).

2.6.2.5. Out-of-Core Solvers. In some practical problems, it is necessary to solve exceedingly large systems of equations. When the storage requirements of the resulting matrices exceed the capacity of the computer memory, it becomes necessary to resort to the utilization of peripheral storage devices. In this approach, the matrix is partitioned into entries and a minimum number of entries are retained in the computer memory at any given instant. The remaining elements of the coefficient matrix are retained in peripheral storage, such as a disk. The most effective program of this kind combines the coefficient generation process with the solution scheme. Stand-alone equation-solving codes are not generally available.

2.6.3. *Iterative Solution of Matrix Equations*

So far we have described variations of Gaussian elimination methods for solving the matrix equation

$$Ax = b \,. \tag{2.6.50}$$

We have seen that for a large system the amount of work is proportional to n^3 for full matrix solvers and is proportional to nm^2 for band matrix solvers. We would wonder whether there are more efficient or reliable methods for solving a large system of equations. The answer is yes, but we will need to overcome a certain prejudice - the prejudice in favor of the "exactness" of the Gaussian methods as compared to the "approximate" methods to be discussed here. Once and for all, one needs to be clear that "exact" answers don't exist in a computer even for methods that are formally exact. Hence, it is perfectly plausible that a formally approximate method may yield faster and more accurate results than an exact one if the method could yield a convergent solution. The principle of all iteration solution schemes is to make an initial guess at $x^{(0)}$, then use a recurrence formula to generate new approximations $x^{(1)}, x^{(2)}, \ldots$, that converge to x.

An iteration of the degree m is defined as a function of

$$x^{k+1} = F_k(A,b,x^{(k)},x^{(k-1)},\ldots,x^{(k-m+1)})$$

The iteration is said to be stationary if F_k is independent of k, that is, for any iteration, the functional form is the same. The iteration is said to be linear if F_k is a linear function of $x^{(k)}, x^{(k-1)}, \ldots, x^{(k-m+1)}$.

In this book only the basic iterative methods will be discussed. After a sequence of solutions is obtained from the basic iterative methods, acceleration schemes can be applied to the basic iterates to increase the rate of convergency. The most widely used acceleration schemes include Chebyshev acceleration, conjugate acceleration, and Lanczos acceleration. Descriptions of these acceleration schemes can be found in Hageman and Young (1981) and are constantly evolving. A basic iteration method is any of the linear stationary methods of the first degree in the form of

$$x^{(k+1)} = Gx^{(k)} + v, \ k = 1, 2, \ldots, \tag{2.6.51}$$

where G is the real n x n iteration matrix for the method and v is an associated known vector. The method is of the first degree since $x^{(k+1)}$ is dependent explicitly only on $x^{(k)}$ and not on $x^{(k-1)}, \ldots, x^{(0)}$. The method is linear since neither G nor v depends on $x^{(k)}$, and it is stationary since neither G nor v depends on k.

The principle of basic iterative methods is the matrix splitting. Let

$$A = Q + A - Q \quad \text{then}$$

$$Ax = b$$

$$(Q + A - Q)x = b$$

$$Qx^{(k+1)} = (Q - A)x^{(k)} + b$$

$$x^{(k+1)} = (I - Q^{-1}A)x^{(k)} + Q^{-1}b,$$

which leads to

$$G = I - Q^{-1}A, \ v = Q^{-1}b, \tag{2.6.52}$$

where I is the identity matrix and Q is some nonsingular matrix. Such a matrix Q is called a *splitting matrix*. The assumption of Eq. (2.6.52), together with the fact that A is nonsingular, implies that if $\hat{x}$ is the solution of the related system

$$(I - G)\hat{x} = v \tag{2.6.53}$$

if and only if $\hat{x}$ is also the unique solution to

$$\hat{x} = A^{-1}b. \tag{2.6.54}$$

One of the most important criteria for choosing an iteration method is whether the method will yield a convergent solution. To answer this question, the general requirements for the convergency of iteration methods must be examined. To do this, it is necessary to introduce some definitions. First, the eigenvalues, λ's, of a n × n matrix B are the real or complex numbers determined from

$$\det(B - \lambda kI) = 0 \tag{2.6.55}$$

For example, if B is given by

$$B = \begin{bmatrix} 1 & 0 & 0 \\ -1 & 2 & -1 \\ 0 & -1 & 1 \end{bmatrix}, \tag{2.6.56}$$

then the eigenvalues are given by

$$\begin{vmatrix} 1-\lambda & 0 & 0 \\ -1 & 2-\lambda & -1 \\ 0 & -1 & 1-\lambda \end{vmatrix} = 0 . \qquad (2.6.57)$$

The solution of Eq. (2.6.57) yields

$$\lambda_1 = \frac{(3-\sqrt{5})}{2},\ \lambda_2 = 1,\ \lambda_3 = \frac{(3+\sqrt{5})}{2} . \qquad (2.6.58)$$

The spectral radius S(B) of the n × n matrix B is defined as the maximum of the moduli of the eigenvalues of B, i.e., $\{\lambda_i, i = 1, 2, ..., n\}$ is the set of eigenvalues of B, then

$$S(B) = \max_{i \in n} |\lambda_i| . \qquad (2.6.59)$$

For example, $S(B) = (3+\sqrt{5})/2$ for the matrix B given by Eq. (2.6.56).

With the above definition, we can state that a necessary and sufficient condition for any basic iteration method given by Eq. (2.6.51) to converge is that the spectral radius of the iteration matrix G be less than 1, i.e.,

$$S(G) < 1 . \qquad (2.6.60)$$

The next important consideration for choosing an iteration method is speed of convergence. To measure the speed of convergence of the basic iteration methods, let the error vector $e^{(k)}$ be defined by

$$e^{(k)} = x^{(k)} - \hat{x} . \qquad (2.6.61)$$

Using Eq. (2.6.51) together with the fact that $\hat{x}$ is also satisfies the related equation Eq. (2.6.53), we have

$$e^{(k)} = Ge^{(k-1)} = \dots = G^k e^{(0)} . \qquad (2.6.62)$$

Then, for any vector norm and corresponding matrix norm, we have

$$||e^{(k)}|| \le ||G^k||\ ||e^{(0)}|| . \qquad (2.6.63)$$

Thus $||G^k||$ gives a measure by which the norm of the error has been reduced after k iterations. We define the average rate of convergence of Eq. (2.6.51) by

$$R_k(G) = -\log||G^k||/k . \qquad (2.6.64)$$

It can be shown that if S(G) < 1, then

$$\lim_{k\to\infty} \left(||G^k||\right)^{1/k} = S(G) . \qquad (2.6.65)$$

Hence we are lead to define the asymptotic rate of convergence by

$$R_\infty(G) = \lim_{k\to\infty} R_k(G) = -\log S(G) \qquad (2.6.66)$$

We remark that whereas $R_k(G)$ depends on which norm is used, $R_\infty(G)$ is independent of which norm is used. Frequently we shall refer to $R_\infty(G)$ as the rate of convergency.

When acceleration methods are not used, the basic iteration methods we choose should satisfy Eq. (2.6.60) to guarantee the convergence of the solution. However, when acceleration methods are used, the basic iteration method (based on which acceleration is made) is not necessarily convergent. Normally, when acceleration methods are used, it is sufficient that the basic iteration method be "symmetrizable" in the following sense.

Definition. The iterative method given by Eq. (2.6.51) is symmetrizable if for some non-singular matrix W the matrix $W(I - G)W^{-1}$ is symmetric positive definite (SPD). Such a matrix W is called a *symmetrization matrix*. An iterative method that is not symmetrizable is nonsymmetrizable. A matrix B is SPD if B is symmetric and if the inner product $(u, Au) > 0$ for any nonzero vector u.

It can be shown that if the basic iterative method given in Eq. (2.6.51) is symmetrizable, then the eigenvalues of G are real and the algebraically largest eigenvalues M(G) of G is less than 1. Many iterative methods are symmetrizable. For example, the basic iterative method Eq. (2.6.51) is symmetrizable whenever A and the splitting matrix Q in Eq. (2.6.52) are SPD.

We remark here that the symmetrization property need not imply convergence. If the iterative method is symmetrizable, then the eigenvalues are less than 1 but not necessarily less than 1 in absolute value. Hence, the convergence condition Eq. (2.6.61) need not be satisfied. However, the resulting property that the eigenvalues are less than 1 can be used to derive the so-called extrapolated method based on (2.6.51), which is convergent whenever the basic method is symmetrizable.

The extrapolated method applied to Eq. (2.6.51) is defined by

$$x^{(k+1)} = \gamma\left(Gx^{(k)} + v\right) + (1 - \gamma)Ix^{(k)} = Hx^{(k)} + \gamma v\ , \qquad (2.6.67)$$

where

$$H = \gamma G + (1 - \gamma)I\ . \qquad (2.6.68)$$

Here γ is a parameter that is often referred to as the *extrapolation factor*. If the iterative method is symmetrizable, then the optimum value $\hat{\gamma}$ of γ, in the sense of minimizing S(H), is given by

$$\hat{\gamma} = \frac{2}{2 - M(G) - m(G)}\ , \qquad (2.6.69)$$

where m(G) and M(G) are the smallest and largest eigenvalues of G, respectively. Moreover, it has been shown that

$$S(H) = \frac{M(G) - m(G)}{2 - M(G) - m(G)} < 1\ . \qquad (2.6.70)$$

Thus, the optimum extrapolated method, which is given by

$$x^{(k+1)} = [\hat{\gamma}G + (1-\hat{\gamma})I]x^{(k)} + \hat{\gamma}v\ , \qquad (2.6.71)$$

is a convergent iteration-method.

2.6.4. Most Commonly Encountered Basic Iterative Methods

In this section we describe three most commonly encountered basic iteration methods: the Jacobi method, the Gauss-Seidel method, and the successive relaxation method. We shall assume that the matrix in Eq. (2.6.50) is symmetric and positive definite (SPD). For each method, we describe a computational procedure for carrying out the iterations and discuss convergence properties. For use throughout this section, we express the matrix A as the matrix sum

$$A = D - L - U , \tag{2.6.72}$$

where D is a block diagonal matrix, L is the block lower triangular matrix, and U is the block upper triangular matrix. As we will see, the Jacobi, Gauss-Seidel, and successive relaxation methods can be defined uniquely in terms of these D, L, and U.

2.6.4.1. The Jacobi Method. Substitution of Eq. (2.6.72) into Eq. (2.6.71) yields

$$Dx - Lx - Ux = b . \tag{2.6.73}$$

The Jacobi method is defined by

$$Dx^{(k+1)} = (L+U)x^{(k)} + b , \tag{2.6.74}$$

which can be written as

$$x^{(k+1)} = Bx^{(k)} + v , \tag{2.6.75}$$

where B is the Jacobi iteration matrix defined by

$$B = D^{-1}(L+U) = I - D^{-1}A \tag{2.6.76}$$

and

$$v = D^{-1}b . \tag{2.6.77}$$

The splitting matrix for the Jacobi method is

$$Q = D . \tag{2.6.78}$$

The Jacobi method is convergent if and only if S(B) < 1. It has been shown that S(B) < 1 if the SPD matrix A is irreducible with weak diagonal dominance.

For $n > 1$, a $n \times n$ matrix A is reducible, if there exists a $n \times n$ permutation matrix P such that

$$PAP^T = \begin{bmatrix} A_{11} & A_{12} \\ 0 & A_{22} \end{bmatrix} , \tag{2.6.79}$$

where A_{11} is an $r \times r$ submatrix, A_{22} is an $(n-r) \times (n-r)$ submatrix. A permutation matrix P contains exactly one value of unity and (n-1) zeros in each row and each column. If no such permutation matrix exists, then A is irreducible. The question of whether or not a matrix is irreducible can easily be answer using a geometrical interpretation. In this approach n points $P_1, P_2, ..., P_n$ in a plane are associated with an $n \times n$ matrix A. For every nonzero a_{ij} of the matrix, we connect the point P_i to P_j. A diagonal entry a_{ii} requires a path which joins P_i to itself; such a path is called a *loop*. If a graph constructed according to this procedure generates a path from every point P_i to every point P_j, then the matrix is irreducible.

As an example, consider the matrices and their associated directed paths as shown in Figure 2.6.8. Because the points P_i and P_j can be connected by a directed path in the second, third, and fourth instances, these matrices are irreducible. In the first instance, however, there is no directed path from P_2 to P_1; thus this matrix is reducible.

$$\begin{bmatrix} 1 & 2 \\ 0 & 3 \end{bmatrix}$$

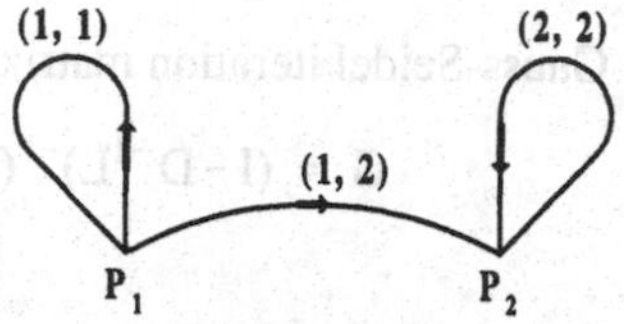

$$\begin{bmatrix} 1 & 2 \\ 4 & 3 \end{bmatrix}$$

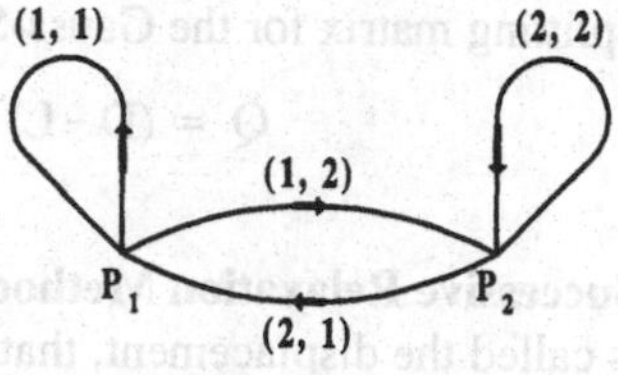

$$\begin{bmatrix} 12 & 1 & 1 \\ 1 & 4 & 1 \\ 1 & 1 & 4 \end{bmatrix}$$

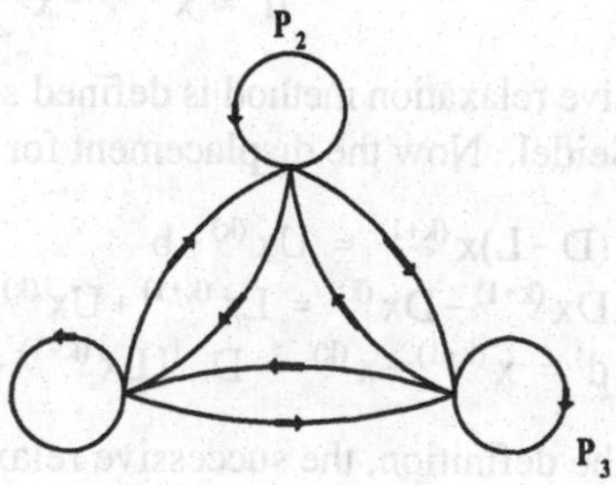

$$\begin{bmatrix} 2 & 0 & 1 \\ 0 & 1 & 5 \\ 4 & 6 & -5 \end{bmatrix}$$

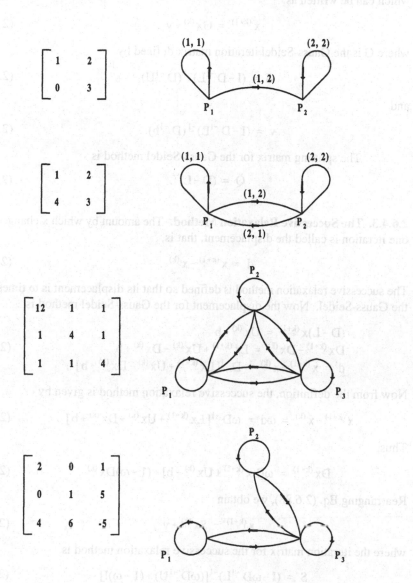

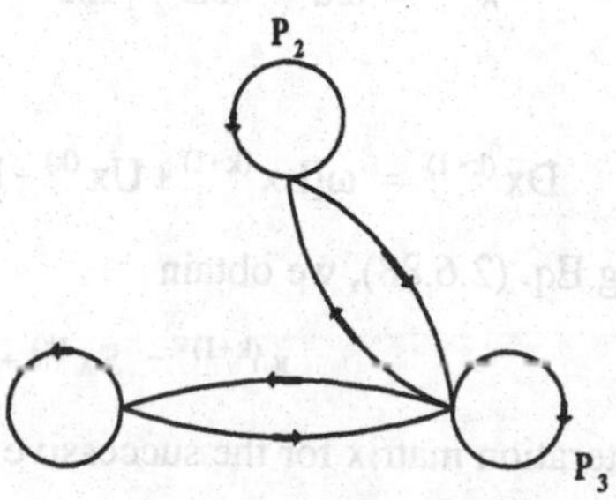

Fig. 2.6.8 Matrix Reducibility and Connecting Path.

2.6.4.2. The Gauss-Seidel Method. The Gauss-Seidel method is defined

$$\begin{aligned} (D-L)x^{(k+1)} &= Ux^{(k)}+b \\ D(I-D^{-1}L)x^{(k+1)} &= Ux^{(k)}+b \\ (I-D^{-1}L)x^{(k+1)} &= D^{-1}Ux^{(k)}+D^{-1}b \\ x^{(k+1)} &= (I-D^{-1}L)^{-1}(D^{-1}U)x^{(k)}+(I-D^{-1}L)^{-1}D^{-1}b\ , \end{aligned} \tag{2.6.80}$$

which can be written as

$$x^{(k+1)} = Gx^{(k)}+v\ , \tag{2.6.81}$$

where G is the Gauss-Seidel iteration matrix defined by

$$G = (I-D^{-1}L)^{-1}(D^{-1}U) \tag{2.6.82}$$

and

$$v = (I-D^{-1}L)^{-1}(D^{-1}b)\ . \tag{2.6.83}$$

The splitting matrix for the Gauss-Seidel method is

$$Q = (D-L)\ . \tag{2.6.84}$$

2.6.4.3. The Successive Relaxation Method. The amount by which x changes during one iteration is called the displacement, that is,

$$d = x^{(k+1)}-x^{(k)}\ . \tag{2.6.85}$$

The successive relaxation method is defined so that its displacement is ω times that of the Gauss-Seidel. Now the displacement for the Gauss-Seidel method is

$$\begin{aligned} (D-L)x^{(k+1)} &= Ux^{(k)}+b \\ Dx^{(k+1)}-Dx^{(k)} &= Lx^{(k+1)}+Ux^{(k)}-Dx^{(k)}+b \\ d = x^{(k+1)}-x^{(k)} &= D^{-1}[Lx^{(k+1)}+Ux^{(k)}-Dx^{(k)}+b]\ . \end{aligned} \tag{2.6.86}$$

Now from the definition, the successive relaxation method is given by

$$x^{(k+1)}-x^{(k)} = \omega d = \omega D^{-1}[Lx^{(k+1)}+Ux^{(k)}-Dx^{(k)}+b]\ . \tag{2.6.87}$$

Thus,

$$Dx^{(k+1)} = \omega[Lx^{(k+1)}+Ux^{(k)}+b]+(1-\omega)Dx^{(k)}\ . \tag{2.6.88}$$

Rearranging Eq. (2.6.88), we obtain

$$x^{(k+1)} = Sx^{(k)}+v\ , \tag{2.6.89}$$

where the iteration matrix for the successive relaxation method is

$$S = (I-\omega D^{-1}L)^{-1}[(\omega D^{-1}U)+(1-\omega)I] \tag{2.6.90}$$

and

$$v = (I - \omega D^{-1}L)^{-1}(\omega D^{-1}b) . \qquad (2.6.91)$$

The splitting matrix for the successive relaxation method is

$$Q = (\omega^{-1}D - L) . \qquad (2.6.92)$$

There are two other well-known basic iteration methods: the RF method and symmetric successive overrelaxation method (SSOR). These two methods are only briefly discussed here. The RF method is based on a variant of Richardson and is defined by

$$x^{(k+1)} = (I - A)x^{(k)} + b . \qquad (2.6.93)$$

The symmetric successive over-relaxation method consists of forward SOR and backward SOR. In other words, one performs the SOR in a certain order, and at the end of computation one performs another SOR in reverse order.

2.6.5. *Preconditioning Conjugate Gradient Methods*

In the early 1950s Hestenes and Stiefel (1952) presented a new iterative method for solving systems of linear algebraic equations. This new method was known as the *conjugate gradient method* (CG) which is a nonstationary and nonlinear iteration of degree one. The CG method, though an iterative method, converges to the true solution of the linear system in a finite number of iterations in the absence of rounding errors. Because of this and many other interesting properties, the CG method attracted considerable attention in the numerical analysis community when it was first presented. However, for various unknown reasons, the method was not widely used, and little was heard about it for many years. Beginning in the mid-1960s there was a strong resurgence of interest in the CG method (Daniel, 1965, 1967; Reid, 1971, 1972; Bartels and Daniel, 1974; Axelsson, 1974; O'Leary, 1975; Chandra et al., 1977; Concus and Golub, 1976; Congus et al., 1976). The application of CG methods to subsurface flow and transport problems has been explored since the early 1980s (Gambolati, 1980; Gambolati and Volpi, 1980a, 1980b; Gambolati and Perdon, 1984; Kaasschieter, 1988; Gambolati et al., 1990; Kuiper, 1987; Hill, 1990).

The classical conjugate gradient method of Hestenes and Stiefel (1952) can be considered as a modification of the method of the *steepest-descent* (SD) method. To derive the steepest-decent method, we assume that A is a symmetric positive definite (SPD) matrix and consider the real valued quadratic function

$$F(x) = \frac{1}{2}x^{T}Ax - b^{T}x . \qquad (2.6.94)$$

Equation (2.6.94) can be written as

$$F(x) = \frac{1}{2}x^TAx - \frac{1}{2}x^TA(A^{-1}b) - \frac{1}{2}x^TA(A^{-1}b)$$

$$= \frac{1}{2}x^TAx - \frac{1}{2}x^TA(A^{-1}b) - \frac{1}{2}(A^{-1}b)^TAx \qquad (2.6.95)$$

$$= \frac{1}{2}\left(x - A^{-1}b\right)^TA\left(x - A^{-1}b\right) - \frac{1}{2}b^TA^{-1}b \, .$$

Since A is symmetric positive definite (SPD), the first term on the right-hand side of Eq. (2.6.95) is always positive. Thus, the function F(x) would attain its minimum value if $(x - A^{-1}b) = 0$; that is, solving the system Ax = b is equivalent to minimizing the quadratic function F(x). Now the downhill gradient (the direction of the steepest decent) is given by

$$-\nabla F(x) = b - Ax \, . \qquad (2.6.96)$$

The direction of the vector grad F(x) is the direction for which the functional F(x) at the point x has the greatest instantaneous change. If $x^{(k)}$ is some approximation to Ax = b, then in the method of steepest decent, we obtain an improved approximation $x^{(k+1)}$ by moving in the direction of grad $F(x^{(k)})$ to a point where $F(x^{(k+1)})$ is minimal. In other words,

$$x^{(k+1)} = x^{(k)} + \alpha_k\left[-\nabla F\left(x^{(k)}\right)\right] , \qquad (2.6.97)$$

where α_k is to be chosen to minimize $F(x^{(k+1)})$. Substituting Eq. (2.6.96) into Eq. (2.6.97), we obtain

$$x^{(k+1)} = x^{(k)} + \alpha_k r^{(k)} , \qquad (2.6.98)$$

where

$$r^{(k)} = b - Ax^{(k)} \qquad (2.6.99)$$

is the residual of the k-th iteration. Substituting Eq. (2.6.98) into Eq. (2.6.94), we have

$$F\left(x^{k+1}\right) = \frac{1}{2}\left(x^{(k)} + \alpha_k r^{(k)}\right)^TA\left(x^{(k)} + \alpha_k r^{(k)}\right) - \qquad (2.6.100)$$

$$b^T\left(x^{(k)} + \alpha_k r^{(k)}\right) .$$

Taking the derivative of Eq. (2.6.100) with respect to α_k to minimize $F(x^{(k+1)})$, we obtain

$$\alpha_k = \frac{\left(r^{(k)}\right)^T\left(r^{(k)}\right)}{\left(r^{(k)}\right)^T A\left(r^{(k)}\right)} \, . \qquad (2.6.101)$$

To summarize, we can express the procedure of the SD method as

$$x^{(0)} \text{ is chosen arbitrary .} \qquad (2.6.102)$$

For k = 0, 1, 2, ...

$$r^{(k)} = b - Ax^{(k)} \quad \text{or} \quad \left[r^{(k+1)} = r^{(k)} - \alpha_k A r^{(k)}\right] \qquad (2.6.103)$$

$$\alpha_k = \frac{\left(r^{(k)}\right)^T\left(r^{(k)}\right)}{\left(r^{(k)}\right)^T A\left(r^{(k)}\right)} \qquad (2.6.104)$$

$$x^{(k+1)} = x^{(k)} + \alpha_k r^{(k)} . \qquad (2.6.105)$$

The SD method looks great, but, for ill-conditioned matrices A, the convergence rate of the method can be very slow (Luenberger, 1973). However, by choosing our direction vectors differently, we obtain the *conjugate gradient* (CG) method, which has been shown to give the solution in at most N iterations in the absence of round-off error (Hestenes and Stiefel, 1952). Instead of improving the solution along the vector $r^{(k)}$, we will improve the solution along a "direction" vector $p^{(k)}$ of our choice; that is, the improved approximation is given by

$$x^{(k+1)} = x^{(k)} + \alpha_k p^{(k)} \quad \text{for} \quad k \geq 0 . \qquad (2.6.106)$$

As before, the minimization of the quadratic function $F(x^{(k+1)})$ would yield

$$\alpha_k = \frac{\left(p^{(k)}\right)^T\left(r^{(k)}\right)}{\left(p^{(k)}\right)^T A\left(p^{(k)}\right)} \quad \text{for } k \geq 0 . \qquad (2.6.107)$$

The question is what "direction" vector do we choose. In the CG method, we choose the "direction vector" $p^{(k)}$ by the following recursive formula:

$$p^{(k)} = r^{(k)} + \beta_k p^{(k-1)} \quad \text{fork} \geq 1 \qquad (2.6.108)$$

and

$$p^{(0)} = r^{(0)} , \qquad (2.6.109)$$

where β_k is chosen such that $p^{(k)}$ is A-orthogonal or A-conjugate to $p^{(k-1)}$ for k greater than or equal to 1. Evidently, this requirement of orthogonality gives

$$\beta_k = -\frac{\left(r^{(k)}\right)^T A\left(p^{(k-1)}\right)}{\left(p^{(k-1)}\right)^T A\left(p^{(k-1)}\right)} \quad \text{for } k \geq 1 . \qquad (2.6.110)$$

It is easy to derive a recursive relationship between residuals

$$r^{(k+1)} = b - Ax^{(k+1)} = b - A\left(x^{(k)} + \alpha_k p^{(k)}\right) = r^{(k)} - \alpha_k A p^{(k)} . \qquad (2.6.111)$$

In summary, the above discussion leads to the following CG algorithm:

$$x^{(0)} \text{ is arbitrary, } r^{(0)} = b - Ax^{(0)} . \qquad (2.6.112)$$

Starting with k = 0 and continuing with k = 1, 2, ... , we perform the following steps:

$$p^{(k)} = r^{(k)} \quad \text{if } k = 0 \tag{2.6.113}$$

$$\beta_k = -\frac{\left(r^{(k)}\right)^T A\left(p^{(k-1)}\right)}{\left(p^{(k-1)}\right)^T A\left(p^{(k-1)}\right)} \left[\text{or } \frac{\left(r^{(k)}\right)^T\left(r^{(k)}\right)}{\left(r^{(k-1)}\right)^T\left(r^{(k-1)}\right)}\right] \quad \text{if } k \geq 1 \tag{2.6.114}$$

$$p^{(k)} = r^{(k)} + \beta_k p^{(k-1)} \quad \text{if } k \geq 1 \tag{2.6.115}$$

$$\alpha_k = \frac{\left(p^{(k)}\right)^T\left(r^{(k)}\right)}{\left(p^{(k)}\right)^T A\left(p^{(k)}\right)} \left[\text{or } \frac{\left(r^{(k)}\right)^T\left(r^{(k)}\right)}{\left(p^{(k)}\right)^T A\left(r^{(k)}\right)}\right] \tag{2.6.116}$$

$$x^{(k+1)} = x^{(k)} + \alpha_k p^{(k)} \tag{2.6.117}$$

$$r^{(k+1)} = r^{(k)} - \alpha_k A p^{(k)} . \tag{2.6.118}$$

Hestenes and Stiefel (1952) have shown that the "direction vector" $p^{(k)}$ defined by Eq. (2.6.115) using (2.6.114) is also A-orthogonal to the "direction vectors" $p^{(k-1)}$, $p^{(k-2)}$, ..., $p^{(0)}$; that is, in general, we have

$$\left(p^{(k)}\right)^T A p^{(i)} = 0 \quad \text{for } i = 0, 1, \ldots, k-1 . \tag{2.6.119}$$

It was also shown (Hestenes and Stiefel, 1952) that the residuals $r^{(k)}$ are mutually orthogonal; that is, they satisfy the relationship

$$\left(r^{(k+1)}\right)^T\left(r^{(i)}\right) = 0 \quad \text{for } i = 0, 1, 2, \ldots, k . \tag{2.6.120}$$

The N-th residual is to be orthogonal to $r^{(0)}$, $r^{(1)}$, ..., r(N-1). Hence at most $r^{(N)}$ has to be zero; that is, $r^{(s)} = 0$ for some $s \geq N$. Thus, the conjugate gradient method converges, in the absence of round-off error, in at most N iterations.

Because of the round-off error, the residuals are only locally orthogonal, and this seriously affects the theoretical properties of the CG method. The last residual $r^{(N)}$ is usually different from zero, and the equations given by Eqs. (2.6.112) through (2.6.118) have to be repeatedly used beyond N to get satisfactory results. However, if the matrix A is sparse, the CG scheme may be greatly accelerated. This result, which has seen an extensive research since the earlier 1980s, has revalued the accelerated CG methods that appear to be one of today's most efficient techniques to solve large, sparse, symmetric and positive definite systems.

If the eigenvalues of A are clustered around few (possibly one) fundamental values, the CG method converges rapidly (Gambolati and Perdon, 1984). In the finite element matrices arising in subsurface flow modeling, the eigenvalues (which are all real positive values) are usually evenly distributed between the minimum and maximum values. Thus, the idea of accelerating the CG method is to transform the matrix equation into an equivalent system such that the modified matrix would have eigenvalues clustered around a few fundamental values. To carry out this idea, let us rewrite the matrix equation (2.6.50) as

$$By = c\ , \tag{2.6.121}$$

where

$$\begin{aligned} B &= W^{-1}AW^{-1} \\ y &= Wx \\ c &= W^{-1}b\ . \end{aligned} \tag{2.6.122}$$

In Eq. (2.6.122), W^{-1} is an auxiliary matrix that aims at producing a matrix B whose eigenvalues fall in the vicinity of one. It is noted that B is also positive definite. Now applying the CG method to Eq. (2.6.121) and restoring to the original variables, we obtain the preconditioned conjugate gradient algorithm as

$$x^{(0)} \text{ is arbitrary, } r^{(0)} = b - Ax^{(0)}\ . \tag{2.6.123}$$

Starting with k = 0 and continuing with k = 1, 2, ... , we perform the following steps:

$$z_k = Q^{-1} r_k \tag{2.6.124}$$

$$p^{(k)} = z^{(k)} \qquad \text{if } k = 0 \tag{2.6.125}$$

$$\beta_k = -\frac{\left(z^{(k)}\right)^T A\left(p^{(k-1)}\right)}{\left(p^{(k-1)}\right)^T A\left(p^{(k-1)}\right)} \left[\text{or } \frac{\left(z^{(k)}\right)^T\left(r^{(k)}\right)}{\left(z^{(k-1)}\right)^T\left(r^{(k-1)}\right)}\right] \qquad \text{if } k \geq 1 \tag{2.6.126}$$

$$p^{(k)} = z^{(k)} + \beta_k p^{(k-1)} \qquad \text{if } k \geq 1 \tag{2.6.127}$$

$$\alpha_k = \frac{\left(p^{(k)}\right)^T\left(r^{(k)}\right)}{\left(p^{(k)}\right)^T A\left(p^{(k)}\right)} \left[\text{or } \frac{\left(z^{(k)}\right)^T\left(r^{(k)}\right)}{\left(p^{(k)}\right)^T A\left(p^{(k)}\right)}\right] \tag{2.6.128}$$

$$x^{(k+1)} = x^{(k)} + \alpha_k p^{(k)} \tag{2.6.129}$$

$$r^{(k+1)} = r^{(k)} - \alpha_k A p^{(k)}\ , \tag{2.6.130}$$

where

$$Q^{-1} = W^{-1} W^{-1} \tag{2.6.131}$$

is called the preconditioning matrix and we recall Q is the splitting matrix.

2.6.6. MultiGrid Methods

Classical iteration or relaxation schemes presented in the earlier sections suffer some disabling limitations. The convergence rates of those classical iteration schemes dependent on the number of equations to be solved. These limitations have a severe implication in that large three-dimensional problems involving millions of grid points and several degrees of freedom (unknowns) at each grid point would demand computational time greater than real time for the change of physical phenomena. Ideally, one would like to have an iterative scheme that would yield a convergent rate independent of the number of equations to be solved. In other words, an optimal iterative scheme should consume a computational effort proportional only to the change of a real physical system. Numerical stalling resulting from the slow convergency, because of the large number of equations needed to be solved, should not be permitted. Multigrid methods evolved form the attempts to correct these limitations. These attempts have been largely successful; used in the multigrid setting, relaxation or iteration methods are competitive with optimal direct solution methods such as the fast Fourier transform or the method of cyclic reduction. However, these optimal direct solution methods are rather specialized and can be applied primarily to systems arising from separable self-adjoined boundary-value problems. On the other hand, the multigrid relaxation methods have more generality and a wider range of application than the specialized optimal direct solution methods. Multigrid methods have been applied successfully to a variety of problems (Brandt, 1984). But first we must begin with the basics.

2.6.6.1. Problems with Basic Iteration Methods. We have presented three basic iteration methods and some non-stationary, nonlinear iterative methods. There are many more variations on the ones we have presented on iterative solutions for the linear system of algebraic equations. These schemes work very well for the first several iterations. Inevitably, however, the convergence slows and the entire scheme appears to stall. A simple explanation for this phenomenon can be found: the rapid decrease in error during the early iterations is due to the efficient elimination of the oscillatory modes of the error. Once the oscillatory modes have been removed, the iteration is much less effective in reducing the remaining smooth components (Briggs, 1987). Many relaxation schemes possess this property of eliminating the oscillatory modes and leaving the smooth modes. We shall call this property the *smoothing property*. It is a serious limitation of these methods. However, it can be overcome and the remedy is one of the pathways to multigrid.

The subject of iterative methods constitutes a large and important domain of classical numerical analysis. It is filled with very elegant mathematics from both linear algebra and analysis (Varga, 1962; Young, 1971; Hageman and Young, 1981). However, esoteric iterative methods are not required for the development of multigrid. Many of the most effective multigrid methods are built on the simple basic relaxation schemes presented earlier. For this reason, we content ourselves with a few basic methods and ask how to develop them into far more powerful methods.

2.6.6.2. Elements of Multigrid Methods. Let use accept the fact that many *basic iteration* methods possess the smoothing property. This property makes these methods very effective at eliminating the high-frequency or oscillatory components of the error, while leaving the low-frequency or smooth components relatively unchanged. The immediate issue is whether these methods can be adapted in some way to make them effective on all error components.

One way to improve a relaxation scheme, at least in the early stages, is to use a good initial guess. A well-known technique for obtaining an improved initial guess is to perform some preliminary iterations on a coarse grid and then use the resulting approximation as an initial guess on the original fine grid. Relaxation on a coarse grid is less expensive since there are fewer unknowns to be solved. Also, since convergence factor behaves like $1 - O(h^2)$, the coarse grid will have a marginally improved convergence rate. The use of coarse grids to generate improved initial guesses is the basis of a strategy called *nested iteration.*

Intuitively and geometrically, we see that smooth modes on a fine grid look less smooth on a coarse grid. This suggests that when relaxation begins to stall, signaling the predominance of smooth error modes, it is advisable to move to a coarser grid, on which those smooth errors appear oscillatory and relaxation will be more effective. The oscillatory error on the coarse grid can be reduced with smoothing and returned to the fine grid for the correction of the approximated solution. This strategy is called the *coarse grid correction* and it requires a mechanics of *intergrid transfer.* For this strategy to work, we must recall that to solve the original equation is equivalent to solving a corresponding residual equation. In other words, relaxation on the original equation $Au = f$ with an arbitrary initial guess v is equivalent to relaxing on the residual equation $Ae = r$ with the specific initial guess $e = 0$.

From the above discussion, it is obvious that the essential elements of the multigrid methods are (1) a basic relaxation scheme that can smooth short wave errors, (2) intergrid transfer, (3) a coarse grid correction strategy to remove errors of long wave components, and (4) a nested strategy to provide an initial guess. The first order of business in multigrid methods is thus the construction of a basic relaxation scheme, which is problem dependent. For a Laplace type of equation, a symmetrical Gauss-Seidel iteration method is an excellent smoother.

Regarding intergrid transfer, we consider only the case in which the coarse grid has twice the grid spacing of the next finer grid. This is a nearly universal practice, since there seems to be no advantage in using grid spacings with ratios other than 2. If we denote a grid of size h the fine grid, the grid of size 2h would be the coarse grid. The transfer of any quantity from the coarse grid 2h to the fine grid h constitutes the first class of intergrid transfer. It is a common procedure in numerical analysis and is generally called *interpolation* or *prolongation.* Many interpolation methods could be used. Fortunately, for most multigrid purposes, the simplest of these are quite effective. For this reason, we will consider only linear interpolation. The linear operator will be denoted I^h_{2h}. It takes coarse grid vectors and produces fine grid vectors according to the rule

$$I^h_{2h} v^{2h} = v^h , \tag{2.6.132}$$

where

$$v^h_{2j} = v^{2h}_j , \quad v^h_{2j+1} = \frac{1}{2}\left(v^{2h}_j + v^{2h}_{j+1}\right), \quad 0 \le j \le \frac{N}{2} - 1 . \tag{2.6.133}$$

Figure 2.6.9 shows graphically the action of I_{2h}. At even-numbered fine-grid points, the values of the vector are transferred directly from R^{2h} to R^h. At odd-numbered fine grid points, the value of v^h is the average of the adjacent coarse grid values.

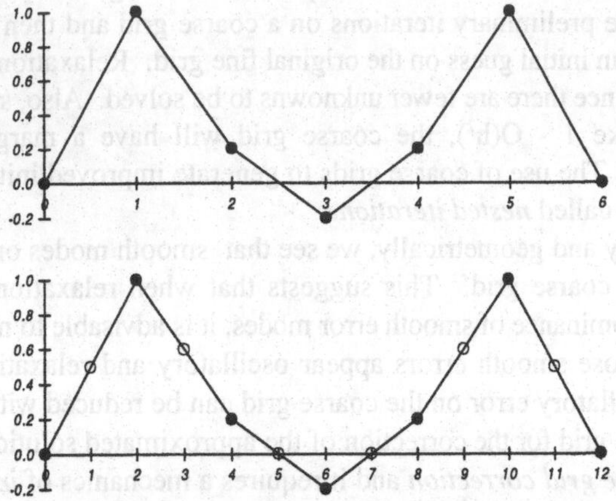

Fig. 2.6.9 Interpolation of a Vector on the Coarse Grid R^{2h} to the Fine Grid R^h.

Writing Eq. (2.6.133) in a matrix form, we note that I^h_{2h} is a linear operator from $R^{N/2-1}$ to R^{N-1}. For the case of N = 8, this operator has the form

$$I^h_{2h} v^{2h} = \frac{1}{2}\begin{bmatrix} 1 & & \\ 2 & & \\ 1 & 1 & \\ & 2 & \\ & 1 & 1 \\ & & 2 \\ & & 1 \end{bmatrix} \begin{Bmatrix} v_1 \\ v_2 \\ v_3 \end{Bmatrix}_{2h} = \begin{Bmatrix} v_1 \\ v_2 \\ v_3 \\ v_4 \\ v_5 \\ v_6 \\ v_7 \end{Bmatrix}_h = v^h. \qquad (2.6.134)$$

How well does this interpolation (prolongation) work? First assume that the "real" error (which is not known exactly) is a smooth vector (indicated by open and close circles in Fig. 2.6.10a) on the fine grid R^h. Assume also that a coarse grid approximation to the error has been determined on R^{2h} and, furthermore, that this approximation is exact at the coarse grid points. When this coarse grid approximation (indicated by close circle in Fig. 2.6.10a) is interpolated to the fine grid, the interpolant is also smooth. Therefore, we expect a relatively good approximation to the fine grid error as shown in Fig. 2.6.10a. By contrast, if the "real" error is oscillatory, even a very good coarse grid approximation may produce an interpolant that is not very accurate, as shown in Fig. 2.6.10b. However, we do not have to worry about the oscillation error on the fine grid since it is readily eliminated by relaxation.

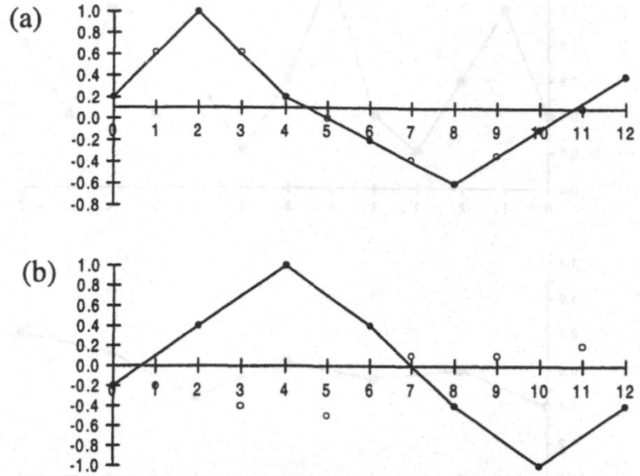

Fig. 2.6.10. Can the Fine Grid Error Be Correctly Obtained from the Interpolation of Coarse Grid Error?

The second class of intergrid transfer functions involves moving vectors from a fine grid to a coarse grid. These are generally called *restriction* operators and are denoted by I_h^{2h}. The most obvious restriction operator is *injection*. It is defined by

$$I_h^{2h} v^h = v^{2h} , \qquad (2.6.135)$$

where

$$v_j^{2h} = v_{2j}^h \qquad 1 \le j \le \frac{N}{2} - 1 . \qquad (2.6.136)$$

In other words, the coarse grid vector simply takes its value directly from the corresponding fine grid point. An alternative restriction operator is call *full weighting* and is defined by

$$I_h^{2h} v^h = v^{2h} , \qquad (2.6.137)$$

where

$$v_j^{2h} = \frac{1}{4}\left(v_{2j-1}^h + 2v_{2j}^h + v_{2j+1}^h\right) \qquad 1 \le j \le \frac{N}{2} - 1 . \qquad (2.6.138)$$

As Figure 2.6.11 shows, the values of the coarse grid vector are a weighted average of values at neighboring fine grid points.

In the discussion that follows, we will use full weighting as a restriction operator. However, in some instances injection may be the better choice. The issue of intergrid transfers, which is an important part of multigrid theory, is discussed at some length by Brandt (Brandt, 1984).

The full weighting operator is a linear operator from R^{N-1} to $R^{N/2-1}$. In the case of N = 8, the full weighting operator has the form

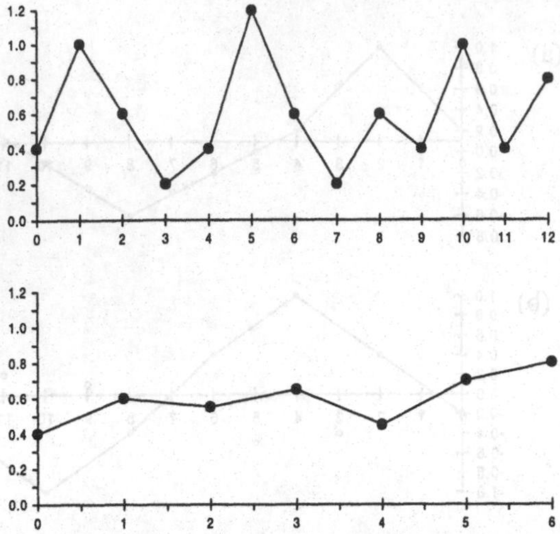

Figure 2.6.11. Restriction by Full Weighting of a Fine Grid Vector to the Coarse Grid.

$$I_h^{2h}\, v^h = \frac{1}{4}\begin{bmatrix} 1 & 2 & 1 & & & & \\ & & 1 & 2 & 1 & & \\ & & & & 1 & 2 & 1 \end{bmatrix}\begin{Bmatrix} v_1 \\ v_2 \\ v_3 \\ v_4 \\ v_5 \\ v_6 \\ v_7 \end{Bmatrix}_h = \begin{Bmatrix} v_1 \\ v_2 \\ v_3 \end{Bmatrix}_{2h} = v^{2h} . \qquad (2.6.139)$$

One reason for our choice of full weighting as a restriction operator is the important fact that

$$I_{2h}^{h} = c\left(I_h^{2h}\right)^T , \qquad (2.6.140)$$

where c is a constant. The fact that the interpolation operator and the full weighting restriction operator are the transpose of each other up to a constant is of importance in more advanced discussion of multigrid methods (Brandt, 1984; Cheng et al., 1998).

We now have a well-defined way to transfer vectors between fine and coarse grids. Therefore, we define the following coarse grid correction scheme.

Coarse Grid Correction Scheme

Relax ν_1 times on	$A^h u^h = f^h$	**on R^h with initial guess v^h.**
Compute	$r^{2h} = I_h^{2h}(f^h - A^h v^h)$.	
Solve	$A^{2h} e^{2h} = r^{2h}$	**on R^{2h}.**
Correct fine grid approximation:	$v^h = v^h + I_{2h}^h e^{2h}$.	
Relax ν_2 times on	$A^h x^h = b^h$	**on R^h with initial guess v^h.**

Several comments are in order. First, notice that when it is important to indicate the grid on which a particular vector or matrix is defined, the superscript h or 2h is used. Second, due to the intergrid transfer operation introduced, all of the quantities in the above procedure are well defined except for A^{2h}. We still have not defined the coarse grid representation of the original matrix. The simplest way of defining A^{2h} is to treat it as the "R^{2h} version of A^h." A more rigorous construction of A^{2h} out A^h is given by

$$A^{2h} = I_h^{2h} A^h I_{2h}^h . \tag{2.6.141}$$

Finally, the integers ν_1 and ν_2 are parameters that control the number of relaxation sweeps before and after the coarse grid correction. In practice, ν_1 is often 1, 2, or 3.

The above coarse grid correction scheme leaves one looming procedural question: what is the best way to solve the coarse problem

$$A^{2h} e^{2h} = r^{2h}? \tag{2.6.142}$$

The answer to this question may be apparent, particularly to those who think recursively. The coarse grid problem is not much different from the original problem. Therefore, we can apply the coarse grid correction scheme to the residual equation on R^{2h}, which means moving to R^{4h} for the correction step. We can repeat this process on successively coarser grids until a direct solution of the residual equation is possible.

To facilitate the description of this procedure, some economy of notation is desirable. We will call the right-hand side vector of the residual equation f^{2h}, rather than r^{2h}, since it is just another right-hand side vector. Instead of calling the solution of the residual equation e^{2h}, we will use v^{2h} since it is just a solution vector. These changes simplify the notation, but it is still important to remember the meaning of these variables. Here, then, is the coarse grid correction scheme, now imbedded within itself (Fig. 2.6.12). We assume that there are $G > 1$ grids, with the coarsest grid spacing given by Lh, where $L = 2^{G-1}$.

The algorithm telescopes down to the coarsest grid, which can be a single interior grid point, and then works its way back to the finest grid. Figure 2.6.12 shows the schedule for the grids in the order in which they are visited. Because of the pattern of this diagram, this algorithm is called the V-cycle. It is the first true multigrid method.

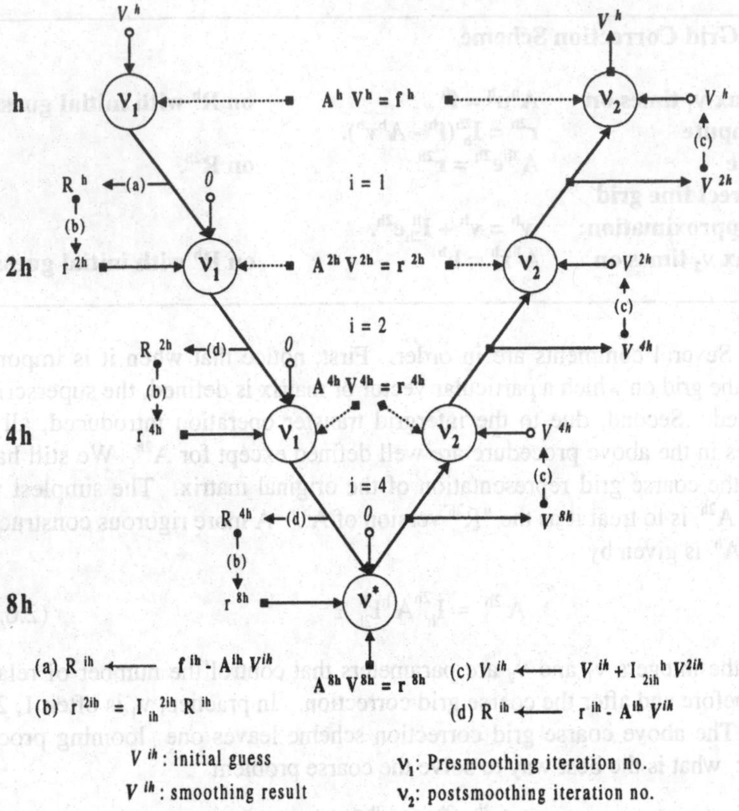

Fig. 2.6.12 Schematic Representation of the V-Cycle Multigrid Method.

$$v^h = MV^h(v^h, f^h)$$

Relax on $A^h u^h = f^h$ ν_1 times with initial guess v^h.
Compute $f^{2h} = I_h^{2h} r^h$.
  Relax on $A^{2h}u^{2h} = f^{2h}$ ν_1 times with initial guess $v^{2h} = 0$.
  Compute $f^{4h} = I_{2h}^{2h} r^h$.
    Relax on $A^{4h}u^{4h} = f^{4h}$ ν_1 times with initial guess $v^{4h} = 0$.
    Compute $f^{8h} = I_{4h}^{8h} r^{4h}$.
      .
      Solve $A^{Lh}u^{Lh} = f^{Lh}$
      .
    Correct $v^{4h} = v^{4h} + I_{8h}^{4h} v^{8h}$
    Relax on $A^{4h}u^{4h} = f^{4h}$ ν_2 times with initial guess v^{4h}.
  Correct $v^{2h} = v^{2h} + I_{4h}^{2h} v^{4h}$
  Relax on $A^{2h}u^{2h} = f^{2h}$ ν_2 times with initial guess v^{2h}.
Correct $v^h = v^h + I_{2h}^{h} v^{2h}$
Relax on $A^h u^h = f^h$ ν_2 times with initial guess v^h.

So far we have developed only the *coarse grid correction* idea. The *nested iteration* idea has yet to be explored. Recall that nested iteration uses coarse grids to obtain improved initial guesses for fine grid problems. In looking at the V-cycle, we might ask how to obtain an informed initial guess for the first fine grid relaxation. Nested iteration would suggest solving a problem on R^{2h}. But how do we obtain a good initial guess for the R^{2h}? Nested iteration sends us to R^{4h}. Clearly we are on another recursive path that leads to the coarsest grid.

The algorithm that joins nested iteration with the V-cycle is called the full multigrid (FMG). Figure 2.6.13 shows the scheduling of grids for FMG. Each V-cycle is preceded by a smaller V-cycle designed to provide the best initial guess possible. As has been demonstrated (Briggs, 1987; Brandt, 1984), the extra work done in these preliminary V-Cycle is not only inexpensive, but generally pays for itself. Full multigrid is the complete knot into which the many threads of the preceding sections are tied. It is a remarkable synthesis of ideas and techniques that individually have been well known and used for a long time. Taken alone, many of these ideas have serious defects. Full multigrid is a technique for integrating them so that they can work together in a way that removes these limitations. The result is a very simple but powerful algorithm.

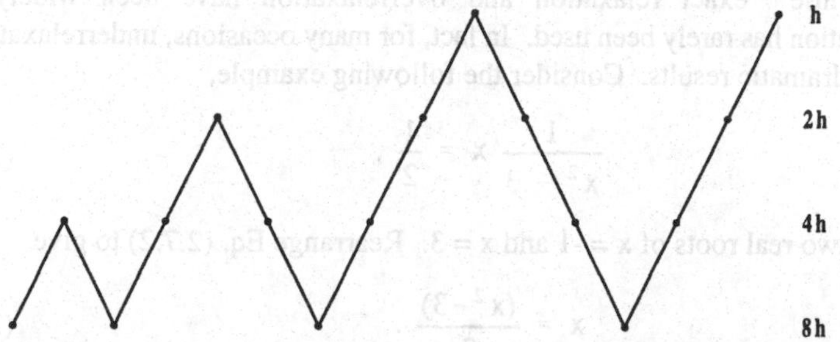

Fig. 2.6.13 Schematic Representation of Full Multigrid Method.

Given in explicit terms, the full multigrid method is described as follows.

Full Multigrid Method

Initialize $f^h, f^{2h}, \ldots; v^h, v^{2h}, \ldots$ **to zero**

Relax on coarsest grid

.

.

$$v^{4h} = v^{4h} + I_{8h}^{4h} v^{8h}$$
$$v^{4h} = MV^{4h}(v^{4h}, f^{4h})$$
$$v^{2h} = v^{2h} + I_{4h}^{2h} v^{4h}$$
$$v^{2h} = MV^{2h}(v^{2h}, f^{2h})$$
$$v^h = v^h + I_{2h}^{h} v^{2h}$$
$$v^h = MV^h(v^h, f^h)$$

2.7. Nonlinear Problems

The matrix equation Eq. (2.6.50) is linear only when the partial differential equation, to which Eq. (2.6.50) is approximated, is linear. To solve the nonlinear matrix equation, two approaches can be taken: (1) the Picard method and (2) the Newton-Ralson method. In the Picard method, an initial estimate is made of the unknown {x}. Using this estimate, we then compute the matrix [A] and solve the linearized matrix equation by the method of linear algebra either with direct elimination or iterative technique. The new estimate is now obtained by the weighted average of the new solution and the previous estimate:

$$x^{(m+1)} = \omega x^{*} + (1-\omega)x^{(m)} , \tag{2.7.1}$$

where $x^{(m+1)}$ is the new estimate, $x^{(m)}$ is the previous estimate, x^{*} is the new solution, and ω is the nonlinear iteration parameter. The procedure is repeated until the new solution x is within a tolerance error. When the parameter is greater than or equal to 0 but less than 1, the iteration is termed underrelaxation. If $\omega = 1$, the iteration is termed exact relaxation. For the cases when ω is greater than 1 but less than or equal to 2, the iteration is termed overrelaxation.

While exact relaxation and overrelaxation have been widely used, underrelaxation has rarely been used. In fact, for many occasions, underrelaxation can yield very dramatic results. Consider the following example,

$$\frac{1}{x^2 - 3}\, x = \frac{1}{2} , \tag{2.7.2}$$

which has two real roots of x = -1 and x = 3. Rearrange Eq. (2.7.2) to give

$$x = \frac{(x^2 - 3)}{2} . \tag{2.7.3}$$

If we take 0 as our initial guess, Table 2.7.1 gives the iterates at various iteration steps with ω = 0.5, 1.0, and 1.5, respectively. It is seen that for this simple example, it would take thousands of iterations to yield a convergent solution for the exact and over relaxation methods, whereas it takes only four iterations for the under-relaxation method to yield the convergent solution. Therefore, when dealing with nonlinear equations, one should not overlook the power of under-relaxation.

For the Newton-Ralson method, we write Eq. (2.6.50) in the following form:

$$y = Ax - b = 0 . \tag{2.7.4}$$

Tayler expansion of Eq. (2.7.4) about the previous iterate yields

$$y^{(k)} + \frac{dy}{dx}\left(x^{(k+1)} - x^{(k)}\right) = 0 , \tag{2.7.5}$$

where $y^{(k)}$ is the value of y evaluated at $x^{(k)}$, $x^{(k)}$ is the value of x from a previous iteration, and $x^{(k+1)}$ is the value of x at a new iteration. Written in matrix notation, Eq. (2.7.5) becomes

Table 2.7.1. Iterates as Function of Iteration Number.

M	OMEGA = 0.5	OMEGA = 1.0	OMEGA = 1.5	NEWTON - RALSON
-	-------------	-------------	-------------	-------------
1	-7.500000D-01	-1.500000D+00	-2.250000D+00	-1.500000D+00
2	-9.843750D-01	-3.750000D-01	2.671875D+00	-1.050000D+00
3	-9.999390D-01	-1.429688D+00	1.768250D+00	-1.000610D+00
4	-1.000000D+00	-4.779968D-01	-7.890950D-01	-1.000000D+00
5	-1.000000D+00	-1.385759D+00	-1.388449D+00	-1.000000D+00
6	-1.000000D+00	-5.398353D-01	-1.099317D-01	-1.000000D+00
7	-1.000000D+00	-2.185970D+00	-1.000000D+00	-1.000000D+00
8	-1.000000D+00	-5.829507D-01	2.426835D+00	-1.000000D+00
9	-1.000000D+00	-1.330084D+00	9.537281D-01	-1.000000D+00
10	-1.000000D+00	-6.154380D-01	-2.044666D+00	-1.000000D+00
100	-1.000000D+00	-8.598072D-01	7.194317D-01	-1.000000D+00
200	-1.000000D+00	-8.999545D-01	6.518091D-02	-1.000000D+00
300	-1.000000D+00	-9.180970D-01	-2.259790D+00	-1.000000D+00
400	-1.000000D+00	-9.289929D-01	2.951971D+00	-1.000000D+00
500	-1.000000D+00	-9.364563D-01	-1.057596D+00	-1.000000D+00
600	-1.000000D+00	-9.419776D-01	2.857897D+00	-1.000000D+00
700	-1.000000D+00	-9.462749D-01	-2.027587D+00	-1.000000D+00
800	-1.000000D+00	-9.497417D-01	6.085009D-01	-1.000000D+00
900	-1.000000D+00	-9.526154D-01	-1.388257D+00	-1.000000D+00
1000	-1.000000D+00	-9.550472D-01	-1.895280D+00	-1.000000D+00

$$Z^{(k)}\left(X^{(k)} - X^{(k+1)}\right) = Y^{(k)}\,, \qquad (2.7.6)$$

where Y is the residue, Z is the Jacobian of Y with respect to X. Thus, the solution of Eq. (2.7.6) involves the following steps. First, knowing $X^{(k)}$, one computes the residual $Y^{(k)}$ via Eq. (2.7.3). Second, one computes the Jacobian $Z^{(k)}$. Third, one solves Eq. (2.7.5) to obtain the value of ΔX (where $\Delta X = X^{(k)} - X^{(k+1)}$). Finally, one obtains the new iterate by

$$X^{(k+1)} = \Delta X - X^{(k)}\,. \qquad (2.7.7)$$

The above steps are repeated until a convergent solution is obtained. The question is, what is the Jacobian? It can be verified that the Jacobian is given by the following formula, if we assume b is independent of x:

$$z_{ij} = a_{ij} + \sum_{k=1}^{n} \frac{\partial a_{ik}}{\partial x_j} x_k\,. \qquad (2.7.8)$$

For the example given in Eq. (2.7.2), the Newton-Ralson algorithm can be written as

$$x^{(k+1)} = x^{(k)} - \frac{\left(x^{(k)}\right)^2 - 2x^{(k)} - 3}{2x^{(k)} - 2}\,. \qquad (2.7.9)$$

The iteration solution by the Newton-Ralson method is also given in Table 2.7.1. It is seen that it also takes only four iterations to yield a convergent solution.

2.8. References

Axelsson, O. 1974. On preconditioning and convergence acceleration in sparse matrix problems, CERN 74-10. European Organization for Nuclear Research (CERN), Data Handling Division.

Bartels, R. and J. W. Daniel. 1974. A conjugategradient approach to nonlinear ellipic boundary value problems in irregular regions. Proceeding of the Conference on Numerical Solution of Differential Equations, Dundee, Scotland, 1973. New York: Springer-Verlag.

Brandt, A. 1984. Multigrid Techniques: 1984 Guide with Applications to Fluid Dynamics. Department of Applied Mathematics, The Weizmann Institute of Science, Rehovot, 76100, Israel.

Brebbia, C. A., J. C. F. Telles, and L. C. Wrobel. 1984. Boundary Element Techniques. New York: Springger-Verlag.

Briggs, W. L., 1987. A Multigrid Tutorial. Philadelphia: Society for Industrial and Applied Mathematics.

Carrano, C. S. Jr. and G. T. Yeh. 1995. A fourier analysis of dynamic optimization of the Petrov-Galerkin finite element method. Int. J. Num. Methods Engg. 38:4123-4155.

Chandra, R., S. C. Eisenstat, and M. H. Schultz. 1977. The modified conjugate residual method for partial differential equations. In R. Vichnevetsky, ed. Advances in Computer Methods for Partial Differential Equations. Vol. II. Minneapollis: IMACS.

Cheng, H. P., G. T. Yeh, M. H., Li, J. Xu, and R. Carsel. 1998. A study of incorporating the multigrid method into the three-dimensional finite element discretization. Intl. J. Numer. Methods Engg., 41:499-526.

Concus, P. and G. H. Golub. 1976. A generalized conjugate gradient method for nonsymetric systems of linear equations. Rep. Stan-CS-76-646, Computer Science Department, Stanford University.

Concus, P., G. H. Golub, and D. P. O'Leary. 1976. A generalized conjugate gradient method for the numerical solution of ellipic partial differential equations. In J. R. Bunch and D. R. Rose, eds. Sparse Matrix Computation (pp. 309-332). New York: Academic Press.

Daniel, J. W. 1965. The conjugate gradient method for linear and nonlinear operator equations. Doctoral thesis, Stanford University.

Daniel, J. W. 1967. The conjugate gradient method for linear and nonlinear operator equations. SIAM J. Numer. Anal. 4:10-26.

Duff, I. S. 1977. MA28 - A set of Fortran Subroutines for Sparse Unsymmetric Linear Equations. AERE-R 8730. Harwell, Oxfordshire: AERE.

Finlayson, B. A. 1972. Existence of variational principles for the Navier-Stokes equation. Phys. Fluids. 15(6): 963-967.

Finalyson, B. A. and L. E. Scriven. 1967. On the search for variational principles. J. Heat Mass Transfer. 19:799-821.

Forsythe, G. E. and W. R. Wasow. 1960. Finite-Difference Methods for Partial Differential Equations. New York: John Wiley & Sons Inc.

Gambolati, G. 1980. Fast solution to finite element flow equations by new iteration and modified conjugate gradient method. Intl. J. Numer. Methods Engg., 15:661-675.

Gambolati, G. and G. Volpi. 1980a. Analysis of performance of the modified conjugate gradient method for solution of sparse linear sets of finite element equations. Third International Conference on Finite Elements in Flow Problems, Banff, Canada,

1980.

Gambolati, G. and G. Volpi. 1980b. An improved iterative scheme for refining the solution of ill-conditioned systems. Information Processing. 80:729-734.

Gambolati, G. and A. M. Perdon. 1984. The conjugate gradients in subsurface flow and land subsidence modeling. In J. Bear and M. Y. Corapcioglu, eds. Fundamentals of Transport Phenomena in Porous Media. NATO-ASI Series (Vol. 82, pp. 953-983). The Hague: Martinus Nijoff.

Gambolati, G., G Galeati, and S. P. Neuman. 1990. A Eulerian-Lagrangian finite element model for coupled groundwater transport. In G. Gambolati, A. Rinaldo, C. A. Brebbia, W. G. Gray, and G. F. Pinder, eds. Computational Methods in Subsurface Hydrology. (pp. 341-348). Eighth International Conference on Computational Methods in Water Resources, Venice, Italy, June 11-15, 1990. Southampton: Computational Mechanics Publications.

Gosman, A. D. and K. Y. M. Lai. 1982. Finite difference and other approximations for the transport and Navier-Stokes equations. Presented at the IAHR Symposium on Refined Modeling of Flows, Paris.

Gottlieb, D. 1984. Spectral methods for compressible flow problems. ICASE Report No. 84-29, June 22, 1984. Institute for Computer Applications in Science and Engineering, NASA Langley Research Center, Hampton, VA.

Hageman, L. A. and D. M. Young. 1981. Applied Iterative Methods. New York: Academic Press.

Hassan, A. A. 1974. Mathematical modeling of water quality for water resources management. Vol. 1. Development of the Groundwater Quality Model. District Report, Department of Water Resources, Southern District, State of California.

Hayes, L., G. F. Pinder, and M. Celia. 1981. Alternating-direction collocation for rectangular regions. Comp. Meth. Appl. Meth. Eng. 17:265-277.

Hestenes, M. R. and E. L. Stiefel. 1952. Methods of conjugate gradients for solving linear system. Nat. Bur. Std. J. Res. 49:409-436.

Hilderbrand, F. B. 1968. Finite-Difference Equations and Simulations. Englewood Cliffs, NJ: Prentice-Hall.

Hill, M. C. 1990. Preconditioned Conjugate-Gradient 2 (PCG2), A Computer Program for Solving Ground-Water Flow Equations. USGS Water-Resources Investigations Report 90-4048. Denver: U.S. Geological Survey.

Hinton, E. and D. R. J. Owen. 1977. Finite Element Programming. New York:

Academic Press.

Hood, P. 1976. Frontal solution program for unsymmetric matrices. Intl. J. Numer. Methods Engg., 10:379-400.

Huebner, K. H. 1975. The Finite Element Method for Engineers. New York: John Wiley & Sons, Inc.

Hussaini, M. Y., C. L. Streett, and T. A. Zang. 1983. Spectral methods for partial differential equations, ICASE Report No. 83-46, August 29. Institute for Computer Applications in Science and Engineering, NASA Langley Research Center, Hampton, VA.

Huyakorn, P. S. and G. F. Pinder. 1977. A pressure enthalpy finite element model for simulating hydrothermal reservoir reservoirs. Present4ed at the Second International Symposium on Computer Methods for Partial Differential Equations. Leigh University, Bethlehem, Pennsylvania, June 22-24.

Irons, B. M. 1970. A frontal solution program. Intl. J. Numer. Methods Engg., 2:5-32, 1970.

Jameson, A., W. Smith, and E. Turkel, 1981. Numerical solution of the Euler equations by finite volume methods using Runge-Kutta time-stepping scheme. AIAA Paper 81-1259, June.

Kaasschieter, E. F. 1988. Guidelines for the use of preconditioned conjugate gradients in solving discretized potential flow problems. In M. A. Celia, L. A. Ferrand, C. A. Brebbia, W. G. Gray, and G. F. Pinder,, eds. Computational Methods in Water Resources Vol. 2, Numerical Methods for Transport and Hydrological Processes, (147-152). Amsterdam: Elsevier, Amsterdam.

Konikow L. F. and J. D. Bredhoeft. 1978. Computer Model of Two-Dimensional Solute Transport and Dispersion in Groundwater, (Bk. 7, Ch. C2), Techniques of Water Resources Investigations. Reston, Va: USGS.

Kuiper, L. K. 1987. Computer Program for Solving Ground-Water Flow Equations by the Preconditioned Conjugate Gradient Methods. USGS Water-Resources Investigation Report 87-4091. Austin, Texas: USGS.

Livesley, R. K. 1983. Finite Elements: An Introduction for Engineers. Cambridge: Cambridge University Press.

Luenberger, D. G. 1973. Introduction to Linear and Nonlinear Programming. Reading, MA: Addison-Wesley.

Melosh, R. J. and R. M. Bamford. 1969. Efficient solution of load-deflection equations. J. Struct.Div. ASCE, 95:661-676.

Mikhlin, S. G. 1965. The Problem of the Minimum of a Quadratic Functional, San Francisco, CA: Holden-Day. (English translation of the 1957 edition)

Mikhlin, S. G. 1964. Variational Methods in Mathematical Physics, New York: Macmillan Company. (English translation of the 1957 editions)

Mikhlin, S. G. and K. Smolistsky. 1967. Approximate Method for the Solution of Differential and Integral Equations, Amsterdam: Elsevier.

Mondkar, R. J. and G. H. Powell. 1974. Large capacity equation solver for structural analyses. Comp. Struc., 4:531-548.

Narasimahn, T. N. and P. A. Weatherspoon. 1976. An integrated finite element method for analyzing fluid flow in porous media. Water Resources Research, 12(1):57-64.

O'Leary, D. P. 1975. Hybrid conjugate gradient algorithms. Doctoral thesis, Stanford University.

Orlob, G.T. and P. C. Woods. 1967. Water quality management in irrigation system. J. Irrigation and Drainage Div., ASCE, 93:49-66.

Orssag, S. A. 1980. Spectral methods in complex geometries. J. Comput. Physics, 37:70-92.

Peric, M. 1985. A finite volume method for the prediction of three-dimensional fluid flow in complex cucts. Ph.D. dissertation. Mechanical Engineering Department, Imperial College, London.

Pinder, G. F. and W. G. Gray. 1977. Finite Element Simulation in Surface and Subsurface Hydrology. New York: Academic Press.

Reid, J. K. 1971. On the method of conjugate graidents for the solution of large sparse systems of linear equations. Proceeding of the Conference on Large Sparse Sets of Linear Equations. (pp. 231-254). New York: Academic Press.

Reid, J. K. 1972. The use of conjugate gradients for systems of linear equations possessing Property A. SIAM J. Numer. Anal., 9:325-332.

Roache, P. T. 1976. Computational Fluid Dynamics. Albuquerque, NM: Hermosa.

Sagan, H. 1961. Boundary and Eigenvalue Problems in Mathematical Physics. New York: John Wiley & Sons, Inc.

Salvadori, M. G. and M. L. Baron. 1961. Numerical Methods in Engineering. Englewood Cliffs, NJ: Prentice-Hall.

Shapiro, A. M. and G. F. Pinder. 1981. Solution of immiscible displacement in porous media using the collocation finite element method. In K. P. Holz. et al., eds. Finite Element in Water Resources (pp. 9.61-9.70). Berlin: Springer-Verlag

Shoup, T. E. 1978. A Practical Guide to Computer Methods for Engineers. Englewood Cliffs, N J: Prentice-Hall.

Spalding, D. B. 1972. A novel finite difference formulation for differential equations involving both first and second derivatives. Intl. J. Numer. Methods Engg., 4:551-559.

Tanji, K. K. 1970. A computer analysis on the leaching of boron from stratified soil column. Soil Sciences, 110:44-51.

Tewarson, R. P. 1973. Sparse Matrices. New York: Academic Press.

Thacker. W. C. 1977. Irregular grid finite-difference technique: simulation of oscillations in shallow circular basin. J. Physical Oceanography, 7:284-292.

Thacker. W. C. 1977. Irregular grid finite-difference technique: simulation of oscillations in shallow circular basin. J. Physical Oceanography, 7:284-292.

Thacker, W. C., A. Gonzaleg, and G. E. Putland. 1980. A method of automating the construction of irregular computational grids for storm surge forecast models . J. Comput. Phys. 37:371-387.

Tonti, E. 1969. Variational formulation of nonlinear differential equations, I, II, Bull. Acad. Roy Belg. (Classe Sci.) (5), 55:262-278.

Vainberg, M. M. 1964. Variational Methods for the Study of Nonlinear Operators. San Francisco, CA: Holden-Day.

Varga, R. 1962. Matrix Iterative Analysis. Englewood Cliffs, NJ: Prentice-Hall.

Wang, H. and M. P. Anderson. 1982. Introduction to Groundwater Modeling: Finite Difference and Finite Element Methods. San Francisco: Freeman.

Westerink, J. M., M. E. Canterkin, and D. Shea. 1988. Non-diffusive N+2 degree upwinding for the finite element solution of the time dependent transport equation. In M. A. Celia, et al. eds. Development in Water Sci. 36, Vol. 2, Numerical Methods in Water Resources (pp. 57-62). Comput. Mechanics Publ., Elsevier.

Wilkinson, J. H. 1965. The Algebraic Eigenvalue Problem. London: Oxford

University Press.

Yeh, G. T. 1981a. Numerical solution of Navier-Stokes equations with an integrated compartment method (ICM), Intl. J. Numeri. Methods Fluids, 1:207-223.

Yeh, G. T. 1981b. ICM: An Integrated Compartment Method for Numerically Solving Partial Differential Equations. ORNL-5684. Oak Ridge, TN: Oak Ridge National Laboratory.

Yeh, G. T. 1983. Solution of groundwater flow equations using an orthogonal finite element scheme. In G. Mesnard, ed., Proceeding of International 83 Summer Conference, Modeling and Simulation, Vol. 4. (pp. 329-351). Tassin, France: AMSE Press.

Yeh, G. T. 1985. An orthogonal-upstream finite element approach to modeling aquifer contaminant transport. Water Resources Research, 22(6):952-964.

Young, D. 1971. Iterative Solution of Large Linear Systems. New York: Academic Press.

Zang, T. A., C. Streett, and M. Y. Hussanini. 1989. Spectral methods for CFD. ICASE Report No. 89-13, February 17, 1989. Hampton, VA: Institute for Computer Applications in Science and Engineering, NASA Langley Research Center.

Zienkiewicz, O. C. 1977. The Finite Element Method. 3rd Ed. New York: McGraw-Hill Book Company.

gradient/concentration gradient varying along the Neumann boundary; (6) treat time-dependent total fluxes distributed over the Cauchy boundary; (7) automatically determine variable boundary conditions of evaporation, infiltration, or seepage on the soil-air interface; (8) include river boundary conditions to deal with stream-subsurface flow interactions; (9) provide two options of treating the mass matrix - consistent and lumping; (10) include the off-diagonal hydraulic conductivity/dispersion diffusion coefficient tensor components in the governing equations for dealing with cases when the coordinate system does not coincide with the principal directions of the hydraulic conductivity/dispersion coefficient tensor; (11) give three options for estimating the nonlinear matrix; (12) include several options for solving the linearized matrix equations: direct band matrix solution, block iterations, successive point iterations, and four preconditioned conjugate gradient methods; (13) use mixtures of various element shapes (quadrilateral and triangular elements in 2-D and hexahedral and tetrahedral elements as well as triangular prisms in 3-D) for ease of discretizating complicated geometries; (14) automatically reset time step size when boundary conditions or source/sinks change abruptly; and (15) check the mass balance computation over the entire region for every time step.

3.1. Governing Equations of Single-Phase Flow and Thermal and Solute Transport

3.1.1. Mass Balance Equations

The mass balance equation for the fluid is expressed in integral form as

$$\frac{D_{V_s}}{Dt}\int_{v} S n_e \rho_f dv = -\int_{\Gamma} \mathbf{n}\cdot(S n_e \rho_f)\mathbf{V}_{fs} d\Gamma + \int_{v} \rho_f^* q dv - \int_{v} \rho_f n_e S m_{f\to s} dv \,, \tag{3.1.1}$$

where t is time, $D_{V_s}()/D_t$ is the material derivative of () with respect to time relative to the solid having a velocity of $\mathbf{V}_s$, v is the material volume containing constant amount of solid media (L^3), S is the degree of saturation (L^3/L^3), n_e is effective porosity (L^3/L^3), Γ is the surface enclosing the material volume v (L^2), $\mathbf{n}$ is the outward unit vector normal to the surface Γ, $\mathbf{V}_{fs}$ is the fluid velocity relative to the solid (L/T), ρ_f^* is the density of source/sink fluid (M/L^3), q is the artificial source/sink per unit volume of the medium ($(L^3/T)/L^3$), and $m_{f\to s}$ is the loss rate of fluid mass per unit fluid mass to the solid phase due to phase change and/or interface diffusion [(M/LM)/T]. By the Reynolds transport theorem (Owczarek, 1964), Eq. (3.1.1) can be written as

$$\begin{aligned}\int_{v} \frac{\partial S n_e \rho_f}{\partial t} dv + \int_{\Gamma} \mathbf{n}\cdot(S n_e \rho_f \mathbf{V}_s) d\Gamma + \int_{\Gamma} \mathbf{n}\cdot(S n_e \rho_f \mathbf{V}_{fs}) d\Gamma \\ = \int_{v} \rho_f^* q dv - \int_{v} \rho_f n_e S m_{f\to s} dv \,,\end{aligned} \tag{3.1.2}$$

where $\mathbf{V}_s$ is the velocity of the deformable surface Γ (L/T).

Applying the Gaussian divergence theorem to Eq. (3.1.2) and using the fact that v is arbitrary, one can obtain the following mass balance equation for the fluid in differential form:

$$\frac{\partial S n_e \rho_f}{\partial t} + \nabla\cdot\left(S n_e \rho_f \mathbf{V}_s\right) + \nabla\cdot\left(S n_e \rho_f \mathbf{V}_{fs}\right) = \rho_f^* q - \rho_f n_e S m_{f\to s} \ . \qquad (3.1.3)$$

The first term of Eq. (3.1.3) represents rate of change of fluid mass per unit medium volume, the second term represents the net flux of fluid mass per unit medium volume due to deformation, the third term represents the net flux of fluid mass per unit medium volume, the fourth term represents the artificial rate per unit medium volume, and the fifth term represents the production rate of fluid mass per unit medium volume due to phase change and/or interface diffusion.

Similarly, applying the mass balance principle to the solid mass, we obtain

$$\frac{\partial (1 - n_e)\rho_s}{\partial t} + \nabla\cdot\left[\left(1 - n_e\right)\rho_s \mathbf{V}_s\right] = -\rho_b m_{s\to f} \ , \qquad (3.1.4)$$

where ρ_s is the solid density (M/L^3), $\rho_b = (1-n_e)\rho_s$ is the bulk density of the media, (M^3/L), and $m_{s\text{-->}f}$ is the loss rate of solid mass per solid mass to the fluid phase due to phase change and/or interface diffusion [(M/M)/T]. Eqs. (3.1.3) and (3.1.4) are subject to the following constraint:

$$\rho_f n_e S m_{f\to s} + \rho_b m_{s\to f} = 0 \ . \qquad (3.1.5)$$

3.1.2. Momentum Balance Equations

The macroscopic balance equation of momentum for the fluid is expressed in integral from as

$$\frac{D_{\mathbf{V}_s}}{Dt}\int_v S n_e \rho_f \mathbf{V}_f \, dv = -\int_\Gamma \mathbf{n}\cdot(S n_e \rho_f \mathbf{V}_f)\mathbf{V}_{fs} d\Gamma + \int_v S n_e \rho_f \mathbf{g} dv + \int_\Gamma \mathbf{n}\cdot\sigma_f d\Gamma + \int_v S n_e \rho_f \mathbf{F}_f dv + \int_v \rho_f^* q \mathbf{V}^* \, dv - \int_v \rho_f n_e S \mathbf{M}_{f\to s} dv \ , \qquad (3.1.6)$$

where $\mathbf{V}_f = \mathbf{V}_{fs} + \mathbf{V}_s$ is the fluid velocity (L/T), σ_f is the stress tensor exerted on the fluid ($ML/(T^2L^2)$), $\mathbf{F}_f$ is the frictional force per unit fluid mass exerted on the fluid (L/T^2), **g** is the gravitational force per unit fluid volume (L/T^2), $\mathbf{V}^*$ is the velocity of the artificial source/sink, and $\mathbf{M}_{f\text{-->}s}$ is the rate of loss of fluid momentum per unit fluid mass to the solid phase, ($ML/(MT^2)$). The fluid stress tensor is made up of the effect of pressure and shear stress as

$$\sigma_f = - S n_e p \delta + \tau_f \ , \qquad (3.1.7)$$

where p is the pressure, δ is the delta tensor, and τ_f is the shear stress tensor. By the Reynolds transport theorem and Eq. (3.1.7), Eq. (3.1.6) can be written as

$$\int_v \frac{\partial S n_e \rho_f \mathbf{V}_f}{\partial t} dv + \int_\Gamma \mathbf{n}\cdot(S n_e \rho_f \mathbf{V}_f \mathbf{V}_s)\, d\Gamma + \int_\Gamma \mathbf{n}\cdot(S n_e \rho_f \mathbf{V}_f \mathbf{V}_{fs})\, d\Gamma$$

$$= \int_v S n_e \rho_f \mathbf{g}\, dv - \int_\Gamma \mathbf{n}\cdot(n_e S p \delta)\, d\Gamma + \int_\Gamma \mathbf{n}\cdot\tau_f\, d\Gamma \quad . \tag{3.1.8}$$

$$+ \int_v S n_e \rho_f \mathbf{F}_f dv + \int_v \rho_f^* q \mathbf{V}^* \, dv - \int_v \rho_f n_e S \mathbf{M}_{f\text{-}s} dv \, .$$

Applying the Gaussian divergence theorem to Eq. (3.1.8), denoting $\theta = Sn_e$, and noting $\nabla(\theta p \delta) = \nabla(\theta p)$, we have

$$\frac{\partial \rho_f \theta \mathbf{V}_f}{\partial t} + \nabla\cdot\left(\rho_f \theta \mathbf{V}_f \mathbf{V}_f\right) = \rho_f \theta \mathbf{g} - \nabla(\theta p) + \tag{3.1.9}$$
$$\nabla\cdot\tau_f + \rho_f \theta \mathbf{F}_f + \rho_f^* q \mathbf{V}^* - \rho_f n_e S \mathbf{M}_{f\text{-}s} \, ,$$

where θ is the moisture content. The first term of Eq. (3.1.9) represents the rate of change of momentum, the second term is the net momentum flux due to advection, the third term is the gravitational force, the fourth term is the pressure force, the fifth term is the force due to shear stress, the sixth term is the frictional force between the solid and fluid, the seventh term is the momentum due to artificial source/sink, and the last term is the production rate of the fluid momentum from the solid to liquid phase.

An equation for the balance of solid momentum similar to Eq. (3.1.9) can be obtained as

$$\frac{\partial \rho_b \mathbf{V}_s}{\partial t} + \nabla\cdot\left(\rho_b \mathbf{V}_s \mathbf{V}_s\right) = \rho_b \mathbf{g} + \nabla\cdot\sigma_s + \rho_b \mathbf{F}_s - \rho_b \mathbf{M}_{s\text{-}f} \, , \tag{3.1.10}$$

where $\rho_b = (1-n_e)\rho_s$ is the bulk density of the media (M/L^3), $\mathbf{M}_{s\text{-->}f}$ is the loss rate of the solid momentum per unit solid mass to the fluid phase ($ML/(MT^2)$), σ_s is the stress tensor exerted on the solid ($ML/(T^2L^2)$), and $\mathbf{F}_s$ is the frictional force per unit solid mass exerted on the solid (L/T^2). Equations (3.1.9) and (3.1.10) are subject to the following constraints:

$$\rho_f \theta \mathbf{F}_f + \rho_b \mathbf{F}_s = 0 \, . \tag{3.1.11}$$

The loss rates of fluid momentum and solid momentum, $\mathbf{M}_{f\text{-->}s}$ and $\mathbf{M}_{s\text{-->}f}$, respectively, are given by

$$\mathbf{M}_{f\text{-}s} = m_{f\text{->}s} \frac{1}{2}\left[\left(1+\mathrm{sign}(m_{f\text{-}s})\right)\mathbf{V}_f + \left(1-\mathrm{sign}(m_{f\text{-}s})\right)\mathbf{V}_s\right] \tag{3.1.12}$$

$$\mathbf{M}_{s\text{-}f} = m_{s\text{->}f} \frac{1}{2}\left[\left(1+\mathrm{sign}(m_{s\text{-}f})\right)\mathbf{V}_s + \left(1-\mathrm{sign}(m_{s\text{-}f})\right)\mathbf{V}_f\right] , \tag{3.1.13}$$

which results in the following constraint:

$$\rho_f \theta M_{f\to s} + \rho_b M_{s\to f} = 0 . \qquad (3.1.14)$$

In retrospect, the above constraint could have been posed by a simple physical argument.

3.1.3. Energy Transport Equations

We will assume that the temperature in the liquid water is in equilibrium with that in the media matrix. With this assumption, we can derive the energy equation for the combined fluid and solid phase rather than for the fluid and solid phases separately. The conservation of energy is expressed in integral form as

$$\frac{D_{V_s}}{Dt}\int_v (Sn_e\rho_f E_f + \rho_b E_s)dv = -\int_\Gamma \mathbf{n}\cdot(Sn_e\rho_f E_f \mathbf{V}_{fs})d\Gamma +$$

$$\int_v \rho_f Sn_e \mathbf{g}\cdot\mathbf{V}_f\, dv + \int_v \rho_b\, \mathbf{g}\cdot\mathbf{V}_s\, dv + \int_v \rho_f Sn_e \mathbf{F}_f\cdot\mathbf{V}_f dv + \int_v \rho_b \mathbf{F}_s\cdot\mathbf{V}_s dv \qquad (3.1.15)$$

$$\int_\Gamma \mathbf{n}\cdot\sigma_f\cdot\mathbf{V}_f d\Gamma + \int_\Gamma \mathbf{n}\cdot\sigma_s\cdot\mathbf{V}_s d\Gamma - \int_\Gamma \mathbf{J}_h\cdot\mathbf{n} d\Gamma + \int_v \rho^* q E_f^* dv ,$$

in which

$$E_f = U_f + \frac{V_f^2}{2} ; \quad E_s = U_s + \frac{V_s^2}{2} , \qquad (3.1.16)$$

where E_f is the sum of internal energy and kinetic energy for the fluid phase (L^2/T^2), E_s is the sum of internal energy and kinetic energy for the solid phase (L^2/T^2), U_f is the internal energy of the fluid per unit fluid mass (L^2/T^2), U_s is the internal energy of the solid per unit solid mass (L^2/T^2), $\mathbf{J}_h$ is the heat flux due to conduction and radiation (M/L^3), and E_f^* is the sum of internal energy and kinetic energy per unit source/sink fluid mass (L^2/T^3).

By the Reynolds transport theorem, Eq. (3.1.15) can be written as

$$\int_v \frac{\partial(\rho_f n_e SE_f + \rho_b E_s)}{\partial t} dv + \int_\Gamma \mathbf{n}\cdot[(S\rho_f n_e E_f + \rho_b E_s)\mathbf{V}_s] d\Gamma + \int_\Gamma \mathbf{n}\cdot(Sn_e\rho_f E_f \mathbf{V}_{fs}) d\Gamma$$

$$= \int_v \rho_f Sn_e \mathbf{g}\cdot\mathbf{V}_f dv + \int_v \rho_b \mathbf{g}\cdot\mathbf{V}_s dv + \int_v \rho_f Sn_e \mathbf{F}_f\cdot\mathbf{V}_f dv + \int_v \rho_b \mathbf{F}_s\cdot\mathbf{V}_s dv \qquad (3.1.17)$$

$$+ \int_\Gamma \mathbf{n}\cdot\sigma_f\cdot\mathbf{V}_f d\Gamma + \int_\Gamma \mathbf{n}\cdot\sigma_s\cdot\mathbf{V}_s d\Gamma - \int_\Gamma \mathbf{J}_h\cdot\mathbf{n} d\Gamma + \int_v \rho_f^* q E_f^* dv .$$

Applying the Gaussian divergence theorem to Eq. (3.1.17) and using the fact that υ is arbitrary, we obtain the following equation of conservation of energy in differential form:

$$\frac{\partial(\rho_f n_e S E_f + \rho_b E_s)}{\partial t} + \nabla\cdot[(\rho_f n_e S E_f + \rho_b E_s)\mathbf{V}_s] +$$

$$\nabla\cdot(S n_e \rho_f E_f \mathbf{V}_{fs}) = \rho_f S n_e \mathbf{g}\cdot\mathbf{V}_f + \rho_f S n_e \mathbf{F}_f\cdot\mathbf{V}_f + \rho_b \mathbf{F}_s\cdot\mathbf{V}_s + \tag{3.1.18}$$

$$\rho_b \mathbf{g}\cdot\mathbf{V}_s + \nabla\cdot(\mathbf{V}_f\cdot\sigma_f) + \nabla\cdot(\mathbf{V}_s\cdot\sigma_s) - \nabla\cdot\mathbf{J}_h + \rho_f^* q E_f^* \ .$$

Recall that the equations for the conservation of fluid and solid momentum are given by a slightly modified Eq. (3.1.8) and (3.1.9) as follows:

$$\frac{\partial \rho_f \theta \mathbf{V}_f}{\partial t} + \nabla\cdot(\rho_f \theta \mathbf{V}_f \mathbf{V}_f)$$

$$= \rho_f \theta \mathbf{g} + \nabla\cdot\sigma_f + \rho_f \theta \mathbf{F}_f + \rho_f^* q \mathbf{V}^* - \rho_f \theta \mathbf{M}_{f\to s} \tag{3.1.19}$$

and

$$\frac{\partial \rho_b \mathbf{V}_s}{\partial t} + \nabla\cdot(\rho_b \mathbf{V}_s \mathbf{V}_s) = \rho_b \mathbf{g} + \nabla\cdot\sigma_s + \rho_b \mathbf{F}_s - \rho_b \mathbf{M}_{s\to f} \ . \tag{3.1.20}$$

Taking the inner product of Eq. (3.1.19) with $\mathbf{V}_f$ and of Eq. (3.1.20) with $\mathbf{V}_s$, subtracting the results from Eq. (3.1.18), and using the constraint Eq. (3.1.14), we obtain

$$\frac{\partial(\rho_f n_e S U_f + \rho_b U_s)}{\partial t} + \nabla\cdot[(\rho_f n_e S U_f + \rho_b U_s)\mathbf{V}_s] + \nabla\cdot(S n_e \rho_f U_f \mathbf{V}_{fs})$$

$$= \sigma_f : \nabla\mathbf{V}_f + \sigma_s : \nabla\mathbf{V}_s - \nabla\cdot\mathbf{J}_h + \rho_f^* q(E_f^* - \mathbf{V}^*\cdot\mathbf{V}_f) + \rho_f \theta \mathbf{M}_{f\to s}\cdot\mathbf{V}_{fs} \ . \tag{3.1.21}$$

Using Eqs. (3.1.3) and (3.1.4) and expanding Eq. (3.1.21), we rewrite the equation of thermal energy as follows:

$$\rho_f n_e S\left(\frac{\partial U_f}{\partial t} + \mathbf{V}_f\cdot\nabla U_f\right) + \rho_b\left(\frac{\partial U_s}{\partial t} + \mathbf{V}_s\cdot\nabla U_s\right)$$

$$= \sigma_f : \nabla\mathbf{V}_f + \sigma_s : \nabla\mathbf{V}_s - \nabla\cdot\mathbf{J}_h + \tag{3.1.22}$$

$$\rho_f^* q(E_f^* - \mathbf{V}^*\cdot\mathbf{V}_f - U_f) + (\rho_f \theta m_{f\to s} U_f + \rho_b m_{s\to f} U_b) \ .$$

Equation (3.1.22) is quite general; no assumption has been made regarding the properties of the fluid and the media. For most engineering applications, it is convenient to have the equation of thermal energy (Eq. (3.1.22) in terms of temperature and heat capacity rather than internal energy. We may rewrite the equation in these terms by recognizing that the internal energy may be considered function of specific volume and temperature. Denoting U as internal energy, T as temperature, and V as specific volume, we have

$$dU = \left(\frac{\partial U}{\partial V}\right)_T dV + \left(\frac{\partial U}{\partial T}\right)_V dT . \tag{3.1.23}$$

From the combined first and second laws of thermodynamics (Sears, 1952), we have

$$ds = \frac{1}{T}(dU + pdV) , \tag{3.1.24}$$

where s is the entropy and p is the pressure. Substituting Eq. (3.1.23) into Eq. (3.1.24), we obtain

$$ds = \frac{1}{T}\left(\frac{\partial U}{\partial T}\right)_V dT + \frac{1}{T}\left[p + \left(\frac{\partial U}{\partial V}\right)_T\right] dV . \tag{3.1.25}$$

On the other hand, considering s as function of T and V, we have

$$ds = \left(\frac{\partial s}{\partial T}\right)_V dT + \left(\frac{\partial s}{\partial V}\right)_T dV . \tag{3.1.26}$$

Comparing Eqs. (3.1.23) and (3.1.24), we have

$$\left(\frac{\partial s}{\partial T}\right)_V = \frac{1}{T}\left(\frac{\partial U}{\partial T}\right)_V \quad \text{and} \quad \left(\frac{\partial s}{\partial V}\right)_T = \frac{1}{T}\left[p + \left(\frac{\partial U}{\partial V}\right)_T\right] . \tag{3.1.27}$$

Differentiating the first equation of Eq. (3.1.27) with respect to V and the second one with respect to T, and taking the difference of the resulting equation, we have

$$\left(\frac{\partial U}{\partial V}\right)_T = - p + T\left(\frac{\partial p}{\partial T}\right)_V . \tag{3.1.28}$$

Substituting Eq. (3.1.28) into Eq. (3.1.23) and using the definition of C_v (the heat capacity at constant volume, per unit mass), we obtain

$$dU = \left[-p + T\left(\frac{\partial p}{\partial T}\right)_V\right] dV + C_v dT . \tag{3.1.29}$$

Using the definition of the specific volume $V = 1/\rho$, we obtain, from Eq. (3.1.29), the following relationship between the internal energy and temperature:

$$\rho^2 dU = \left[p - T\left(\frac{\partial p}{\partial T}\right)_V\right] d\rho + \rho^2 C_v dT . \tag{3.1.30}$$

Using Eq. (3.1.30) and assuming that the temperature in the fluid is in equilibrium with that in the media matrix, we can write the thermal energy equation, Eq. (3.1.22), in terms of temperature:

$$\left(n_eS\rho_fC_{vf}+\rho_bC_{vs}\right)\frac{\partial T}{\partial t}+n_eS\rho_fC_{vf}\mathbf{V}_{fs}\cdot\nabla T+\left(n_eS\rho_fC_{vf}+\rho_bC_{vb}\right)\mathbf{V}_s\cdot\nabla T+$$

$$\left[p-T\left(\frac{\partial p}{\partial T}\right)_V\right]\left[\frac{n_eS}{\rho_f}\left(\frac{\partial\rho_f}{\partial t}+\mathbf{V}_f\cdot\nabla\rho_f\right)+\frac{1}{\rho_b}\left(\frac{\partial\rho_b}{\partial t}+\mathbf{V}_s\cdot\nabla\rho_b\right)\right] \qquad (3.1.31)$$

$$=\rho_fqU_f+\sigma_f:\nabla\mathbf{V}_f+\sigma_s:\nabla\mathbf{V}_s-\nabla\cdot\mathbf{J}_h+\rho_f^*q\left(E_f^*-\mathbf{V}^*\cdot\mathbf{V}_f-U_f\right)$$

$$+\left(\rho_f\theta m_{f\to s}U_f+\rho_bm_{s\to f}U_b\right).$$

Using the continuity relationships Eqs. (3.1.3) and (3.1.4), we can rewrite Eq. (3.1.31) as

$$\left(n_eS\rho_fC_{vf}+\rho_bC_{vs}\right)\frac{\partial T}{\partial t}+n_eS\rho_fC_{vf}\mathbf{V}_{fs}\cdot\nabla T+\left(n_eS\rho_fC_{vf}+\rho_bC_{vb}\right)\mathbf{V}_s\cdot\nabla T$$

$$\left[p-T\left(\frac{\partial p}{\partial T}\right)_V\right]\left[\left(\frac{\partial n_eS}{\partial t}+\nabla\cdot\left(n_eS\mathbf{V}_f\right)-\frac{\rho_f^*}{\rho_f}q+\theta m_{f\to s}\right)+\nabla\cdot\mathbf{V}_s+m_{s\to f}\right] \qquad (3.1.32)$$

$$=\sigma_f:\nabla\mathbf{V}_f+\sigma_s:\nabla\mathbf{V}_s-\nabla\cdot\mathbf{J}_h+\rho_f^*q\left(E_f^*-\mathbf{V}^*\cdot\mathbf{V}_f-U_f\right)+\left(\rho_f\theta m_{f\to s}U_f+\rho_bm_{s\to f}U_b\right).$$

Equation (3.1.31) or (3.1.32) is the equation of thermal energy in terms of temperature. It is as general as Eq. (3.1.22) except for the assumption of thermodynamic equilibrium between the fluid and the media matrix, but it is in a more useful form for calculating temperature profiles in the subsurface environment.

3.1.4. Solute Transport Equations

The general transport equation governing the temporal-spatial distribution of any solute can be derived based on the principle of conservation of mass. Let us consider a system of a single solute. Let C_w be the concentration in the dissolved phase in mass per unit liquid mass. The governing equation for C_w can be obtained by applying this principle in integral form as follows:

$$\frac{D_{V_s}}{Dt}\int_v \rho_f\theta C_w\,dv=-\int_\Gamma \mathbf{n}\cdot(\rho_f\theta C_w)\mathbf{V}_{fs}d\Gamma-\int_\Gamma \mathbf{n}\cdot\mathbf{J}\,d\Gamma- \qquad (3.1.33)$$

$$\int_v\rho_f\theta R_{f\to s}dv-\int_v\rho_f\theta\iota^c dv+\int_v M^c dv\,,$$

where v is the material volume containing constant amount of solid media (L^3), C_w is the dissolved concentration of the solute (M/M), Γ is the surface enclosing the material volume v (L^2), **n** is the outward unit vector normal to the surface Γ, **J** is the surface flux of the solute with respect to relative fluid velocity $\mathbf{V}_{fs}$ [$(M/T)/L^2$], $R_{f\to s}$ is the loss rate of the solute per unit fluid mass due to phase change [(M/M)/T)], ι^c is the rate of decay per unit fluid mass, [(M/M)/T], and M^c is the external source/sink rate per unit medium volume [$(M/L^3)/T$].

By the Reynolds transport theorem (Owczarek, 1964), Eq. (3.1.33) can be written as

$$\int_v \frac{\partial \rho_f \theta C_w}{\partial t} dv + \int_\Gamma \mathbf{n}\cdot(\rho_f \theta C_w \mathbf{V}_f)\, d\Gamma + \int_\Gamma \mathbf{n}\cdot\mathbf{J}\, d\Gamma \tag{3.1.34}$$

$$= \int_v \rho_f \theta R_{f\to s} dv - \int_v \rho_f \theta \iota^c dv + \int_v M^c dv \ .$$

Applying the Gaussian divergence theorem to Eq. (3.1.34) and using the fact that v is arbitrary, one can obtain the following continuity equation for the solute in differential form:

$$\frac{\partial \rho_f \theta C_w}{\partial t} + \nabla\cdot(\rho_f \theta C_w \mathbf{V}_f) + \nabla\cdot\mathbf{J} = \rho_f \theta(-R_{f\to s} - \iota^c) + M^c \ . \tag{3.1.35}$$

Eq. (3.1.35) is simply the statement of mass balance over a differential volume. The first term represents the rate of mass accumulation, the second term represents the net rate of mass flux due to advection, the third term is the net mass flux relative to the mean fluid velocity that could be due to dispersion and diffusion, the fourth term is the rate of mass production and reduction due to interfacial chemical reactions and radioactive decay, and the last term is the source/sink term corresponding to artificial injection and or withdrawal.

Let C_s be the concentration (in chemical mass per unit mass of solid phase) of the solute in the solid phases. Let it be further assumed that the solute on the solid surface is not subject to hydrological transport; then its mass balance equation can be obtained by replacing $\rho_f\theta$ with ρ_b and dropping the term associated with the relative fluid velocity and the surface flux term on the left-hand side of Eq. (3.1.35):

$$\frac{\partial \rho_b C_s}{\partial t} + \nabla\cdot(\mathbf{V}_s \rho_b C_s) = \rho_b(-R_{s\to f} - \iota^s) + M^s \ , \tag{3.1.36}$$

where is ρ_b the bulk density, $R_{s\text{-->}f}$ is the loss rate of the solute per unit solid mass due to phase change [(M/M)/T)], ι^s is the rate of decay per unit solid mass [(M/M)/T], and M^c is the external source/sink rate per unit medium volume [$(M/L^3)/T$].

Equations (3.1.35) and (3.1.36) are subject to the following constraints due to the invariance of chemical reactions:

$$\rho_f \theta R_{f\to s} + \rho_b R_{s\to f} = 0 \ . \tag{3.1.37}$$

Equations (3.1.3), (3.1.4), (3.1.9), (3.1.10), (3.1.32), (3.1.35), and (3.1.36) (representing continuity of fluid, continuity of solids, momentum balance of the fluid, momentum balance of the solid, conservation of energy, and mass balance of a chemical, respectively) and the constraints given by Eqs. (3.1.5), (3.1.11), (3.1.14), and (3.1.37) constitute a total of 11 equations. The unknowns involved in these 11 equations are ρ_f, n_e, S, $\mathbf{V}_{fs}$, $m_{f\text{-->}s}$, $\mathbf{V}_s$, ρ_s, $m_{s\text{-->}f}$, p, τ_f, $\mathbf{F}_f$, $\mathbf{M}_{f\text{-->}s}$, σ_s, $\mathbf{F}_s$, $\mathbf{M}_{s\text{-->}f}$, T, C_w, C_s, $R_{f\text{-->}s}$, and $R_{s\text{-->}f}$. The number of unknowns (20) is more than the number of equations (11), hence nine constitutive relationships must be proposed. For this book, we will make several assumptions, as follows.

First, we will assume that the loss of fluid to the solid phase due to interphase change is zero, that is, $m_{f\text{-->}s} = 0$. As a result of this assumption: $m_{s\text{-->}f} = 0$ because of Eq. (3.1.5). In addition, both $\mathbf{M}_{f\text{-->}s}$ and $\mathbf{M}_{s\text{-->}f}$ are zero because of the definition of these two variables as given by Eqs. (3.1.12) and (3.1.13); furthermore, Eq. (3.1.14) will automatically be satisfied. Thus, in one assumption, we have reduced four variables and removed two governing equations, that is, we have reduced the required constitutive relationships by two. It should be noted that it appears that the assumption $m_{f\text{-->}s} = 0$ would be contradictory to the assumption $R_{f\text{-->}s} \neq 0$ if the species C_s is considered as a portion of the solid mass. To resolve this dilemma, we should consider the species C_s as an immobile liquid species, which would allow an assumption $R_{f\text{-->}s} = 0$ that is consistent with the assumption $m_{f\text{-->}s} = 0$.

Second, we will assume that the solid grain is incompressible, although the matrix skeleton is compressible. This assumption would yield

$$\frac{\partial \rho_s}{\partial t} + \mathbf{V}_s \cdot \nabla \rho_s = 0 . \tag{3.1.38}$$

Third, we assume that the fluid is macroscopically nonviscous; that is, $\tau_f = 0$, which reduces the variables by one and hence also reduces the required constitutive relationships by one.

Fourth, in addition to the assumption of thermodynamic equilibrium already made in the derivation of energy equation, we assume that the solid is a thermoelastic material. This assumption results in the expression for the solid stress tensor as (Bear and Corapcioglu, 1981)

$$\sigma_s = \nabla(\mathbf{E}:\nabla \mathbf{U}) - \chi p \delta , \tag{3.1.39}$$

where **U** is the displacement vector, **E** is the elasticity coefficient tensor, and χ is a number between 0 and 1. The displacement vector is related to the solid velocity by

$$\mathbf{V}_s = \frac{\partial \mathbf{U}}{\partial t} . \tag{3.1.40}$$

Equations (3.1.39) and (3.1.40) give an implicit relationship among the variables, S, p, $\mathbf{V}_s$, and σ_s.

Fifth, based on the linear theory and the principles of objectivity and admissibility, the following constitutive relationship can be derived following the work of Hassanizadeh (1986)

$$\begin{aligned} \theta \rho_f \mathbf{F} = p\nabla(\theta) + \pi \cdot \nabla T + \left[\mathbf{A} \cdot \nabla \rho_f + \chi \cdot \nabla C_w\right] + \\ \mathbf{R} \cdot \left(\mathbf{V}_f - \mathbf{V}_s\right) - \rho_f^* q \mathbf{V}^* , \end{aligned} \tag{3.1.41}$$

where π is the material coefficient tensor for heat, **A** is the material coefficient tensor for bulk fluid, χ is the material coefficient tensor for the solute, and **R** is the resistivity tensor for bulk fluid.

Sixth, in general the thermal equation of state can be posed that the fluid density is a function of pressure, temperature, and mole fractions of all chemical species. Here we will assume that the fluid density is a function of pressure, temperature, and mass fraction of the solute as

$$\rho_f = \rho(p,T,C_w) . \tag{3.1.42}$$

Two more constitutive relationships are needed: one would state how the degree of saturation is related to other variables and the other would state how the solute is distributed between phases. These constitutive relationships will be posed when we deal with a particular system. Before leaving this section, we wish to simplify the momentum equations for the fluid and solid.

Substituting Eq. (3.1.41) into Eq. (3.1.9) and neglecting the inertial force and (macroscopic) viscous effects (Polubarinova-Kochina, 1962), we obtain a generalized Darcy's law as

$$\theta \mathbf{V}_{fs} = -\frac{1}{\mu_f}\mathbf{P}\cdot(\nabla p - \rho_f \mathbf{g}) - \boldsymbol{\kappa}\cdot\nabla T - \left[\mathbf{B}\cdot\nabla\rho_f + \mathbf{D}\cdot\nabla C_w\right], \tag{3.1.43}$$

where $\mathbf{P} = (-\mathbf{R})^{-1}\theta^2\mu_f$ is the intrinsic permeability tensor of the media (L^2), μ_f is the dynamic viscosity of the fluid ($ML/T^2/L_2/T$), $\varkappa = (-\mathbf{R})^{-1}\cdot\theta\pi$ is the thermal diffusivity tensor, $\mathbf{B} = (-\mathbf{R})^{-1}\cdot\theta\mathbf{A}$ is the mass diffusivity tensor, and $\mathbf{D} = (-\mathbf{R})^{-1}\cdot\theta\chi$ is the diffusion coefficient tensor for solute.

Adding Eq. (3.1.9) to (3.1.10), using the constitutive relationship Eq. (3.1.39) and the constraint Eq. (3.1.11), and assuming the inertia forces in both fluid and solid phases are small, one can derive a three-dimensional consolidation equation for an elastic medium as (Bear and Corapcioglu, 1981; Verruijt, 1969) as

$$\nabla\cdot\nabla[\mathbf{E}:\nabla\mathbf{U}] = -\nabla\cdot\left(\left[n_e S\rho_f + (1-n_e)\rho_s\right]\mathbf{g}\right) + \nabla^2(Sp) . \tag{3.1.44}$$

Equation (3.1.44) is used to model the deformation or subsidence of the surface media under variably saturated flow conditions.

3.2. Three-Dimensional Saturated Flows of Essentially Incompressible Fluid

In addition to the assumptions that $m_{f\text{->}s} = 0$ and that the solid grain is incompressible, let us further simplify the system by assuming that the pore is completely saturated with one fluid, the temperature field is constant, and the solute concentration will not affect the flow. We will also assume that the medium is elastically isotropic. Under these conditions, Eqs. (3.1.3), (3.1.4), (3.1.43), (3.1.42), (3.1.44), and (3.1.40) are reduced to

$$\frac{\partial \rho_f n_e}{\partial t} + \nabla\cdot(\rho_f n_e \mathbf{V}_s) + \nabla\cdot(n_e \rho_f \mathbf{V}_{fs}) = \rho_f q , \tag{3.2.1}$$

which is continuity equation for the fluid phase;

$$\frac{\partial(1 - n_e)}{\partial t} + \nabla\cdot\left[(1 - n_e)\mathbf{V}_s\right] = 0 \ , \tag{3.2.2}$$

which is the continuity equation for the solid phase with the assumption of incompressible solid grains;

$$\mathbf{V}_{fs} = -\frac{1}{\mu_f n_e}\mathbf{P}\cdot(\nabla p - \rho_f \mathbf{g}) \ , \tag{3.2.3}$$

which is the Darcy's law, a simplified momentum equation;

$$\rho_f = \rho(p) \ , \tag{3.2.4}$$

which is the equation of state;

$$(\lambda_s + 2\mu_s)\nabla^2 e = \nabla^2 p + \nabla\cdot\left[(\rho_f n_e + \rho_b)\mathbf{g}\right] \tag{3.2.5}$$

which is the Biot (1941) three-dimensional consolidation equation; and where ρ_b is the bulk density of the medium (M/L^3), λ_s is the Lamè first constant [$(ML/T^2)/L^2$], μ_s is the Lamè second constant [$(ML/T^2)/L^2$], and $e = \nabla\circ\mathbf{U}$ is the dilatation of the media (dimensionless), and

$$\nabla\cdot\mathbf{V}_s = \frac{\partial}{\partial t}(\nabla\cdot\mathbf{U}) = \frac{\partial e}{\partial t} . \tag{3.2.6}$$

Equations (3.2.1) through (3.2.6) (representing continuity of fluid, continuity of solids, Darcy's law, the equation of state, consolidation of the medium, and the definition of $\mathbf{V}_s$, respectively) contain six variables: ρ_f, n_e, $\mathbf{V}_{fs}$, p, e, and $\mathbf{V}_s$. Hence the number of equations is equal to the number of unknowns. The system is complete, and a mathematical statement is posed. However, we can combine these six equations into a single one to simplify the problem. The simplification is demonstrated below.

Expanding Eqs. (3.2.1) and (3.2.2), we have

$$\rho_f\left[\frac{\partial n_e}{\partial t} + \nabla\cdot(n_e\mathbf{V}_s)\right] + n_e\frac{\partial \rho_f}{\partial t} + (n_e\mathbf{V}_s)\cdot\nabla(\rho_f) = -\nabla\cdot(\rho_f n_e \mathbf{V}_{fs}) + \rho_f n_e q \tag{3.2.7}$$

and

$$\frac{\partial n_e}{\partial t} = -\nabla\cdot(n_e\mathbf{V}_s) + \nabla\cdot\mathbf{V}_s \ , \tag{3.2.8}$$

respectively. Neglecting the second-order term, $(n_e\mathbf{V}_s)\cdot\nabla(\rho_f)$, and substituting Eq. (3.2.8) into (3.2.7), we obtain

$$n_e\frac{\partial \rho_f}{\partial t} + \rho_f\nabla\cdot\mathbf{V}_s = -\nabla\cdot(\rho_f n_e\mathbf{V}_{fs}) + \rho_f n_e q \ . \tag{3.2.9}$$

Now let us assume that the gravitational force is small in comparison with the pressure force in the consolidation of the media; then we can integrate Eq. (3.2.5) to yield the following:

$$\left(\lambda_s + 2\mu_s\right)e = p + f \, , \tag{3.2.10}$$

where f is the integration function. The integration function f must satisfy the Laplace's equation for all time. To simplify the matter further, we will only consider vertical consolidation. Under this condition, it has been shown (Verruijt, 1969) that the integration function f is equal to 0. It then follows from Eq. (3.2.10) that

$$\frac{\partial e}{\partial t} = \alpha \frac{\partial p}{\partial t} \, , \tag{3.2.11}$$

in which

$$\alpha = \frac{1}{\left(\lambda_s + 2\mu_s\right)} \, , \tag{3.2.12}$$

in which α is the coefficient of consolidation of the media. Substituting Eq. (3.2.11) into Eq. (3.2.6), we obtain

$$\nabla \cdot \mathbf{V}_s = \frac{\partial e}{\partial t} = \alpha \frac{\partial p}{\partial t} \, . \tag{3.2.13}$$

Differentiating the equation of state Eq. (3.2.4) with respect to time, we have

$$\frac{\partial \rho_f}{\partial t} = \frac{\partial \rho_f}{\partial p} \frac{\partial p}{\partial t} \, . \tag{3.2.14}$$

Substituting Eqs. (3.2.3) -- Darcy's law, (3.2.11) -- vertical consolidation, and (3.2.14) -- equation of state into Eq. (3.2.9), we have the following:

$$-\left(n_e\beta + \alpha\right)\rho_f \frac{\partial p}{\partial t} = \nabla \cdot \left[\frac{\rho_f}{\mu_f}\mathbf{P} \cdot \left(\nabla p + \rho_f \mathbf{g}\right)\right] + \rho_f n_e q \, , \tag{3.2.15}$$

where

$$-\beta = \frac{1}{\rho_f} \frac{\partial \rho_f}{\partial p} \tag{3.2.16}$$

is the compressibility of the fluid. The assumptions with Eq. (3.2.15) include (1) no mass exchange takes place between the fluid and solid phases, (2) the solid grain is incompressible, (3) there is a thermodynamic equilibrium between the fluid and solid, and (4) initial and viscous forces are much smaller compared to the frictional force (shear force or interface momentum).

When the fluid flow is essentially incompressible, for convenience, we define a pressure head h as

$$h = \int_{P_0}^{P} \frac{dp}{\rho_f g} \, , \tag{3.2.17}$$

in which P_o is the datum pressure $[(ML/T^2)/L^2]$. Taking the gradient of Eq. (3.2.17), we obtain

$$\nabla h = \frac{\nabla p}{\rho_f g} . \tag{3.2.18}$$

Differentiating Eq. (3.2.17) with respect to t, we obtain

$$\frac{\partial h}{\partial t} = \frac{1}{\rho_f g} \frac{\partial p}{\partial t} . \tag{3.2.19}$$

In the meantime, the gravitational field vector can be expressed in terms of a potential as:

$$\mathbf{g} = - g \nabla z , \tag{3.2.20}$$

where **g** is the acceleration of the gravity (L/T^2) and z is the potential head (L).

Substituting Eqs. (3.1.18) through (3.2.20) into Eq. (3.2.15), we obtain

$$S_s \rho_f \frac{\partial h}{\partial t} = \nabla \bullet [\rho_f \mathbf{K} \bullet \nabla H] + \rho_f n_e q , \tag{3.2.21}$$

in which

$$S_s = \alpha' + \beta' n_e, \quad \alpha' = \rho_f g \alpha, \quad \beta' = \rho_f g \beta \tag{3.2.22}$$

$$\mathbf{K} = \frac{\rho_f g}{\mu} \mathbf{P}, \quad H = h + z , \tag{3.2.23}$$

where S_s is the specific storage, α' is the modified compressibility of the media, β' is the modified compressibility of the fluid, K is the hydraulic conductivity, and H is total head.

Expanding the first term on the right-hand side of Eq. (3.2.21) and neglecting the second-order term $[(\mathbf{K} \bullet \nabla H) \bullet (\nabla \rho_f)]$, we have the following governing equation for essentially incompressible (or slightly compressible) fluid flows in saturated media under isothermal conditions:

$$S_s \frac{\partial h}{\partial t} = \nabla \bullet [\mathbf{K} \bullet (\nabla h + \nabla z)] + q, \tag{3.2.24}$$

where q is the source/sink per unit bulk volume.

To completely define the problems of saturated flow under isothermal conditions, Eq. (3.2.24) must be constrained by initial and boundary conditions. This can be done using physical considerations. Three types of boundary conditions can be prescribed for Eq. (3.2.24) when the boundary is given a priori, that is, when the boundary does not move. The first is the Dirichlet boundary on which the pressure is given

$$h = h_d \quad \text{on} \quad B_d(x,y,z,t) = 0 , \tag{3.2.25}$$

where h_d is the specified Dirichlet head, B_d is a part of the boundary $B(x,y,z,t) = 0$, and x, y, z are the spatial Cartesian coordinates. The second type is the Neumann boundary on which the gradient flux is given as

$$- \mathbf{n}\cdot(\mathbf{K}\cdot\nabla h) = q_n \quad \text{on } B_n(x,y,z,t) = 0 , \tag{3.2.26}$$

where q_n is the prescribed Neumann flux and B_n is the part of the boundary B. The third type is the Cauchy (specified flow) boundary on which the total flux is given as

$$- \mathbf{n}\cdot[\mathbf{K}\cdot(\nabla h + \nabla z)] = q_c \quad \text{on } B_c(x,y,z,t) = 0 , \tag{3.2.27}$$

where q_c is the Cauchy flux and B_c is the part of boundary B.

In addition to the above three basic types of boundary conditions, another type of boundary condition can be encountered when the river/stream and aquifer interactions are considered. This type of boundary condition is termed the radiation condition and is expressed mathematically as

$$-\mathbf{n}\cdot\mathbf{K}\cdot(\nabla h + \nabla z) = -\frac{K_R}{b_R}(h_R - h) \quad \text{on } B_r . \tag{3.2.28}$$

K_R is the hydraulic conductivity of the river bottom sediment layer, b_R is the thickness of the river bottom sediment layer, and h_R is the depth of the river bottom measured from the river surface. It should be noted that this type of boundary condition is applicable only when the river is hydraulically connected to the aquifers in the case of saturated flows.

If the boundary B is independent of time, the specification of boundary conditions outlined above is relatively easy. However, for many problems, a portion of the boundary is moving with time and a governing equation is required to specify such a kinematic condition. The kinematic condition written specifically for the subsurface media becomes

$$h = 0 \quad \text{on } B_v \tag{3.2.29}$$

that is, when the free surface reaches the ground surface, the Dirichlet boundary condition with zero pressure head is imposed. On the other hand, when the free surface remains in the medium, the following moving phreatic condition is imposed:

$$S_y\frac{\partial\zeta}{\partial t} + \left(u_s\frac{\partial\zeta}{\partial x} + v_s\frac{\partial\zeta}{\partial y} - w_s\right) = -q_v \quad \text{on } B_v|_{z=\zeta(x,y,t)} \tag{3.2.30}$$

$$h = 0 \quad \text{on } z = \zeta(x,y,t) , \tag{3.2.31}$$

where B_v is the variable moving boundary; q_v is the outward mass flux rate across the moving surface given by $z = \zeta(x,y,t)$; S_y is the specific yield; u_s, v_s, and w_s are the x-, y-, and z-components of Darcy velocity on the free surface; and ζ is the elevation of the free surface from the datum. It should be noted that the term in the parenthesis in Eq. (3.2.27b) is equal to the inward flux normal to the free surface, that is,

$$\mathbf{n}\cdot\mathbf{K}\cdot\nabla(h+z)|_{z=\zeta(x,y,t)} = \left(u_s\frac{\partial\zeta}{\partial x} + v_s\frac{\partial\zeta}{\partial y} - w_s\right) . \tag{3.2.32}$$

It is assumed that an initial distribution of pressure can be described to complete the mathematical statement of the problem

$$h = h_i(x,y,z) \quad \text{in} \quad R , \tag{3.2.33}$$

where R is the region of interest. The initial pressure can be obtained from field measurements or from the simulation of the steady-state version of Eq. (3.2.24).

Equation (3.2.24), along with the boundary and initial conditions Eqs. (3.2.25) through (3.2.33), describes general transient three-dimensional problems in a saturated subsurface under isothermal conditions. For the compressible air flow, the presence of a free-moving boundary condition is unlikely. For essentially incompressible water flow, the presence of a moving boundary condition must be dealt with. Although a number of three-dimensional or quasi three-dimensional saturated water flow models have been developed (Gupta et al., 1975; McDonald and Harbaugh, 1985; Reeves et al., 1986; Kipp, 1987), their applicability to real-world problems under transient conditions is generally limited to confined aquifers because the moving boundary condition Eqs. (3.2.29) through (3.2.33) has not been rigorously included. By non-rigorous we mean that the equations of free-surface involving kinematic boundary conditions, Eqs. (3.2.28) through (3.2.33), are not solved. For example, MODFLOW (McDonald and Harbaugh, 1985) and all its descendants use a *well-term* to approximate a flux boundary condition at the water table. The flux boundary condition is, in turn, an approximation of the true free-surface boundary condition. The danger of such an approach is that it is not always clear when such approximations will faithfully reproduce the physical phenomena that one is attempting to model.

Although it was recognized that the free-surface method is a legitimate approach to the problem (Kipp, 1987), the HST code also neglects the kinematic boundary condition, approximately locating the free-surface by interpolating the heads to satisfy the Dirichlet boundary condition. Kinematics of the free-surface movement are neglected because the "approximation is acceptable when the free-surface movement is small relative to the horizontal interstitial velocity." The heuristic in HST are such that "When designing the grid for a free-surface boundary problem, the uppermost layer of cells must be made thick enough to accommodate the maximum variations in the free-surface location. . . . The free surface may not drop below the lower boundary of the uppermost layer of cells."

The alternative to the nonrigorous approach is to solve the groundwater flow equations with free-surface and seepage boundary conditions. This alternative approach has only recently been explored (Yeh et al., 1994; Knupp, 1996; Diersch and Michels, 1996).

3.3. Finite-Element Modeling of Three-Dimensional Saturated Flow

Finite-element approximation of saturated flows as specified by Eq. (3.2.24) and the associated boundary and initial conditions specified by Eqs. (3.2.25) through (3.2.33) can be made following the nine basic steps given in Chapter 2.

3.3.1. Finite-Element Approximations in Space

Let $N_j(x,y,z)$ be the base function of node point j in a three-dimensional space. The base function, $N_j(x,y,z)$, will have a value of 1.0 at the nodal point j and a value of 0.0 at all other nodal points. Furthermore, $N_j(x,y,z)$ will be given nonzero values only over those elements that have one nodal point coinciding with point j and zero values over all other elements in the domain. Three types of base functions are used in the development of the three-dimensional saturated flow model: linear hexahedral isoparametric elements, linear triangular prisms, and linear tetrahedral elements.

Recall from Chapter 2 that these base functions are given in the following equations. For a linear hexahedral element e (Fig. 3.3.1), the eight basis functions are given by

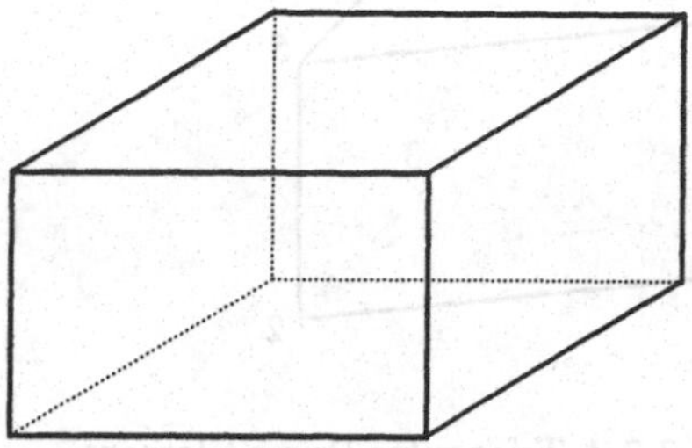

Fig. 3.3.1 A Hexahedral Element.

$$N_j^e = \frac{(1+\xi_j\xi)(1+\eta_j\eta)(1+\zeta_j\zeta)}{8}, \quad j = 1, 2, \ldots, 8, \qquad (3.3.1)$$

where N_j^e is the base function of node j in terms of the local coordinate (ξ,η,ζ) and (ξ_j,η_j,ζ_j) is the coordinate of the j-th node. The transformation from the local coordinate (ξ,η,ζ) to the global coordinate (x,y,z) is achieved by

$$x = \sum_{j=1}^{8} x_j N_j^e(\xi,\eta,\zeta), \; y = \sum_{j=1}^{8} y_j N_j^e(\xi,\eta,\zeta), \text{ and } z = \sum_{j=1}^{8} z_j N_j^e(\xi,\eta,\zeta) . \qquad (3.3.2)$$

Since the base functions are used for coordinate transformation, this element formulation is termed isoparametric.

For a linear triangular-prism element e (Fig. 3.3.2), the six base functions for its six nodes are given by the product of the two-dimensional area coordinate and one-dimensional local coordinate as

$$N_1^e = L_1\frac{(1-\zeta)}{2}, \; N_2^e = L_2\frac{(1-\zeta)}{2}, \; N_3^e = L_3\frac{(1-\zeta)}{2}$$
$$N_4^e = L_1\frac{(1+\zeta)}{2}, \; N_5^e = L_2\frac{(1+\zeta)}{2}, \text{ and } N_6^e = L_3\frac{(1+\zeta)}{2}, \qquad (3.3.3)$$

where N_j^e is the base function of node j in terms of the area coordinate (L_1,L_2,L_3) and the local coordinate ζ. The global coordinate (x,y,z) and the area coordinate (L_1,L_2,L_3) and local coordinate ζ are related by

$$x = \sum_{j=1}^{6} x_j N_j(\zeta,L_1,L_2,L_3), \quad y = \sum_{j=1}^{6} y_j N_j(\zeta,L_1,L_2,L_3), \tag{3.3.4}$$

$$z = \sum_{j=1}^{6} z_j N_j(\zeta,L_1,L_2,L_3), \quad \textit{and} \quad L_1 + L_2 + L_3 = 1 \ . \tag{3.3.5}$$

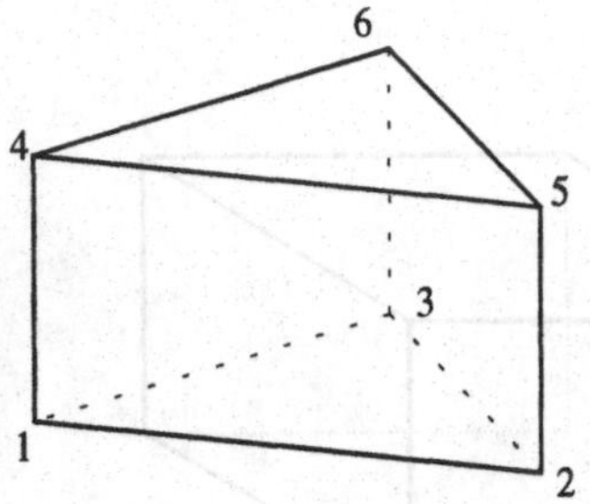

Fig. 3.3.2 A Triangular-Prism Element.

For a linear tetrahedral element e (Fig. 3.3.3), the four base functions for its four nodes are given by the volume coordinate as

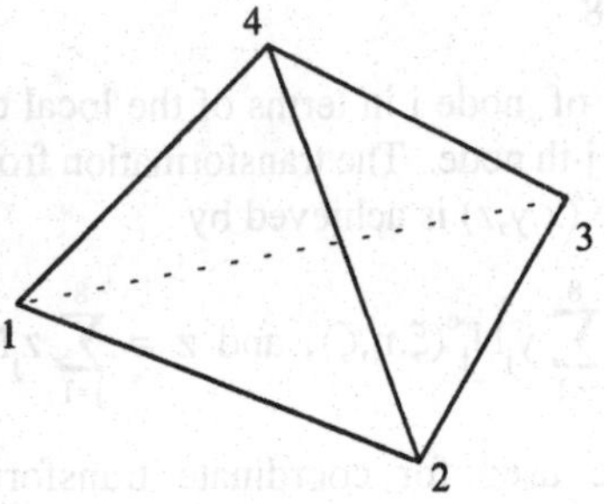

Fig. 3.3.3 A Tetrahedral Element.

$$N_1^e = L_1, \quad N_2^e = L_2, \quad N_3^e = L_3, \quad \text{and} \quad N_4^e = L_4 \ , \tag{3.3.6}$$

where N_j^e is the base function of node j in terms of the volume coordinate (L_1,L_2,L_3,L_4). The global coordinate (x,y,z) and the volume coordinate (L_1,L_2,L_3,L_4) are related by

$$x = \sum_{j=1}^{4} x_j N_j(L_1,L_2,L_3,L_4), \quad y = \sum_{j=1}^{4} y_j N_j(L_1,L_2,L_3,L_4) \ , \tag{3.3.7}$$

$$z = \sum_{j=1}^{4} z_j N_j(L_1, L_2, L_3, L_4), \quad and \quad L_1 + L_2 + L_3 + L_4 = 1 . \qquad (3.3.8)$$

Thus, it is clear that only three of the four volume coordinates, L_1, L_2, L_3, and L_4, can be independent, just as in the original coordinate system, where there are only three independent coordinates, x, y, and z.

Having defined the base functions, we now approximate the pressure head h:

$$h \approx \hat{h} = \sum_{j=1}^{N} h_j(t) N_j(x,y,z) , \qquad (3.3.9)$$

where N is the total number of nodes in the region, and N_j and h_j are the base function and the amplitude of h, respectively, at nodal point j. It is interesting to note that the fundamental distinction between finite-element methods (FEMs) and finite-difference methods (FDMs) is that the former is based directly on approximation of the function, as in Eq. (3.3.9), whereas the latter is based on approximation of derivative (Hilderbrand, 1968).

Since h is only an approximate solution, it will not satisfy the governing equations. Thus, we define a residual R_r for Eq. (3.2.24):

$$R_r = S_s \frac{\partial h}{\partial t} - \nabla \bullet [\mathbf{K} \bullet (\nabla h + \nabla z)] - q . \qquad (3.3.10)$$

According to the principle of the FEM, we choose the coefficient h_j's in Eq. (3.3.9) such that the residual in Eq. (3.3.10) weighted by the set of weighting functions is zero. For the Galerkin finite-element method, we select the set of weighting functions as the same set of base functions. Applying the principle of weighted residual, we obtain

$$\int_R N_i \left\{ S_s \frac{\partial h}{\partial t} - \nabla \cdot [\mathbf{K} \cdot (\nabla h + \nabla z)] - q \right\} dR , \quad i = 1, 2, \ldots, N . \qquad (3.3.11)$$

Substituting Eq. (3.3.9) into Eq. (3.3.11) and integrating by part, we obtain

$$\left[\int_R N_i S_s N_j dR \right] \left\{ \frac{dh_j}{dt} \right\} + \left[\int_R (\nabla N_i) \cdot \mathbf{K} \cdot (\nabla N_j) dR \right] \{h_j\} = \int_R N_i q dR - \int_R (\nabla N_i) \cdot \mathbf{K} \cdot (\nabla z) dR + \int_B \mathbf{n} \cdot \mathbf{K} \cdot (\nabla h + \nabla z) N_i dR , \quad i = 1, 2, \ldots, N \qquad (3.3.12)$$

Equation (3.3.12) written in matrix form is

$$[M] \left\{ \frac{dh}{dt} \right\} + [S]\{h\} = \{G\} + \{Q\} + \{B\} , \qquad (3.3.13)$$

where {dh/dt} and {h} are the column vectors containing the values of dh/dt and h, respectively, at all nodes; [M] is the mass matrix resulting from the storage term; [S] is stiff matrix resulting from the action of conductivity; {G}, {Q}, and {B} are the load vectors from the gravity force, internal source/sink, and boundary conditions, respectively. The matrices, [M] and [S], are given by

$$M_{ij} = \sum_{e \in M_e} \int_{R_e} N_\alpha^e S_r N_\beta^e dR \tag{3.3.14}$$

$$S_{ij} = \sum_{e \in M_e} \int_{R_e} \left[(\nabla N_\alpha^e) \cdot \mathbf{K} \cdot (\nabla N_\beta^e)\right] dR , \tag{3.3.15}$$

where R_e is the region of element e, M_e is the set of elements that have a local side α-β coinciding with the global side i-j, N_α^2 is the α-th local basis function of element e. Similarly the load vectors {G}, {Q} and {B} are given by

$$G_i = -\sum_{e \in M_e} \int_{R_e} (\nabla N_\alpha^e) \cdot \mathbf{K} \cdot \nabla z dR \tag{3.3.16}$$

$$Q_i = \sum_{e \in M_e} \int_{R_e} N_\alpha^e q dR \tag{3.3.17}$$

and

$$B_i = -\sum_{e \in N_{se}} \int_{B_e} N_\alpha^e \cdot [-\mathbf{K} \cdot (\nabla h + \nabla z)] dB , \tag{3.3.18}$$

where B_e is the length of boundary segment e, N_{se} is the set of boundary segments that have a local node α coinciding with the global node i. It should be noted that in applying the weighted-residual finite element method to Eq. (3.2.24), we have used the set of base functions as the set of weighting functions for all terms. Hence the Galerkin finite-element method results.

The reduction of the partial differential equation (PDE), Eq. (3.2.24), to the set of ordinary differential equations (ODE), Eq. (3.3.13), simplifies to the evaluation of integrals on the right-hand side of Eqs. (3.3.14) through (3.3.18) for every element or boundary segment e to yield the element mass matrix $[M^e]$ and stiff matrix $[S^e]$ as well as the element gravity column vector, $\{G^e\}$, source/sink column vector $\{Q^e\}$, and boundary column vector $\{B^e\}$ as

$$M_{\alpha\beta}^e = \int_{R_e} N_\alpha^e S_s N_\beta^e dR \tag{3.3.19}$$

$$S_{\alpha\beta}^e = \int_{R_e} \left(\nabla N_\alpha^e\right) \cdot \mathbf{K} \cdot \left(\nabla N_\beta^e\right) dR \tag{3.3.20}$$

$$G_\alpha^e = -\int_{R_e} \left(\nabla N_\alpha^e\right)\cdot \mathbf{K}\cdot\nabla z dR \tag{3.3.21}$$

$$Q_\alpha^e = \int_{R_e} N_\alpha^e q \, dR \tag{3.3.22}$$

and

$$B_\alpha^e = -\int_{B_e} N_\alpha^e \mathbf{n}\cdot[-\mathbf{K}\cdot(\nabla h + \nabla z)]dB \, , \tag{3.3.23}$$

where the superscript or subscript e denotes the element and α, $\beta = 1, \ldots, 8$ for linear hexahedral elements, α, $\beta = 1, \ldots, 6$ for linear triangular-prism elements, or α, $\beta = 1, \ldots, 4$ for linear tetrahedral elements.

For a hexahedral or triangular-prism element, Eqs. (3.3.19) through (3.3.22) are computed by Gaussian/nodal quadrature (Conte, 1965) because it is not easy to invert Eq. (3.3.2) or (3.3.4) for ξ, η, ζ or L_1, L_2, and ζ in terms of x, y, and z. On the other hand, for a tetrahedral element, the transformation from the global coordinate (x,y,z) to the volume coordinate $(L_1.L_2,L_3,L_4)$ is easily achievable by inverting Eqs. (3.3.7) and (3.3.8). This fact and the integration identity

$$\int_{R_e} L_1^\alpha L_2^\beta L_3^\gamma L_4^\delta dR = 6V \frac{\alpha!\ \beta!\ \gamma!\ \delta!}{(\alpha+\beta+\gamma+\delta+3)!} \tag{3.3.24}$$

make the evaluation of Eqs. (3.3.19) through (3.3.22) for linear triangular element obtainable analytically. However, to simplify programming and make code general, a four-point Gaussian/nodal quadrature is used to numerically integrate Eq. (3.2.24).

The surface integration of Eq. (3.3.23) in a three-dimensional space is not as straightforward as in a two-dimensional space. It requires further elaboration. Any surface integral of a continuous function F(x,y,z) specified on the surface S (Figure 3.3.4) can be reduced to an area integral. Let I represent a surface integral:

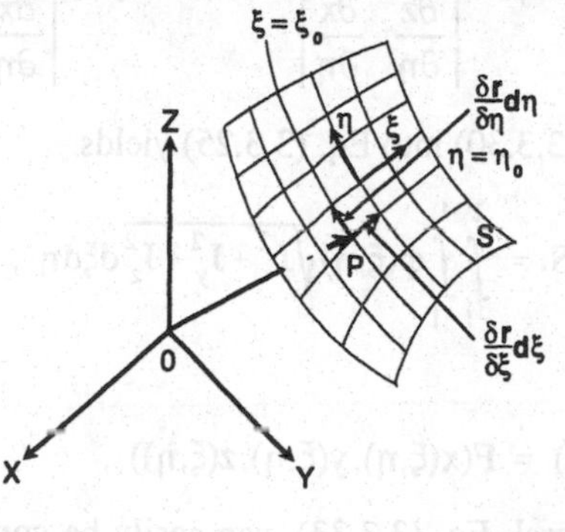

Fig. 3.3.4 A Surface Area and Its Imbedded Local Coordinate.

$$I = \int_S F(x,y,z)dS , \tag{3.3.25}$$

where the surface S is given by the following equation:

$$z = f(x,y) . \tag{3.3.26}$$

Let P be any point on the surface S with coordinates (x,y,z) or (ξ,η) (Figure 3.3.4). Then the vector **r** from O to P is given by

$$\mathbf{r} = x\mathbf{i} + y\mathbf{j} + z\mathbf{k} . \tag{3.3.27}$$

The vectors tangent to the coordinate curves $\xi = \xi_0$ and $\eta = \eta_0$ on the surface S are $\partial r/\partial\eta$ and $\partial r/\partial\xi$, respectively. The area dS is given by

$$dS = \left|\frac{\partial \mathbf{r}}{\partial \xi} \times \frac{\partial \mathbf{r}}{\partial \eta}\right| d\xi d\eta , \tag{3.3.28}$$

where × represents vector multiplication. But

$$\frac{\partial \mathbf{r}}{\partial \xi}\times\frac{\partial \mathbf{r}}{\partial \eta} = \begin{vmatrix} \mathbf{i} & \mathbf{j} & \mathbf{k} \\ \frac{\partial x}{\partial \xi} & \frac{\partial y}{\partial \xi} & \frac{\partial z}{\partial \xi} \\ \frac{\partial x}{\partial \eta} & \frac{\partial y}{\partial \eta} & \frac{\partial z}{\partial \eta} \end{vmatrix} \tag{3.3.29}$$

so that

$$dS = \sqrt{J_x^2 + J_y^2 + J_z^2}\, d\xi d\eta , \tag{3.3.30}$$

where

$$J_x = \begin{vmatrix} \frac{\partial y}{\partial \xi} & \frac{\partial z}{\partial \xi} \\ \frac{\partial y}{\partial \eta} & \frac{\partial z}{\partial \eta} \end{vmatrix}, \quad J_y = \begin{vmatrix} \frac{\partial z}{\partial \xi} & \frac{\partial x}{\partial \xi} \\ \frac{\partial z}{\partial \eta} & \frac{\partial x}{\partial \eta} \end{vmatrix}, \quad \text{and } J_z = \begin{vmatrix} \frac{\partial x}{\partial \xi} & \frac{\partial y}{\partial \xi} \\ \frac{\partial x}{\partial \eta} & \frac{\partial y}{\partial \eta} \end{vmatrix} . \tag{3.3.31}$$

Substituting Eq. (3.3.30) into Eq. (3.3.25) yields

$$\int_S F(x,y,z)\,dS = \int_{-1}^{1}\int_{-1}^{1} \phi(\xi,\eta)\sqrt{J_x^2 + J_y^2 + J_z^2}\, d\xi d\eta , \tag{3.3.32}$$

where

$$\phi(\xi,\eta) = F(x(\xi,\eta), y(\xi,\eta), z(\xi,\eta)) . \tag{3.3.33}$$

The surface integral, Eq. (3.3.33), can easily be computed with either the Gaussian or the nodal quadrature. The above consideration of surface integrals is used for quadrilateral surface cases. For triangular surface cases, the implementation is much simpler. The surface area can be calculated based on the global coordinates given on the nodes of the triangular surface. Then it is straightforward to compute the surface integral Eq. (3.3.25) directly with a Gaussian/nodal quadrature. Using Eq. (3.3.32), we can easily perform the integration of Eq. (3.3.23) for element matrices resulting from boundary conditions.

With the element matrices $[M^e]$ and $[S^e]$ and the element column vectors $\{G^e\}$ and $\{Q^e\}$ computed, the global matrices [M] and [S] and the global column vectors {G} and {Q} are then assembled element by element.

3.3.2. Mass Lumping

Referring to [M], one may recall that this is a unit matrix if the finite-difference formulation is used in spatial discretization. Hence, by proper scaling, the mass matrix can be reduced to the finite-difference equivalent by lumping (Clough, 1971). In many cases, the lumped mass matrix would result in a better solution, in particular, if it is used in conjunction with the central or backward-difference time marching (Yeh and Ward, 1980). Under such circumstances, it is preferred to the consistent mass matrix (mass matrix without lumping). Therefore, options are provided for the lumping of the matrix [M]. More explicitly, [M] will be lumped according to

$$M_{ij} = \sum_{e \in M_e} \left(\sum_{\beta=1}^{4} \int_{R_e} N_\alpha^e S_s N_\beta^e \, dR \right) \quad \text{if} \quad j = i$$

$$M_{ij} = 0 \quad \text{if} \quad j \neq i . \tag{3.3.34}$$

3.3.3. Finite-Difference Approximation in Time

Next, we derive a matrix equation by integrating Eq. (3.3.13). For the time integration of Eq. (3.3.13), the load vector {B} will be ignored. This load vector will be discussed in the next section on the numerical implementation of boundary conditions. An important advantage of finite-element approximation over the finite-difference approximation is its inherent ability to handle complex boundaries and obtain the normal derivatives therein. In the time dimension, such advantages are not evident. Thus, finite difference methods are typically used in the approximation of the time derivative. Three time-marching methods are adopted in the present water-flow model. In the first one, the central or Crank-Nicolson formulation may be written as

$$\frac{[M]}{\Delta t}\left(\{h\}_{t+\Delta t} - \{h\}_t\right) + \frac{1}{2}[S]\left(\{h\}_{t+\Delta t} + \{h\}_t\right) = \{G\} + \{Q\} , \tag{3.3.35}$$

where [M], [S], {G}, and {Q} are evaluated at $t + \Delta t/2$.

In the second method, the backward difference formulation may be written as

$$\frac{[M]}{\Delta t}\left(\{h\}_{t+\Delta t} - \{h\}_t\right) + [S]\left(\{h\}_{t+\Delta t} + \{h\}_t\right) = \{G\} + \{Q\} , \qquad (3.3.36)$$

where [M], [S], {G}, and {Q} are evaluated at $t + \Delta t$. In the third optional method, the values of unknown variables are assumed to vary linearly with time during the time interval, Δt. In this middifference method, the recurrence formula is written as

$$\left(\frac{2}{\Delta t}[M] + [S]\right)\{h\}_{t+\Delta t/2} - \frac{2}{\Delta t}[M]\{h\}_t = \{G\} + \{Q\} \qquad (3.3.37)$$

and

$$\{h\}_{t+\Delta t} = 2\{h\}_{t+\Delta t/2} - \{h\}_t , \qquad (3.3.38)$$

where [M], [S], {G}, and {Q} are evaluated at $(t + \Delta t/2)$.

Equations (3.3.35) through (3.3.37) can be written as a matrix equation

$$[A]\{h\} = \{b\} , \qquad (3.3.39)$$

where [A] is the matrix, {h} is the unknown vector to be found and represents the values of discretized pressure-head field at new time, and {b} is the load vector. Take for example, Eq. (3.3.35); [A] and {b} represent the following:

$$[A] = \frac{[M]}{\Delta t} + [S], \quad \textit{and} \quad \{b\} = \frac{[M]}{\Delta t}\{h\}_t + \{G\} + \{Q\} , \qquad (3.3.40)$$

where $\{h\}_t$ is the vector of the discretized pressure-head field at previous time.

3.3.4. Numerical Implementation of Boundary Conditions

The matrix equation, Eq. (3.3.39), is singular. To make the problem uniquely defined, boundary conditions [Eqs. (3.2.25) through (3.2.32)] must be implemented. For the Cauchy boundary condition given by Eq. (3.2.27), we simply substitute Eq. (3.2.27) into Eq. (3.3.23) to yield a boundary-element column vector $\{B_c^e\}$ for a Cauchy segment:

$$\left\{B_c^e\right\} = \left\{q_c^e\right\}, \qquad (3.3.41)$$

where $\{q_c^e\}$ is the Cauchy boundary flux vector given by

$$q_{c\alpha}^e = -\int_{B_e} N_\alpha^e q_c \, dB, \qquad \alpha = 1, 2, 3, \text{ or } 4 . \qquad (3.3.42)$$

This Cauchy boundary flux vector represents the normal fluxes through the three or four nodal points of the element surface B_e on B_c.

For the Neumann boundary condition given by Eq. (3.2.26), we substitute Eq. (3.2.26) into Eq. (3.3.23) to yield a boundary-element column vector $\{B_n^e\}$ for a Neumann segment:

$$\{B_n^e\} = \{q_n^e\}, \tag{3.3.43}$$

where $\{q_n^e\}$ is the Neumann boundary flux vector given by

$$q_{n\alpha}^e = \int_{B_e} \left(N_\alpha^e \mathbf{n}\cdot\mathbf{K}\cdot\nabla z - N_\alpha^e q_n\right) dB \quad \alpha = 1, 2, 3, \text{ or } 4, \tag{3.3.44}$$

which is independent of the pressure head and represents the fluxes through the three or four nodal points of the element surface B_e on B_n.

For the river boundary condition, we substitute Eq. (3.2.28) into Eq. (3.3.23) to yield a boundary-element matrix, $[B_r^e]$, and a boundary-element column vector, $[B_r^e]$, for a given river segment

$$\left[B_r^e\right] = \left[q_r^e\right], \quad \left\{B_r^e\right\} = \left\{q_r^e\right\}, \tag{3.3.45}$$

where $[q_r^e]$ and $\{q_r^e\}$ are the contrition to the boundary-element matrix and column vector, respectively, due to the river boundary condition and are given by

$$q_{r\alpha\beta}^e = \int_{B_e} N_\alpha^e \frac{K_R}{b_R} N_\beta^e dB, \quad q_{r\alpha}^e = \int_{B_e} N_\alpha^e \frac{K_R}{b_R} dB. \tag{3.3.46}$$

While the implementation of Cauchy, Neumann, and river boundary conditions is straightforward, the implementation of the moving boundary condition given by Eqs. (3.2.29) through (3.2.32) requires an elaboration. Eq. (3.2.30) states that rate of increase of water stored per unit area due to moving free surface plus the inward normal flux from the inside world equals the inward normal flux from the outside world. On the other hand Eq. (3.2.31) states that on the moving free surface the pressure head, h = 0, that is, on the free surface, the total head H (= h + z) is equal to the free surface elevation ζ. Hence, substituting Eqs. (3.2.30) and (3.2.32) into Eq. (3.3.23) and using the fact on the free surface ζ = H, we obtain

$$\int_{z=\zeta(x,y,t)} \mathbf{n}\cdot\mathbf{K}\cdot\nabla H dB = \int_{z=\zeta(x,y,t)} \left(-q_v - s_y\frac{\partial H}{\partial t}\right) dB = \int_{z=\zeta(x,y,t)} \left(-q_v - s_y\frac{\partial h}{\partial t}\right) dB. \tag{3.3.47}$$

Eq. (3.4.47) indicates that the boundary conditions on the free surface contribute to both the global load vector and coefficient matrix as

$$\left\{B_v^e\right\} = \left\{q_v^e\right\}, \quad \left[B_v^e\right] = \left[q_v^e\right], \tag{3.3.48}$$

where $\{q_v^e\}$ and $[q_v^e]$ are the moving-boundary vector and matrix, given by

$$q_{v\alpha}^e = -\int_{B_e} N_\alpha^e q_v dB \quad \text{and} \quad q_{v\alpha\beta}^e = \int_{B_e} N_\alpha^e s_y N_\beta^e dB. \tag{3.3.49}$$

The moving boundary vector represents the normal fluxes through the three or four nodal points of the element surface B_e on B_v, and the moving boundary matrix represents the contribution to the coefficient matrix through the three or four nodal points of the element surface B_e on B_v.

Assembling over all Neumann, Cauchy, river, and variably-moving boundary element-sides, we obtain the boundary column vector {B} as

$$\{B\} = \{q_b\} \; , \tag{3.3.50}$$

in which

$$\{q_b\} = \sum_{e \in N_{ne}} \{q_n^e\} + \sum_{e \in N_{ce}} \{q_c^e\} + \sum_{e \in N_{ve}} \{q_v^e\} + \sum_{e \in N_{re}} \{q_r^e\} \; , \tag{3.3.51}$$

where N_{ce}, N_{ne}, and N_{ve} are the number of Cauchy, Neumann, and variably moving boundary sides, respectively. Similarly, assembling over all the variably moving boundary element-sides, we obtain the boundary matrix [B] as

$$[B] = [q_b], \quad [q_b] = \sum_{e \in N_{ve}} [q_v^e] + \sum_{e \in N_{re}} [q_r^e] \tag{3.3.52}$$

This boundary matrix and boundary vector should be used to modify the coefficient matrix [A] and the load vector {b} in Eq. (3.3.39).

Finally, at nodes where Dirichlet boundary conditions are applied, an identity equation is generated for each node and included in the matrix equation of Eq. (3.3.39). The detailed method of applying this type of boundary conditions can be found elsewhere (Wang and Conner, 1975). After incorporating the boundary conditions described in this subsection, Eq. (3.3.39) becomes

$$[C]\{h\} = \{R\} \; . \tag{3.3.53}$$

3.3.5. Solution Techniques

Eq. (3.3.53) is a linear matrix equation. The nonlinearity of the problem lies in the fact that the free surface is not a priori. This free surface should be located such that Eq. (3.2.31) is satisfied. Thus, the algorithm to locate the free surface proceeds in eight steps as follows:

1. Set iteration counter $k = 0$.
2. Assume the initial position of the free surface ζ^k.
3. Assemble the finite element equation.
4. Implement moving free surface condition given by Eq. (3.2.27) along with all other types of boundary conditions at fixed boundaries.
5. Solve the resulting matrix equation for the pressure head h at all nodes on and below the free surface.
6. For all nodes on the moving free surface, check if total head $H = \zeta^k$
7. If yes, stop. If not go to Step 8.
8. $\zeta^{K+1} = \zeta^k + \omega h$; Update iteration number $k = k+1$; Go to Step 3.

where ω is a relaxation parameter, which is normally taken to be greater than 0 and less than 1.0.

For the solution of the linearized equations, six optional solvers are provided: (1) successive block iteration methods; (2) successive point iteration methods; and (3) four preconditioned conjugate gradient methods using the polynomials, incomplete Cholesky decomposition, modified incomplete Cholesky decomposition, and symmetric successive overrelaxation as the preconditioners, respectively. The successive block and point iteration methods can be found in any standard text book on applied linear algebra. The computational algorithm for preconditioned conjugate gradient methods is the same as that in Section 2.6.5

3.3.6. *Computation of Mass Balance*

One of the most important aspects in numerical modeling of subsurface flow is to check the mass balance over the whole region. The error in mass balance provides a crude index on the accuracy and convergence of numerical computations. The mass balance over a region R enclosed by the boundary B(x,y,z) = 0 can be obtained by integrating Eq. (3.2.24)

$$F_V = \int_R \left(S_s \frac{\partial h}{\partial t} - q \right) dR \quad \textit{and} \quad F_B = \int_B F_n dB \,, \tag{3.3.54}$$

where F_V represents the net volume-increasing rate of fluid in the region, F_B is the net volume-flow rate through the entire boundary out from the region, and F_n is the outward normal flux. In fact, F_n can be defined as

$$F_n = -\, \mathbf{n} \cdot \mathbf{K} \cdot (\nabla h + \nabla z) \,. \tag{3.3.55}$$

Having obtained the pressure-head field by solving Eq. (3.3.53), one can integrate two equations in Eq. (3.3.54) independently. If the solution for h is free of error, one would expect the sum of two integrals to be equal to zero. Here the integral of the first equation in Eq. (3.3.54) is broken into two parts:

$$F_h = \int_R S_s \frac{\partial h}{\partial t} dR, \quad F_s = \int_R q \, dR \,, \tag{3.3.56}$$

where F_h and F_s represent the flow rates due to pressure-head change and artificial sources, respectively. Similarly, the integral of the second equation in Eq. (3.3.54) is broken into five parts:

$$F_d = \int_{B_d} F_n \, dB, \quad F_c = \int_{B_c} F_n \, dB, \quad F_n = \int_{B_n} F_n \, dB \tag{3.3.57}$$

$$F_v = \int_{B_d} F_n \, dB, \quad F_r = \int_{B_r} F_n \, dB, \quad F_L = \int_{B-B_d-B_c-B_n} F_n \, dB \,, \tag{3.3.58}$$

where F_d, F_c, F_n, F_v, F_r and F_L represent fluxes through the Dirichlet boundary B_d, the

Cauchy boundary B_c, the Neumann boundary B_n, the variably-moving boundary B_v, the river boundary B_r, and the unspecified boundary $B\text{-}B_d\text{-}B_c\text{-}B_n\text{-}B_v$.

For an exact solution, the sum of the net outgoing flux F_B across the entire boundary and the total mass increase rate F_V should be equal to zero. In addition, F_L should theoretically be equal to zero. However, in any practical simulation, F_B plus F_V will not be equal to zero, and F_L will be nonzero. Nevertheless, the mass balance computation should provide a means to check the numerical scheme and the consistency in the computer code.

3.3.7. Example Problems

A computer program has been designed to solve Eq. (3.2.23) along with the boundary and initial conditions, Eqs. (3.2.24) through (3.2.28). The operation and construction of this computer program can be found elsewhere (Yeh et al., 1994). This computer code acronym 3DFEWA, along with its documentation, is contained in the attached floppy diskette in this book. To verify 3DFEWA, three example problems dealing with an unconfined aquifer are presented. The first one is a steady-state problem from recharge over a parallel plate. The second problem is a transient problem from recharge over a vertical column. The third problem is the same as the first problem except for the fact that it is a transient problem.

3.3.7.1. Problem 1: Steady-State Two-Dimensional Drainage Problem. This example is selected to represent the simulation of a two-dimensional drainage problem in a unconfined aquifer with 3DFEWA. The region of interest is bounded on the left and right by parallel drains fully penetrating the medium, on the bottom by an impervious aquifuge, and on the top by an air-soil interface (Fig. 3.3.5). The distance between the two drains is 20 m (Fig. 3.3.5). The medium is assumed to have a saturated hydraulic conductivity of 0.01 m/h, and a porosity and specific yield of 0.25.

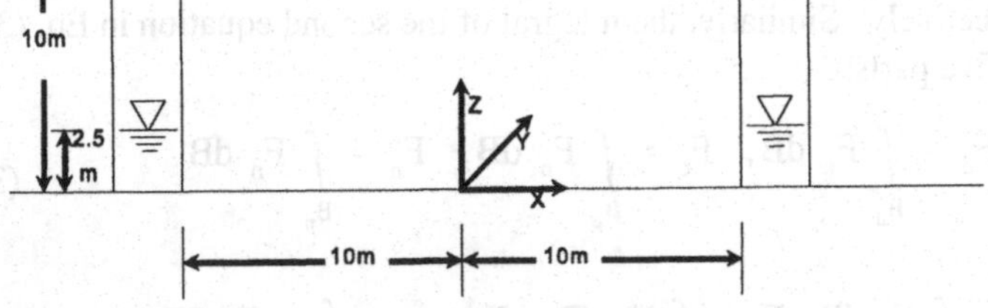

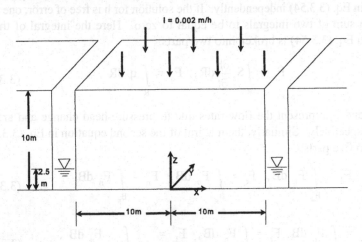

Fig. 3.3.5 Problem Sketch of Problem 1.

Because of the symmetry, the region for numerical simulation will be taken as $0 < x < 10$ m and $0 < z < 10$ m, and 10 m wide along the y-direction will be assumed. The boundary conditions are given as follows: no flux is imposed on the left ($x = 0$), front ($y = 0$), back ($y = 10$), and bottom ($z = 0$) sides of the region; pressure head is assumed to vary from zero at the water surface ($z = 2.5$) to 2.5 m at the bottom ($z = 0$) on the right side ($x = 10$); and variable conditions are applied to the free surface that is time dependent and varies anywhere between the bottom and top surfaces. The right side above the water surface is specified as a seepage face that is represented as a Dirichlet boundary condition with negative profile type. Steady-state solutions are sought for three cases with rainfall rates equal to 0.0, 2×10^{-3}, and 4×10^{-3} m/h, respectively. A preinitial condition is set, $h = 10 - z$.

The region of interest is discretized with 10 x 1 x 10 x 4 = 400 triangular prism elements with element size = 0.5(1x5) x 1 m^3, resulting in 11 x 2 x 11 + 11 x 10 = 352 node points (Fig. 3.3.6). For 3DFEWA simulation, each of the three vertical planes will be considered a subregion. Thus, the total of the three subregions, each with 121, 121, and 110 node points, respectively, is used for the subregional block iteration simulation.

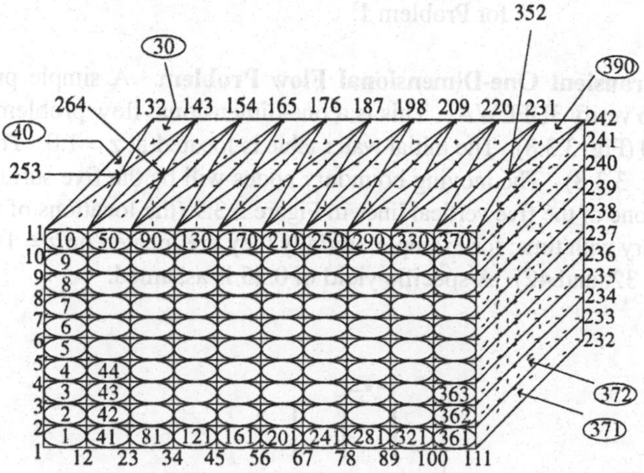

Fig. 3.3.6 Finite Element Grid for Problem 1.

The pressure head tolerance is 2×10^{-4} m for cycle iteration to locate water table and is 2×10^{-3} m for block iteration. The relaxation factors for both the nonlinear iteration and block iteration are set equal to 0.5. With the above descriptions, the input data can be prepared according to the instructions in Appendix A of 3DFEWA (Yeh et al., 1994). Executing 3DFEWA with the input data set, we obtain the results of the free surface location as shown in Figure 3.3.7.

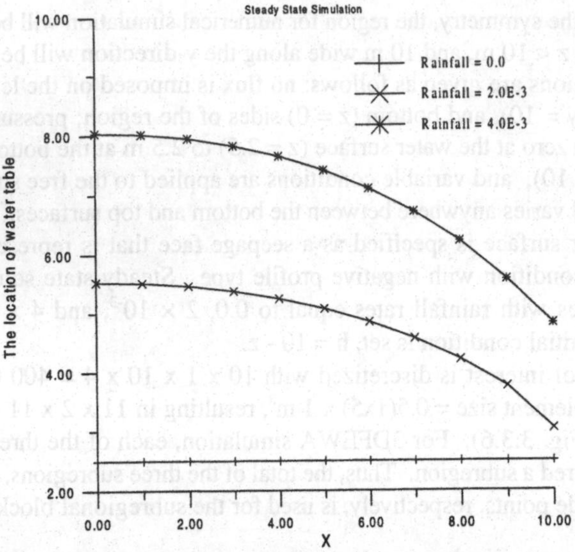

Fig. 3.3.7 Water Tables at Various Times for Problem 1.

Problem 2: Transient One-Dimensional Flow Problem. A simple problem is presented here to verify 3DFEWA. This is a one-dimensional flow problem between $z = 0$ and $z = 9.0$ (Fig. 3.3.8). The initial water table is located at $z = 1.0$. The rainfall rate is 0.25 (Fig. 3.3.8). The moving boundary nodes will be the five surface nodes, each located on one of the five vertical lines in Figure 3.3.9 (the locations of these five surface nodes vary with time; for example, initially they are located on $z = 1.0$, that is, nodes 2, 12, 22, 32, and 42). A specific yield of 0.25 is assumed.

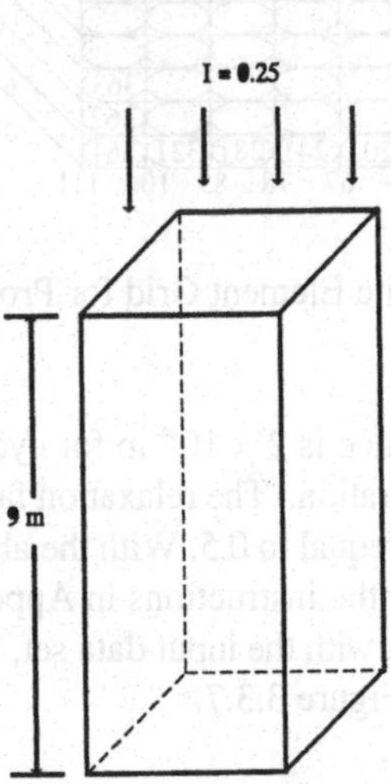

Fig. 3.3.8 Problem Definition and Sketch for Problem 2.

The region is divided into (1 x 1 x 9) x 4 = 36 triangular prism elements resulting in (1 x 1 x 10) x 4 + 10 = 50 nodes (Fig. 3.3.9). The subregion block iteration method is used to solve the resulting matrix equation. Three subregions are used: the first subregion consists of nodes on the front face, the second region is made up of nodes in the back face, and the third subregion is composed of nodes in the center vertical line. A time-step size of 0.5 is used and a 10 time-step simulation is made.

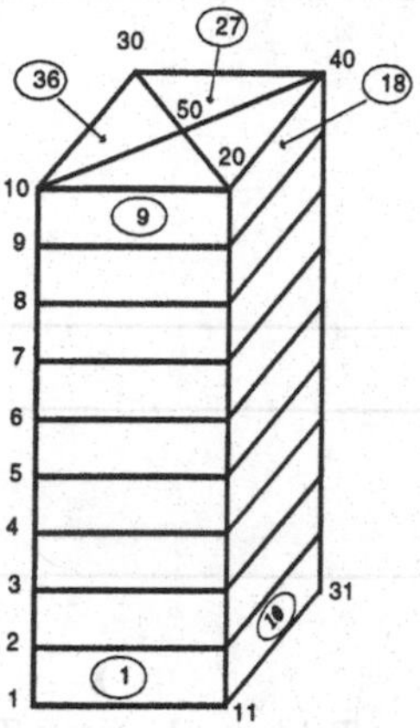

Fig. 3.3.9 Finite Element Discretization for Problem 2.

With the above descriptions, the input data can be prepared according to the instructions in Appendix A of 3DFEWA. Executing 3DFEWA, we obtain the simulation results. Figure 3.3.10 depicts the water table as function of time from the simulation of 3DFEWA. It clearly demonstrates that the moving rate of the water table is 1.0 as expected. This rate is equal to the infiltration rate divided by the specific yield (0.25/0.25 = 1.0). Hence 3DFEWA yielded an almost perfect solution for this simple one-dimensional problem of transient moving free surfaces.

Problem 3: Transient Two-Dimensional Drainage Problem. This example is selected to represent the transient simulation of a two-dimensional drainage problem in an unconfined aquifer with 3DFEWA. The region of interest is bounded on the left and right by parallel drains fully penetrating the medium, on the bottom by an impervious aquifuge, and on the top by an air-soil interface as in Problem 1 (Fig. 3.3.5). The distance between the two drains is 20 m. The medium is assumed to have a saturated hydraulic conductivity of 0.01 m/h, and a porosity and specific yield of 0.25.

The boundary conditions are given as follows: no flux is imposed on the left (x = 0), front (y = 0), back (y = 10), and bottom (z = 0) sides of the region; pressure head is assumed to vary from zero at the water surface (z = 2.75) to 2.75 m at the bottom (z = 0) on the right side (x = 10); and variable conditions are applied to the free surface, which is time dependent and varies anywhere between the bottom and top

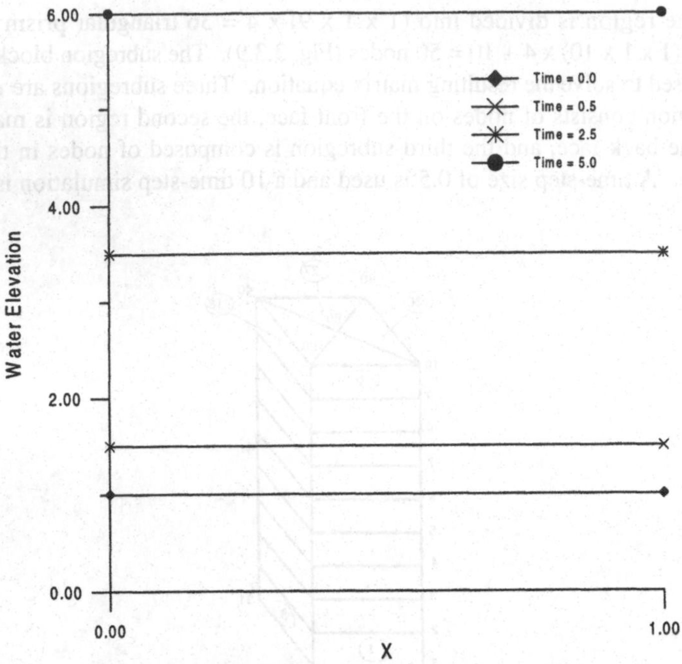

Fig. 3.3.10 Water Tables at Various Times for Problem 2.

surfaces. The right side above the water surface is specified as a seepage face that is represented as a Dirichlet boundary condition with negative profile type. Fluxes on the top side of the variable boundary are assumed equal to 0.002 m/h. A transient state solution will be sought. A preinitial condition is set, $h = 2.75 - z$. The initial condition is obtained by simulating steady-state problem without any infiltration.

Because of the symmetry, for numerical simulation, the region for numerical simulation will be taken as $0 < x < 10$ m and $0 < z < 4$ m, and 10 m wide along the y-direction will be assumed. The region of interest is discretized in a similar fashion as in Problem 1, except that the region is divided into (10 x 1 x 16) x 4 = 640 triangular prism elements resulting in 11 x 17 x 2 + 10 x 17 = 544 nodes. The element size in the vertical direction is 0.25. The rectangle that forms four triangles has the size of 1 by 10. A time-step size of 2.0 was used and a 50 time-step simulation was made to illustrate how to use 3DFEWA.

With the above descriptions, the input data can be prepared according to the instructions in Appendix A of 3DFEWA whose documentation is included in the attached floppy diskette. The input data can be obtained by a minor modification of that in Problem 1. Figure 3.3.11 depicts the water table as a function of time from the simulation of 3DFEWA. It is seen that at times greater than 60, the water table spans the entire element for some elements, and it spans a partial element for some other elements.

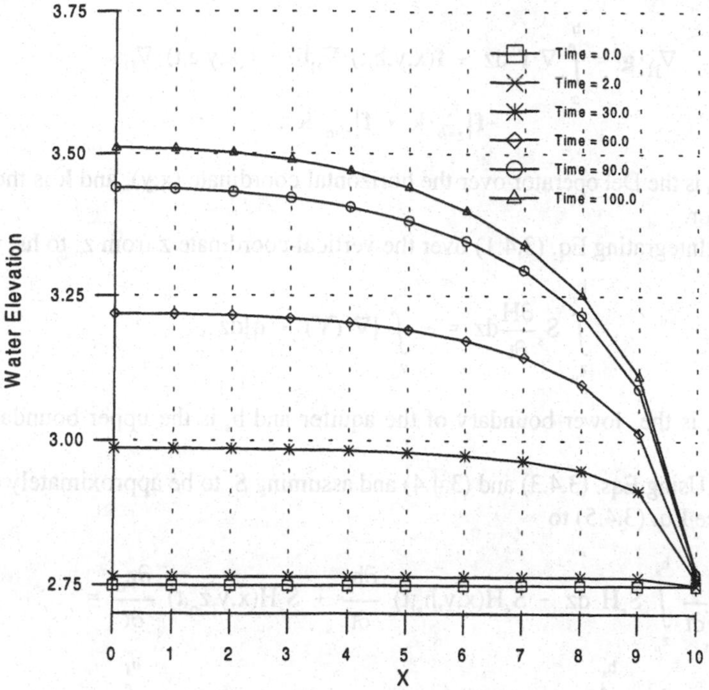

Fig. 3.3.11 Water Tables at Various Times for Problem 3.

3.4. Vertically Integrated Flows of Essentially Incompressible Fluids

Let us write Eq. (3.2.23) in the following form

$$S_s \frac{\partial H}{\partial t} = - \nabla \cdot (\mathbf{V}) + q; \quad \mathbf{V} = - \mathbf{K} \cdot (\nabla h + \nabla z) \; . \tag{3.4.1}$$

The Leibenitz rule states that if

$$g(x,y,t) = \int_{a(x,y,t)}^{b(x,y,t)} f(x,y,z,t) dz \; , \tag{3.4.2}$$

then

$$\frac{\partial g}{\partial t} = \int_a^b \frac{\partial f}{\partial t} dz + f(x,y,b,t) \frac{\partial b}{\partial t} - f(x,y,a,t) \frac{\partial a}{\partial t} \; . \tag{3.4.3}$$

Similarly, if both g and f are vector functions, we can obtain

$$\nabla_H \cdot \mathbf{g} = \int_a^b \nabla\cdot\mathbf{f}\,dz + \mathbf{f}(x,y,b,t)\cdot\nabla_H b - \mathbf{f}(x,y,a,t)\cdot\nabla_H a$$
$$-\mathbf{f}|_{z=b}\cdot\mathbf{k} + \mathbf{f}|_{z=a}\cdot\mathbf{k}\,, \tag{3.4.4}$$

where ∇_H is the Del operator over the horizontal coordinate (x,y), and **k** is the vertical unit vector.

Integrating Eq. (3.4.1) over the vertical coordinate z from z_o to hs, we have

$$\int_{z_o}^{h_s} S_s \frac{\partial H}{\partial t} dz = - \int_{z_o}^{h_s} [\nabla\cdot(\mathbf{V}) + q]dz\,, \tag{3.4.5}$$

where z_o is the lower boundary of the aquifer and h_s is the upper boundary of the aquifer.

Using Eqs. (3.4.3) and (3.4.4) and assuming S_s to be approximately constant, we reduce Eq. (3.4.5) to

$$\frac{\partial}{\partial t}\int_{z_o}^{h_s} S_s H\,dz - S_s H(x,y,h_s,t)\frac{\partial h_s}{\partial t} + S_s H(x,y,z_o,t)\frac{\partial z_o}{\partial t} =$$
$$-\nabla_H \cdot \int_{z_o}^{h_s} \mathbf{V}dz + \mathbf{V}|_{z=H_s}\cdot(\nabla_H h_s) - \mathbf{V}|_{z=z_o}\cdot(\nabla_h z_o) + \int_{z_o}^{h_s} q dz \tag{3.4.6}$$
$$- \mathbf{V}|_{z=h_s}\cdot\mathbf{k} + \mathbf{V}|_{z=z_o}\cdot\mathbf{k}\,.$$

The kinematic condition for a free surface

$$z - h_s(x,y,t,) = 0 \tag{3.4.7}$$

can be written as

$$S_y \frac{\partial h_s}{\partial t} = - \mathbf{V}|_{z=h_s}\cdot(\nabla_H h_s) + \mathbf{V}|_{z=h_s}\cdot\mathbf{k} - \mathbf{I}|_{z=h_s}\,, \tag{3.4.8}$$

and for a fixed surface

$$z - z_o(x,y) = 0 \tag{3.4.9}$$

can be written as

$$0 = - \mathbf{V}|_{z=z_o}\cdot(\nabla_h z_o) + \mathbf{V}|_{z=z_o}\cdot\mathbf{k} - \mathbf{I}|_{z=z_o}\,, \tag{3.4.10}$$

where S_y is the specific yield and I evaluated at x = h_s is the specific infiltration.

Substituting Eqs. (3.4.7) through (3.4.10) into Eq. (3.4.6) and assuming $H(x,y,h_s,t)$ to be approximately equal to the average value of H(x,y,z,t), we obtain the following equation for the confined aquifer:

$$bS_s \frac{\partial \Phi}{\partial t} = \nabla_H \cdot (b\mathbf{K} \cdot \nabla_H \Phi) - Q_U - Q_L + W \tag{3.4.11}$$

and the following equation for unconfined aquifer:

$$(S_y + bS_s) \frac{\partial \Phi}{\partial t} = \nabla_H \cdot (b\mathbf{K} \cdot \nabla_H \Phi) - Q_U - Q_L + W \, , \tag{3.4.12}$$

where

$$\Phi = \frac{1}{b} \int_{z_o}^{h_s} H \, dz, \quad \mathbf{K} \cdot \nabla_H \Phi = \frac{1}{b} \int_{z_o}^{h_s} \mathbf{V} \, dz, \quad W = \int_{z_o}^{h_s} q \, dz \tag{3.4.13}$$

$$Q_U = - I|_{z=h_s}, \quad Q_L = - I|_{z=z_o}, \quad and \quad b = h_s - z_o \, . \tag{3.4.14}$$

It should be noted that in deriving Eq. (3.4.12), we have assumed that $\Phi \approx h_s$.

Eqs. (3.4.11) and (3.4.12) can be rewritten as

$$S \frac{\partial \Phi}{\partial t} = \nabla_H \cdot (\mathbf{T} \cdot \nabla_H \Phi) - Q_U - Q_L + W \, , \tag{3.4.15}$$

in which

$$\mathbf{T} = b\mathbf{K} \quad and \quad S = bS_s + s_y \tag{3.4.16}$$

are the transmissivity tensor and generalized storage coefficient, respectively. In the case of confined aquifer, T is independent of Φ, whereas in the case of unconfined aquifers, T is a function of Φ resulting in a nonlinear equation. In Eq. (3.4.12), Q_U is the infiltration rate from precipitation, Q_L is the leakage from the bottom confining aquitard, and S is given by Eq. (3.4.16) for unconfined aquifers; whereas Q_U and Q_L are the leakage from the top and bottom confining aquitards, respectively, and S_y is equal to zero. If we assume that the storage in the confining aquitards can be neglected, then the Q_U and Q_L can be written as

$$Q_U = K_U(\Phi - \Phi_U) \quad and \quad Q_L = K_L(\Phi - \Phi_L) \, , \tag{3.4.17}$$

where K_U and K_L are the leakage coefficients of the upper and lower confining aquitards, respectively, and Φ_U and Φ_L are the hydraulic head in the aquifers above the upper confining aquitard and below the lower confining aquitard, respectively. If the storage of the confining aquitards is not negligible, the equations to relate Q_U and Q_L to the Φ can be found elsewhere (Trescott et al., 1976).

Two types of source/sink are considered here: one is the distributed element source/sink and the other is the point (well) source/sink. Thus, W is formulated as

$$W = S_W + \sum_{i=1} S_i \delta(x - x_i) \delta(y - y_i) \, , \tag{3.4.18}$$

where S_w is the distributed source/sink function, S_i is the point source/sink strength at the i-th point source/sink, (x_i, y_i) is the coordinate of the i-th point source/sink, and δ is a delta function.

Substituting Eqs. (3.4.17) and (3.4.18) into (3.4.15), we have

$$S\frac{\partial \Phi}{\partial t} = \nabla_H \cdot (\mathbf{T} \cdot \nabla_H \Phi) - K_U(\Phi - \Phi_U) - K_L(\Phi - \Phi_l) + S_w + \sum_{i=1} S_i \delta(x - x_i)\delta(y - y_i) . \tag{3.4.19}$$

The initial and boundary conditions for Eq. (3.4.19) can be obtained by integrating Eqs. (3.2.33) and (3.2.25) through (3.2.28). Starting with Eq. (3.2.33), we first add a potential head z to both sides of the equation and integrate it vertically to yield

$$\Phi = \Phi_i(x,y) \quad \text{in} \quad R , \tag{3.4.20}$$

where Φ_i is the prescribed initial condition, which can be defined or obtained by solving the steady state version of Eq. (3.4.19). Similarly, the integration of Eq. (3.2.25) will yield the Dirichlet boundary condition as follows:

$$\Phi = \Phi_d(x,y,t) \quad \text{on} \quad B_d(x,y) = 0 , \tag{3.4.21}$$

where Φ_d is the given Dirichlet hydraulic head. However, because of the elimination of z dependence, the integration of Eqs. (3.4.11) and (3.4.12) will render the identical condition

$$-\mathbf{n} \cdot \mathbf{T} \cdot \nabla \Phi = q_f(x,y,t) \quad \text{on} \quad B_f(x,y) = 0 , \tag{3.4.22}$$

where $q_f(x,y,t)$ is the prescribed flux resulting from the vertical integration of either q_r or q_c and B_f is the union of B_n and B_c.

Equations (3.4.19) through (3.4.22) constitute a general mathematical statement of the physical problem of flow in aquifers. Analytical solutions for this general system do not exist. Numerical algorithms must be devised to solve the problem. Finite-difference approximations have been widely reported in the literature. A finite element model taken from FEWA: A Finite Element Model of Water Flows through Aquifers by G. T. Yeh and D. D. Huff (Yeh and Huff, 1983) will be presented in this book.

3.5. Finite-Element Modeling of Groundwater Flows (FEWA)

Finite-element approximation of groundwater flows is specified by Eq. (3.4.19) and the associated initial and boundary conditions specified by Eqs. (3.4.20) through (3.4.22) can be made following the nine basic steps given Chapter 2.

3.5.1. Finite-Element Approximations in Space

Let $N_j(x,y)$ be the base function of node point j in the two-dimensional space. The base function, $N_j(x,y)$, will have a value of 1.0 at the nodal point j and a value of 0.0 at all other nodal points. Furthermore, $N_j(x,y)$ will be given nonzero values only over those elements that have one nodal point coinciding with point j and zero values over all other elements in the domain. Two types of base functions are used in the development of groundwater flow model: linear quadrilateral isoparametric elements and linear triangular elements. We now approximate the hydraulic head ϕ by

$$\phi \approx \phi = \sum_{j=1}^{N} \phi_j N_j(x,y) \, , \tag{3.5.1}$$

where N is the total number of nodes in the region, and N_j and ϕ_j are the base function and the amplitude of ϕ, respectively, at nodal point j.

Since ϕ is only an approximate solution, it will not satisfy the governing equations. Thus, we define a residual R_r by

$$R_r = S\frac{\partial \phi}{\partial t} - \nabla_H \cdot (\mathbf{T} \cdot \nabla_H \phi) + K_U(\phi - \phi_U) + K_L(\phi - \phi_L) - S_w - \sum_{i=1} S_i \delta(x - x_i)\delta(y - y_i) \, . \tag{3.5.2}$$

According to the principle of the FEM, we choose the coefficients ϕ_j in Eq. (3.5.1) such that the residual in Eq. (3.5.2) weighted by the set of weighting functions is zero. Thus, we have

$$\int_{R_e} N_i \left\{ S\frac{\partial \phi}{\partial t} - \nabla_H \cdot \mathbf{T} \cdot \nabla_H \phi) + K_U(\phi - \phi_U) - K_L(\phi - \phi_L) \right\} dR = \int_{R_e} N_i \left\{ S_w + \sum_{k=1} S_k \delta(x - x_k)\delta(y - y_k) \right\} dR \, . \tag{3.5.3}$$

Integration of Eq. (3.5.3) by parts yields the following :

$$\left[\int_R N_i S N_j \, dR \right] \left\{ \frac{\partial \phi_j}{\partial t} \right\} + \left[\int_R (\nabla N_i) \cdot \mathbf{T} \cdot (\nabla N_j) dR \right] \{\phi_j\} + \left[\int_R N_i (K_U + K_L) N_j \, dR \right] \{\phi_j\} = \int_R N_i (K_U \phi_U + K_L \phi_L + S_w) dR + \sum_{k=1} S_k \delta_{ik} + \int_B N_i \mathbf{n} \cdot \mathbf{T} \cdot (\nabla_H \phi) dB, \quad i \in N \, . \tag{3.5.4}$$

In deriving Eq. (3.5.4), we have implicitly assumed that the point source/sink coincide with nodal points. Equation (3.5.4) written in matrix form is

$$[M]\left\{ \frac{\partial \phi}{\partial t} \right\} + [S]\{\phi\} = \{Q\} + \{B\} \, , \tag{3.5.5}$$

where {dϕ/dt} and {ϕ} are the column vectors containing the values of dϕ/dt and ϕ, respectively, at all nodes; [M] is the mass matrix resulting from the storage term; [S] is the stiff matrix resulting from the transmisivity term; {Q} and {B} are the load vectors from the internal source/sink as well as leakage, and boundary conditions, respectively. The matrices, [M] and [S], are given by

$$M_{ij} = \sum_{e \in M_e} \int_{R_e} N_\alpha^e S N_\beta^e dR \tag{3.5.6}$$

$$S_{ij} = \sum_{e \in M_e} \int_{R_e} \left[(\nabla_H N_\alpha^e) \cdot \mathbf{T} \cdot (\nabla_H N_\beta^e) + N_\alpha^e (K_U + K_L) N_\beta^e \right] dR \ , \tag{3.5.7}$$

where R_e is the region of element e, M_e is the set of elements that have a local side $\alpha-\beta$ coinciding with the global side i-j, and N_α^e is the α-th local basis function of element e. Similarly, the load vectors {Q} and {B} are given by

$$Q_i = \sum_{e \in M_e} \int_{R_e} N_\alpha^e (S_w + K_U \phi_U + K_L \phi_L) dR + \sum_{k=1} S_k \delta_{ki} \tag{3.5.8}$$

and

$$B_i = -\sum_{e \in N_{se}} \int_{B_e} N_\alpha^e \mathbf{n} \cdot [-\mathbf{T} \cdot (\nabla_H \phi)] dB \ , \tag{3.5.9}$$

where B_e is the length of boundary segment e and N_{se} is the set of boundary segments that have a local node α coinciding with the global node i. It should be noted that in applying the weighted-residual finite element method to Eq. (3.4.19), we have used the set of base functions as the set of weighting functions for all terms. Hence the Galerkin finite-element method results.

The reduction of the partial differential equation (PDE), Eq. (3.4.19), to the set of ordinary differential equations (ODE), Eq. (3.5.5), simplifies to the evaluation of integrals on the right-hand side of Eqs. (3.5.6) through (3.5.9) for every element or boundary segment e to yield the element mass matrix $[M^e]$ and stiff matrix $[S^e]$ as well as the element source/sink column vector $\{Q^e\}$ and boundary column vector $\{B^e\}$ as

$$M_{\alpha\beta}^e = \int_{R_e} N_\alpha^e S N_\beta^e dR \tag{3.5.10}$$

$$S_{\alpha\beta}^e = \int_{R_e} \left[(\nabla_H N_\alpha^e) \cdot \mathbf{T} \cdot (\nabla_H N_\beta^e) + N_\alpha^e (K_U + K_L) N_\beta^e \right] dR \tag{3.5.11}$$

$$Q_\alpha^e = \int_{R_e} N_\alpha^e (S_w + K_U \phi_U + K_L \phi_L) dR \tag{3.5.12}$$

and

$$B_\alpha = -\int_{B_e} N_\alpha^e \mathbf{n} \cdot [-\mathbf{T} \cdot (\nabla_H \phi)] dB \ , \tag{3.5.13}$$

where the superscript or subscript e denotes the element and α, β = 1, 2, 3, or 4 for linear quadrilateral elements or α, β = 1, 2, or 3 for linear triangular elements.

For a quadrilateral element, Eqs. (3.5.10) through (3.5.13) are computed by

Gaussian quadrature (Conte, 1965). On the other hand, for a triangular element, the integration of Eqs. (3.5.10) through (3.5.13) is performed analytically using the integration identity,

$$\int_{R_e} L_1^{\,n} L_2^{\,m} L_3^{\,k} dR = 2A \frac{n!m!k!}{(m+n+k+2)!} \ . \qquad (3.5.14)$$

With the element matrices $[M^e]$ and $[S^e]$ and the element column vector $\{Q^e\}$ computed, the global matrices [M] and [S] and the global column vector {Q} are then assembled element by element.

3.5.2. *Mass Lumping*

Referring to [M], one may recall that this is a unit matrix if the finite difference formulation is used in spatial discretization. Hence, by proper scaling, the mass matrix can be reduced to the finite-difference equivalent by lumping (Clough, 1971). In many cases, the lumped mass matrix would result in a better solution, in particular, if it is used in conjunction with the central or backward-difference time marching (Yeh and Ward, 1980). Under such circumstances, it is preferred to the consistent mass matrix (mass matrix without lumping). Therefore, options are provided for the lumping of the matrix [M]. More explicitly, [M] will be lumped according to

$$M_{ij} = \sum_{e \in M_e} \left(\sum_{\beta=1}^{4} \int_{R_e} N_\alpha^{\,e} S N_\beta^{\,e} dR \right) \text{ if } j = i,\ M_{ij} = 0 \text{ if } j \neq i \ . \qquad (3.5.15)$$

3.5.3. *Finite-Difference Approximation in Time*

Next, we derive a matrix equation by integrating Eq. (3.5.5). For the time integration of Eq. (3.5.5), the load vector {B} will be ignored. This load vector will be discussed in the next section on the numerical implementation of boundary conditions. An important advantage of finite-element approximation over the finite-difference approximation is the inherent ability to handle complex boundaries and obtain the normal derivatives therein. In the time dimension, such advantages are not evident. Thus, finite-difference methods are typically used in the approximation of the time derivative. Three time-marching methods are adopted in the present water-flow model. In the first one, the central or Crank-Nicolson formulation may be written as

$$\frac{[M]}{\Delta t}\left(\{\phi\}_{t+\Delta t} - \{\phi\}_t\right) + \frac{1}{2}[S]\left(\{\phi\}_{t+\Delta t} + \{\phi\}_t\right) = \{Q\} \ , \qquad (3.5.16)$$

where [M], [S], and {Q} are evaluated at $t+\Delta t/2$

In the second method, the backward-difference formulation may be written as

$$\frac{[M]}{\Delta t}\left(\{\phi\}_{t+\Delta t}-\{\phi\}_{t}\right)+[S]\{\phi\}_{t+\Delta t}=\{Q\}\ , \tag{3.5.17}$$

where [M], [S], and {Q} are evaluated at $t+\Delta t$. In the third optional method, the values of unknown variables are assumed to vary linearly with time during the time interval, Δt. In this middifference method, the recurrence formula is written as

$$\left(\frac{2}{\Delta t}[M]+[S]\right)\{\phi\}_{t+\Delta t/2}-\frac{2}{\Delta t}[M]\{\phi\}_{t}=\{Q\} \tag{3.5.18}$$

and

$$\{\phi\}_{t+\Delta t}=2\{\phi\}_{t+\Delta t/2}-\{\phi\}_{t}\ , \tag{3.5.19}$$

where [M], [S], and {Q} are evaluated at $(t+\Delta t/2)$.

Equations (3.5.16) through (3.5.18) can be written as a matrix equation

$$[A]\{\phi\}=\{b\}\ , \tag{3.5.20}$$

where [A] is the matrix, $\{\phi\}$ is the unknown vector to be found and represents the values of discretized pressure field at new time, and {b} is the load vector. Take for example, Eq. (3.5.16); [A] and {b} represent the following:

$$[A]=\frac{[M]}{\Delta t}+[S]\quad \textit{and}\quad \{b\}=\frac{[M]}{\Delta t}\{\phi\}_{t}+\{Q\}\ , \tag{3.5.21}$$

where $\{\phi\}_t$ is the vector of the discretized pressure field at previous time.

3.5.4. Numerical Implementation of Boundary Conditions

The matrix equation, Eq. (3.5.20), is singular. To make the problem uniquely defined, boundary conditions [Eqs. (3.5.20) and (3.5.21)] must be implemented. For the flux boundary condition given by Eq. (3.5.21), we simply substitute Eq. (3.5.21) into Eq. (3.5.13) to yield a boundary-element column vector $\{B_c^e\}$ for a Cauchy segment:

$$\left\{B_c^e\right\}=\left\{q_f^e\right\}\ , \tag{3.5.22}$$

where $\{q_f^e\}$ is the boundary flux vector given by

$$q_{f\alpha}^{e}=-\int_{B_e}N_\alpha^e q_f dB\ ,\qquad \alpha=1\ \text{or}\ 2\ . \tag{3.5.23}$$

This vector represents the normal fluxes through the two nodal points of the segment B_e on B_f.

At nodes where Dirichlet boundary conditions are applied, an identity equation is generated for each node and included in the matrices of Eq. (3.5.20). The detailed method of applying this type of boundary condition can be found elsewhere (Wang and Conner, 1975). After incorporating the boundary conditions described in this subsection, Eq. (3.5.20) becomes

$$[C]\{\phi\} = \{R\} . \qquad (3.5.24)$$

3.5.5. Solution of the Matrix Equations

Equation (3.5.24) is in general a banded sparse matrix equation. It may be solved numerically by either direct or iteration methods. In direct methods, a sequence of operation is performed only once. This would result in each solution except for round-off error. In this method, one is concerned with the efficiency and magnitude of round-off error associated with the sequence of operations. On the other hand, in an iterative method, one attempts to research a solution by a process of successive approximations. This involves in making an initial guess, then improving the guess by some iterative process until an error criterion is obtained. Therefore, in this technique, one must be concerned with convergence, and the rate of convergence. The round-off errors tend to be self-corrected.

For practical purposes, the advantages of the direct method are (1) the efficient computation when the bandwidth of the matrix [C] is small, and (2) the fact that no problem of convergency is encountered when the matrix equation is linear or less severe in convergence than iterative methods even when the matrix equation is nonlinear. The disadvantages of direct methods are the excessive requirements on CPU storage and CPU time when a large number of nodes is needed for discretization. On the other hand, the advantages of iteration methods are efficiencies in terms of CPU storage and CPU time when large problems are encountered. Their disadvantages are the requirements that the matrix [C] must be well conditioned to guarantee a convergent solution. Hence, we provide six options to solve the matrix equation so that as wide a range of problems as possible can be dealt with. One is the Gaussian direct elimination method, one is the successive point iteration methods, and the other four are the preconditioned conjugate gradient methods using the polynomials, incomplete Cholesky decomposition, modified incomplete Cholesky decomposition, and symmetric successive overrelaxation as the preconditioners, respectively. The first one should be used when the size of the problems is such that the resulting bandwidth of matrix [C] is not large, say, no greater than 50, whereas the other five alternatives should be used if the bandwidth of the matrix [C] is large.

The matrix equation, Eq. (3.5.24), is nonlinear when the media are unconfined aquifers. To solve it, some type of iterative procedure is required. The approach taken here is to make an initial estimate of the unknown $\{\phi^k\}$. Using this estimate, we then compute the coefficient matrix [C] and solve the linearized matrix equation by the method of linear algebra. The new estimate is now obtained by the weighted average of the new solution and the previous estimate:

$$\{\phi^{(k+1)}\} = \omega\{\phi\} + (1-\omega)\{\phi^k\} , \qquad (3.5.25)$$

where $\{\phi^{(k+1)}\}$ is the new estimate, $\{\phi^k\}$ is the previous estimate, $\{\phi\}$ is the new solution, and ω is the iteration parameter. The procedure is repeated until the new solution $\{\phi\}$ is within a tolerance error. If ω is greater than or equal to 0 but is less than 1, the iteration is underrelaxation. If $\omega = 1$, the method is the exact relaxation.

If ω is greater than 1 but less than or equal to 2, the iteration is termed overrelaxation. The underrelaxation should be used to overcome cases when nonconvergency or the slow convergent rate is due to fluctuation rather than due to "blowup" computations. Overrelaxation should be used to speed up convergent rate when it decreases monotically.

In summary, there are 108 optional numerical schemes available to deal with as wide a range of problems as possible. These are the combinations of (1) two ways of treating the mass matrix (lumping and no-lumping), (2) three ways of approximating the time derivatives (central difference, backward difference, and middifference), (3) six ways of solving the linearized matrix equation (direct method and five iteration methods), and (4) three ways of estimating the coefficient matrix (underrelaxation, exact relaxation, and overrelaxation).

3.5.6. Computer Implementation

A computer program has been designed to solve Eq. (3.5.17) along with the initial and boundary conditions Eqs. (3.4.20) through (3.4.22). The use of this computer program can be found in *FEWA: User's Manual of A Finite Element Model of Water Flow through Aquifers* (Yeh and Huff, 1983). An updated computer code along with its documentation is included in the attached floppy diskette in this book.

3.5.7. Verification Problems

Three examples are given here to illustrate the use of FEWA. The first one deals with confined aquifer and the other two treat unconfined aquifers.

3.5.7.1. Verification Example 1. This problem is a transient simulation in a confined aquifer without sources/sinks or leakage. Consider a confined aquifer with a thickness of b = 10 m (Fig. 3.5.1). Assume the aquifer is filled with a porous medium that has a compressibility of skeleton of $\alpha' = 10^{-5}$ /m and a compressibility of water of $\beta' = 4 \times 10^{-6}$ /m. The hydraulic conductivity of the medium is $K = 4.4 \times 10^{-7}$ m/s, and the effective porosity is $n_e = 0.25$. This confined aquifer would then have a generalized storage coefficient $S = 1.1 \times 10^{-4}$ and a transmissivity of $T = 4.4 \times 10^{-6}$ m^2/s. Let us further assume that no leaking to or from the underlying or overlying aquifer would take place. We will also assume that the internal source/sink is zero. Initially the aquifer is observed to have a hydraulic head $\phi = H = 25$ m (Fig. 3.5.1). If one starts to pump $2Q = 4.4 \times 10^{-6}$ m^3/s of water per unit width perpendicular to the x-z plane at x = 0, what will be the hydraulic head distribution along the x-axis at any time?

The above problem can be described by the following set of equations:

$$S\frac{\partial \phi}{\partial t} = T\frac{\partial^2 \phi}{\partial x^2} \tag{3.5.26}$$

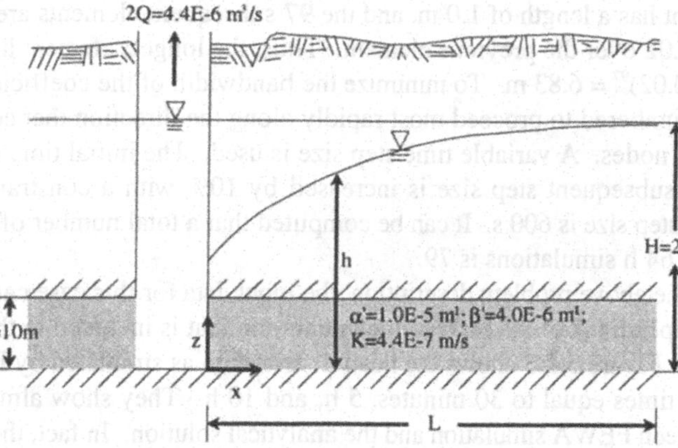

Fig. 3.5.1 Problem Definition and Sketch of Problem 1

$$-T\frac{\partial \phi}{\partial x} = Q \quad \text{at} \quad t > 0 \quad \text{and} \quad x = 0 \tag{3.5.27}$$

$$\phi = H \quad \text{at} \quad t > 0 \quad \text{and} \quad x = L\,. \tag{3.5.28}$$

Equations (3.5.26) through (3.5.29) are in fact the special case of Eqs. (3.4.19) through (3.4.2), respectively. It can be verified that the solution of Eq. (3.5.26) subject to the initial and boundary conditions Eqs. (3.5.27) through (3.5.29) is given by Carslaw and Jaeger (1959) as

$$\phi = H - \frac{2Q}{T}\left\{\left(\frac{Tt}{S\pi}\right)^{1/2} \exp\left(-\frac{Sx^2}{4Tt}\right) - \frac{x}{2}\operatorname{erfc}\left(\frac{\sqrt{S}x}{2\sqrt{Tt}}\right)\right\}. \tag{3.5.29}$$

To compare the FEWA calculation with that obtained by Eq. (3.5.30), a total of approximately 10.64 h real-time simulation is made. With this real time, a length of approximately L = 298.2 m (Fig. 3.5.2) can be considered infinite. To start the simulation, the region is divided into 98 elements and 198 nodes (Fig. 3.5.2). The

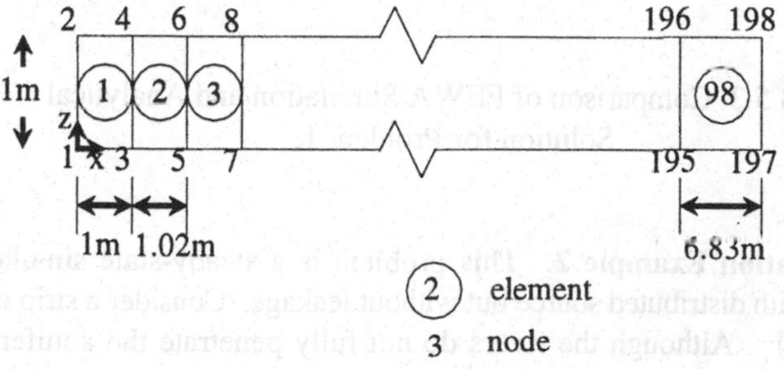

Fig. 3.5.2 Finite-Element Discretization for Problem 1.

smallest element has a length of 1.0 m, and the 97 subsequent elements are increased by a factor of 1.02 over the previous element. Thus, the longest element has a length of $1.0 * (1.0 + 0.02)^{97} = 6.83$ m. To minimize the bandwidth of the coefficient matrix, the nodes are numbered to proceed most rapidly along the direction that contains the least number of nodes. A variable time step size is used. The initial time step size is 60 s and each subsequent step size is increased by 10% with a constraint that the maximum time step size is 600 s. It can be computed that a total number of time steps required for 10.64 h simulations is 79.

With the above problem description, the input data for this case can be coded according to Appendix A of the FEWA documentation that is included in the attached floppy diskette. Figure 3.5.3 shows the head distribution as simulated by FEWA and Eq. (3.5.30) at times equal to 30 minutes, 5 h, and 10 h. They show almost perfect agreement between FEWA simulation and the analytical solution. In fact, the computer printout indicated that the maximum error in the drawdown as predicted by FEWA is less than 0.2%.

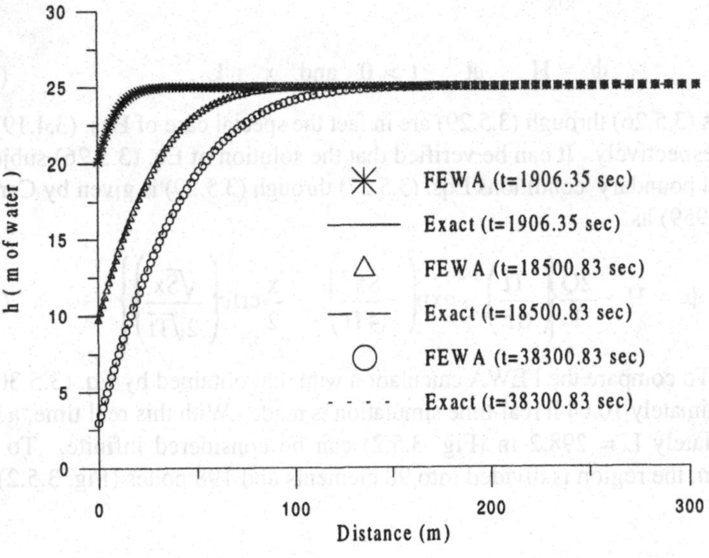

Fig. 3.5.3 Comparison of FEWA Simulation and Analytical Solution for Problem 1.

3.5.7.2. Verification Example 2. This problem is a steady-state simulation in a phreatic aquifer with distributed source but without leakage. Consider a strip of parallel rivers (Fig. 3.5.4). Although the rivers do not fully penetrate the aquifer, we will assume that at $x = 0$ and $x = L$, we have vertical equipotential, $h_0 = 5$ m and $h_L = 10$ m, respectively. The aquifer is assumed to have a compressibility of skeleton of $\alpha' = 0$ and a compressibility of water of 0. The hydraulic conductivity of the medium is $K = 10^{-5}$ m/s, and the effective porosity is $n_e = 0.25$. The specific yield is approximately equal to the effective porosity. A uniform source $S_w = 10^{-8}$ m/s is infiltrating through the ground surface to the aquifer. We will further assume that the bottom of the aquifer is impervious, that is, $K_L = 0$. The question being asked is: what will be the hydraulic head distribution along the x-axis under steady-state conditions?

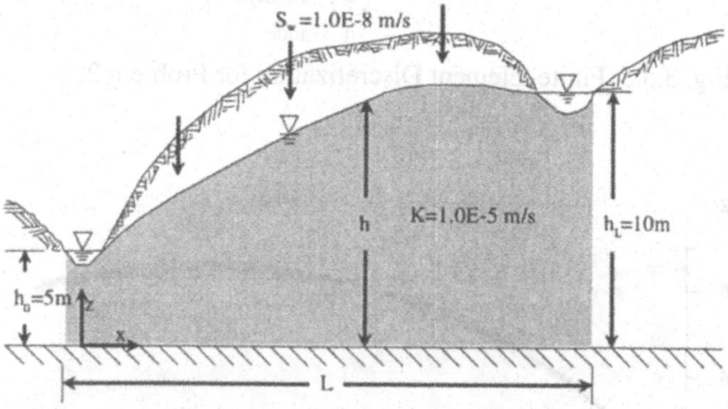

Fig. 3.5.4 Problem Definition and Sketch of Problem 2.

The above problem can be described by the following set of equations:

$$K\frac{d}{dx}\left(\phi\frac{d\phi}{dx}\right) + S_w = 0 \qquad (3.5.30)$$

$$\phi = h_o \quad \text{at} \quad x = 0 \qquad (3.5.31)$$

and

$$\phi = h_L \quad \text{at} \quad x = L = 392 \text{ m} . \qquad (3.5.32)$$

Equations (3.5.31) through (3.5.33) can be verified to have the following solution (Bear 1977):

$$\phi = \left[h_o^2 + \frac{x}{L}\left(h_L^2 - h_o^2\right) + \frac{S_w}{K}x(L-x)\right]^{1/2} . \qquad (3.5.33)$$

For FEWA simulation, the region is divided into 98 elements and 198 nodes (Fig. 3.5.5). Each element has a length of 4 m. With these problem descriptions, the input data for this case can be coded according to Appendix A of the FEWA document. Figure 3.5.6 shows the head distribution as simulated by FEWA and Eq. (3.5.34). They show almost perfect agreement between FEWA simulation and the analytical solution. In fact, the computer printout indicated that the maximum error in the drawdown as predicted by FEWA is less than 0.2%.

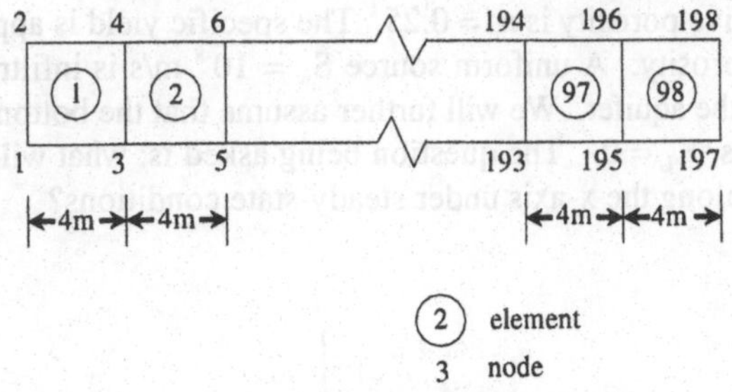

Fig. 3.5.5 Finite-Element Discretization for Problem 2.

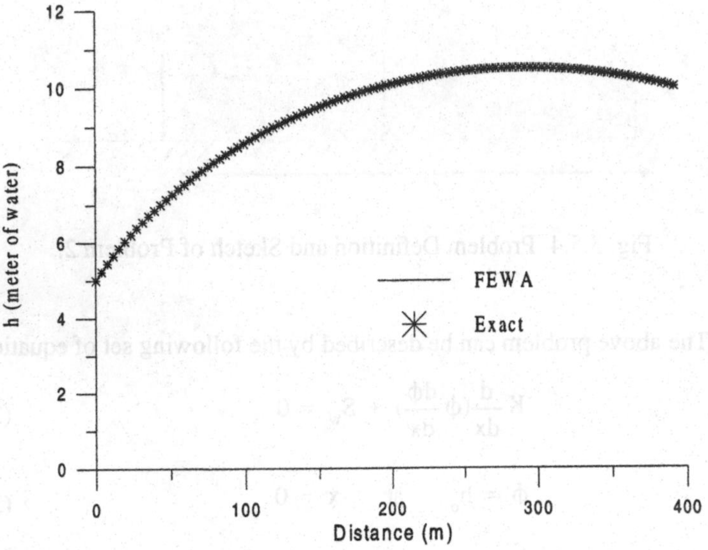

Fig. 3.5.6 Comparison of FEWA Simulation and Analytical Solutions for Problem 2.

3.5.7.3. Verification Example 3. This problem is a steady-state simulation in a phreatic aquifer with distributed source and partial leakage. Consider an aquifer that overlies a thin layer of aquitard (Fig. 3.5.7) that is impervious to the left of x = 0 and is semipervious to the right of x = 0 with a leakage coefficient of $K_L = 10^{-7}\ s^{-1}$. On the ground surface a constant recharge rate of $S_w = 2.5 \times 10^{-7}$ m/s is added to the aquifer. Assume that it is observed that at x = 0, the hydraulic head is constant at $\phi = h_0 = 20$ m in reference to the bottom of the aquifer (Fig. 3.5.7). We will further assume that below the semipervious layer, a suction of $\phi = h_L = -10$ m is maintained. The aquifer is assumed to have a compressibility of skeleton $\alpha' = 0$ and a compressibility of water of $\beta' = 0$. The hydraulic conductivity of the medium is $K = 10^{-5}$ m/s, and the effective porosity is $n_e = 0.25$. The specific yield is approximately equal to the effective porosity. The question being asked is: what will be the hydraulic head distribution along the x-axis under steady-state conditions?

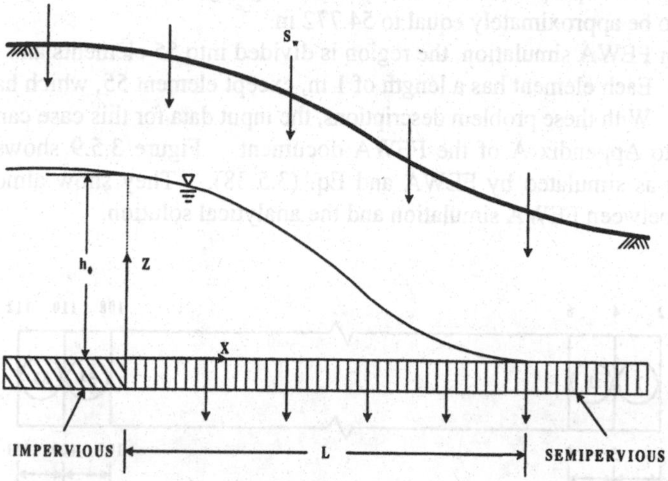

Fig. 3.5.7 Problem Sketch and Definition for Problem 3.

The above problem can be described by the following set of equations:

$$K\frac{d}{dx}\left(\phi\frac{d\phi}{dx}\right) + S_w - K_L(\phi-\phi_L) = 0 \tag{3.5.34}$$

$$\phi = h_o \quad \text{at} \quad x = 0 \tag{3.5.35}$$

and

$$-K\phi\frac{d\phi}{dx} = 0 \quad \text{at} \quad x = L\,, \tag{3.5.36}$$

with the additional requirement that $\phi = 0$ at $x = L$ (Bear, 1977). It can be shown that Eqs. (3.5.35) through (3.5.37) have the following solution (Bear, 1977):

$$\phi = h_o + \frac{Ax^2}{6} - \left(\frac{2Ah_o}{3}+B\right)^{1/2} x\,, \tag{3.5.37}$$

where

$$A = \frac{K_L}{K} \quad \text{and} \quad B = -\frac{K_L\phi_L+S_w}{K}. \tag{3.5.38}$$

It can be shown that with the additional requirement that $h = 0$ at $x = L$, L is given by

$$L = \frac{3}{A}\left\{\left(\frac{2Ah_o}{3}+B\right)^{1/2} - B^{1/2}\right\}. \qquad (3.5.39)$$

With the hydraulic properties and the source strength given in the example, L can be computed to be approximately equal to 54.772 m.

For FEWA simulation, the region is divided into 55 elements and 112 nodes (Fig. 3.5.8). Each element has a length of 1 m, except element 55, which has a length of 0.772 m. With these problem descriptions, the input data for this case can be coded according to Appendix A of the FEWA document. Figure 3.5.9 shows the head distribution as simulated by FEWA and Eq. (3.5.38). They show almost perfect agreement between FEWA simulation and the analytical solution.

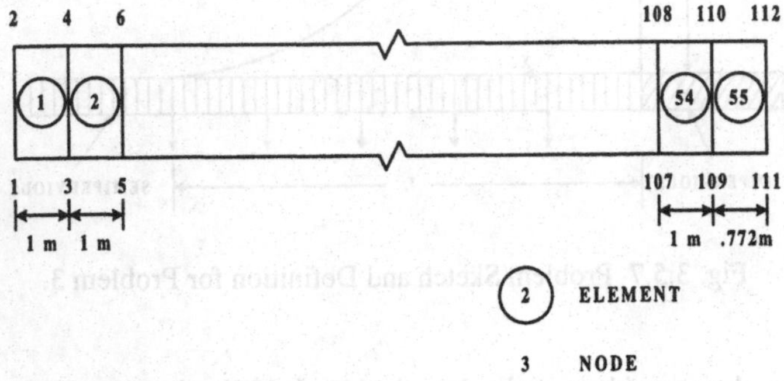

Fig. 3.5.8 Finite-Element Discretization for Problem 3.

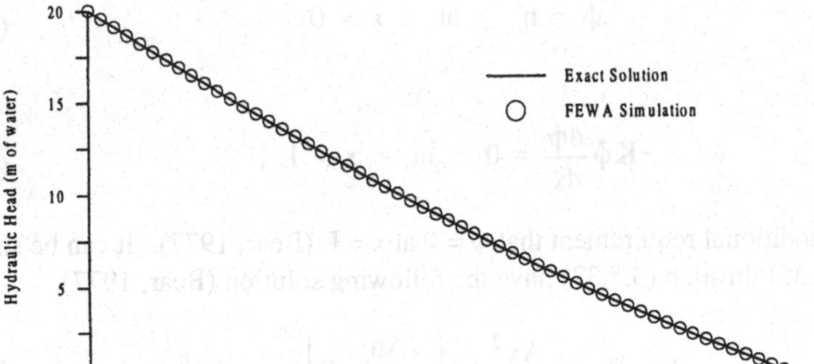

Fig. 3.5.9 Comparison of FEWA Simulation and Exact Solution for Problem 3.

3.6. Convergence, Stability, and Compatibility

This section deals with the conditions that must be satisfied if the solution of the approximating system of algebraic equations is to be a reasonable, accurate approximation to the solution of the corresponding partial differential equations. These conditions are associated with three different but interrelated problems. The first concerns the convergence of the solution of the approximating equations to the solution of differential equations; the second concerns the unstable growth or stable decay of the errors in the arithmetical operations needed to solve the approximating equations; and the third concerns whether the approximating equations approach to the right differential equations.

Let U represent the exact solution of a partial differential equation with independent variables x and t, and u represent the exact solution of approximating equations. The approximating equations are said to be convergent when u tends U as Δx and Δt both tend to 0. Although conditions under which u converges to U have been established for linear elliptic, parabolic, and hyperbolic second-order partial differential equations with solutions satisfying fairly general boundary and initial conditions, they are not yet known for nonlinear equations except for a few particular cases. The difference (U-u) is called the discretization error. The magnitude of the discretization error at any mesh point depends on the finite-sizes of the mesh length and on the number of terms in a polynomial to approximate the derivatives. Thus, the discretization error can be decreased by either approximating the derivatives with a higher order of polynomial or by reducing the mesh size. The former has the disadvantage of involving more pivotal values of u, and the latter leads to an increase in the number of equations to be solved.

The equations that are actually solved are of course the approximating equations. If it were possible to carry out all calculations to an infinite number of decimal places, we would obtain their exact solution u. In practice, however, each calculation is carried out to a finite number of decimal places or significant figures, a procedure that introduces a "round-off" error every time it is used, and the solution actually computed is not u but N (say). N will be called the numerical solution. Generally, a set of approximating equations is stable when the cumulative effect of all the rounding error is negligible. More specifically, if errors $e_{1,1}$, $e_{1,2}$ ..., $e_{i,j}$ are introduced at the mesh points $P_{1,1}$, $P_{1,2}$, ..., $P_{i,j}$, respectively, and $|e_{1,1}|$, $|e_{1,2}|$, ..., $|e_{i,j}|$, are each less than δ, then the approximating equations are stable when the maximum value of (u-N) tends to zero as δ tends to zero and does not increase exponentially with the number of columns or rows of calculation. It is usually not possible to determine the exact magnitude of (u-N) at mesh point (ih,jk) for an arbitrary distribution of errors, but there are standard methods for deriving an estimate of it in special cases, such as an isolated error. For example, consider the following simple parabolic equation:

$$\frac{\partial U}{\partial t} = \frac{\partial^2 U}{\partial x^2} \quad \text{for } 0 < x < 1 \,, \tag{3.6.1}$$

with the initial and boundary conditions given by

$$U = 2x \quad \text{for } 0 < x < 1/2 \quad t = 0 \tag{3.6.2}$$

$$U = 2(1-x) \quad \text{for } 1/2 < x < 1 \quad t = 0 \tag{3.6.3}$$

$$U = 0 \quad \text{at} \quad x = 0 \quad \text{and} \quad x = 1 \quad \text{for all} \quad t\,. \tag{3.6.4}$$

If one approximates the above equations with explicit finite difference and uses $\Delta x = 1/10$ and $\Delta t = 1/200$, then one would have the following approximating equation:

$$u_{i,j+1} = \frac{1}{2}(u_{i-1,j} + u_{i+1,j})\,. \tag{3.6.5}$$

The manner in which errors are propagated when using the above explicit formulae can easily be seen by applying the finite difference equation to the errors themselves. The propagation of an initially isolated error is shown in Table 3.6.1. It is seen that the modulus of the maximum error along each row gradually decreases as j increases, so it would be reasonable to surmise that this scheme is stable.

Table 3.6.1 Propagation of Numerical Errors.

	i=0	1	2	3	4	5	6	7	8	9	10
j=0	0	0	0	0	0	e	0	0	0	0	0
1	0	0	0	0	e/2	0	e/2	0	0	0	0
2	0	0	0	e/4	0	e/2	0	e/4	0	0	0
3	0	0	e/8	0	3e/8	0	3e/8	0	e/8	0	0
4	0	e/16	0	e/4	0	3e/8	0	e/4	0	e/16	0
5	0	0	5e/32	0	5e/16	0	5e/16	0	5e/32	0	0
6	-	-	-	-	-	-	-	-	-	-	-

It should be noted that stability is not associated directly with the solution of the partial differential equation. For example, if we use $\Delta x = 1/10$ and $\Delta t = 1/100$ and approximate Eq. (3.6.1) with explicit scheme, we would have the following computational scheme

$$u_{i,j+1} = u_{i-1,j} - u_{i,j} + u_{i+1,j}\,. \tag{3.6.6}$$

The solution of this finite-difference scheme is given in Table 3.6.2.

Table 3.6.2 Numerical Solution of Diffusion Equation with an Explicit Scheme.

	i = 0	1	2	3	4	5	6
j=0	0.0	0.2	0.4	0.6	0.8	1.0	0.8
1	0.0	0.2	0.4	0.6	0.8	0.6	0.8
2	0.0	0.2	0.4	0.6	0.4	1.0	0.4
3	0.0	0.2	0.4	0.2	1.2	-0.2	1.2
4	0.0	0.2	0.0	1.4	-1.2	2.6	-1.2
5	-	-	-	-	-	-	-

The oscillatory solution of the above scheme is due to nonconvergence, not instability. The numerical solution of the approximating equations is exact; no rounding errors were ever introduced.

Compatibility is a measure of the difference between the differential equation and the approximating equation. Like stability and convergence it can be investigated by a standard method but is much easier to deal with than convergence. Studies to date indicate that compatibility and stability imply convergence, but this relationship has been established for only a limited number of differential equations, such as linear equations with constant coefficients, certain linear hyperbolic and parabolic equations in one-dimension space when approximated by explicit schemes, and a few quasi-linear second-order equations (Richtmyer, 1957). Consider the following equation:

$$\frac{\partial u}{\partial t} - \frac{\partial^2 u}{\partial x^2} = 0 \qquad (3.6.7)$$

and the approximating equation using explicit finite difference:

$$\frac{u_{i,j+1} - u_{i,j}}{k} - \frac{u_{i+1,j} - 2u_{i,j} + u_{i-1,j}}{h^2} = 0 \; . \qquad (3.6.8)$$

Let v(x,t) be any function processing continuous partial derivatives. Then the difference

$$\left(\frac{v_{i,j+1} - v_{i,j}}{k} - \frac{v_{i+1,j} - 2v_{i,j} + v_{i-1,j}}{h^2} \right) - \left(\frac{\partial v}{\partial t} - \frac{\partial^2 v}{\partial x^2} \right)_{i,j} = T \qquad (3.6.9)$$

is called the truncation error at the point (ih,jk) and gives an indication of the error resulting from the replacement of the left hand-side of Eq. (3.6.8) by its finite-difference approximation. If T tends to zero as h and k both tend to zero, the approximating equation is said to be compatible with the differential equation. When v = U, the exact solution of the differential equation, then the term in the second parenthesis in Eq. (3.6.9) is zero, and the expansion of T in powers of h and k provides a measure of the rate at which the value of the difference equation approaches the value of the differential equation at a mesh point.

Expanding $v_{i,j+1}$, $v_{i+1,j}$, and $v_{i-1,j}$ in terms of $v_{i,j}$ and its derivatives, we have

$$v_{i+1,j} = v_{i,j} + h\left(\frac{\partial v}{\partial x}\right)_{i,j} + \frac{h^2}{2!}\left(\frac{\partial^2 v}{\partial x^2}\right)_{i,j} + \dots \tag{3.6.10}$$

etc., and substitution into Eq. (3.6.9) gives

$$T = \frac{1}{2}k\frac{\partial^2 v}{\partial t^2} - \frac{1}{12}h^2\frac{\partial^4 v}{\partial x^4} + \frac{1}{6}k^2\frac{\partial^3 v}{\partial t^3} - \frac{1}{360}h^4\frac{\partial^6 v}{\partial x^6} + \dots \tag{3.6.11}$$

Equation (3.6.11) shows that this explicit difference scheme is compatible with the differential equation because T tends to zero as h and k both tend to zero.

It is sometimes possible to approximate an initial-value type partial differential equation, such as a parabolic or hyperbolic equation, by a finite difference scheme that is stable but that has a solution that converges to the solution of a different differential equation as the mesh sizes tend to zero. Such a difference scheme is said to be incompatible or inconsistent with the partial differential equation. For example, Du Fort and Frankel's explicit scheme to approximate Eq. (3.6.7) is given by

$$\frac{u_{i,j+1} - u_{i,j-1}}{2k} = \frac{u_{i+1,j} - (u_{i,j+1} + u_{i,j-1}) + u_{i-1,j}}{h^2}, \tag{3.6.12}$$

which would result in a truncation error:

$$T = \frac{1}{6}k^2\frac{\partial^3 v}{\partial t^3} - \frac{1}{12}h^2\frac{\partial^4 v}{\partial x^4} + \frac{k^2}{h^2}\frac{\partial^2 v}{\partial t^2} + \cdot \tag{3.6.13}$$

This example is of interest because it shows that the difference equation is compatible with the differential equation only when k/h tends to zero as k and h both tend to zero. If k/h tend to c as k and h both tend to zero, the difference equation approximates the hyperbolic equation

$$\frac{\partial u}{\partial t} + c^2\frac{\partial^2 u}{\partial t^2} - \frac{\partial^2 u}{\partial x^2} = 0 . \tag{3.6.14}$$

3.6.1. Treatment of Convergence

Convergence is more difficult to investigate than stability, and only one example will be considered in detail at this stage. The method below derives a difference equation for the discretization error e, which fortunately, in our example, can be dealt with easily.

Denote the exact solution of the partial differential equation by U and the exact solution of the finite-difference equation by u. Then e = U - u is the discretization error. Consider the equation

$$\frac{\partial U}{\partial t} - \frac{\partial^2 U}{\partial x^2} = 0\ , \tag{3.6.15}$$

where U is known for $0 < x < 1$ when $t = 0$, and at $x = 0$ and 1 when $t > 0$. The simplest explicit finite difference approximation to (3.6.15) is

$$\frac{u_{i,j+1} - u_{i,j}}{k} - \frac{u_{i+1,j} - 2u_{i,j} + u_{i-1,j}}{h^2} = 0\ . \tag{3.6.16}$$

At the mesh points

$$u_{i,j} = U_{i,j} - e_{i,j},\quad u_{i,j+1} = U_{i,j+1} - e_{i,j+1},\quad \text{etc.} \tag{3.6.17}$$

Substitution of Eq. (3.6.17) into Eq. (3.6.16) leads to

$$\begin{aligned} e_{i,j+1} &= re_{i-1,j} + (1-2r)e_{i,j} + re_{i+1,j} + U_{i,j+1} \\ &\quad - U_{i,j} + r(2U_{i,j} - U_{i-1,j} - U_{i+1,j})\ , \end{aligned} \tag{3.6.18}$$

where $r = k/h^2$. By Taylor's theorem, we have

$$U_{i+1,j} = U_{i,j} + h\left(\frac{\partial U}{\partial x}\right)_{i,j} + \frac{h^2}{2!}\left(\frac{\partial^2 U}{\partial x^2}\right)^{(x_i+\theta_1 h, t_j)} \tag{3.6.19}$$

$$U_{i-1,j} = U_{i,j} - h\left(\frac{\partial U}{\partial x}\right)_{i,j} + \frac{h^2}{2!}\left(\frac{\partial^2 U}{\partial x^2}\right)^{(x_i-\theta_2 h, t_j)} \tag{3.6.20}$$

$$U_{i,j+1} = U_{i,j} + k\left(\frac{\partial U}{\partial t}\right)^{(x_i, t_j+\theta_3 k)}\ , \tag{3.6.21}$$

where $0 < \theta_1 < 1$, $0 < \theta_2 < 1$, and $0 < \theta_3 < 1$. Substitution of Eqs. (3.6.19) through (3.6.21) into (3.6.18) gives

$$e_{i,j+1} = re_{i-1,j} + (1-2r)e_{i,j} + re_{i+1,j} + k\left(\frac{\partial U^{(x_i, t_j+\theta_3 k)}}{\partial t} - \frac{\partial^2 U^{(x_i+\theta_4 h, t_j)}}{\partial x^2}\right)\ , \tag{3.6.22}$$

where $-1 < \theta_4 < 1$. This is a difference equation for $e_{i,j}$ which fortunately we need not solve.

Let E_j denote the modulus of the maximum error along the j-th time-row and M is the modulus of the maximum value of the expression in the braces for j = 1 to j. When $r \leq 1/2$, all the coefficients in Eq. (3.6.22) are positive or zero, so

$$
\begin{aligned}
|e_{i,j+1}| &\le r|e_{i-1,j}| + 2(1-2r)|e_{i,j}| + r|e_{i+1,j}| + kM \\
&\le rE_j + (1-2r)E_j + rE_j + kM = E_j + kM
\end{aligned} \tag{3.6.23}
$$

As this is true for all values of i

$$
E_{j+1} \le E_j + kM \le (E_{j-1} + kM) + kM = E_{j-1} + 2kM \tag{3.6.24}
$$

and so on, from which it follows that

$$
E_j \le E_o + jkM = tM \tag{3.6.25}
$$

because the initial values for u and U are the same, that is, $E_0 = 0$. When h tends to zero, $k = rh^2$ also tends to zero and M tends to

$$
\left(\frac{\partial U}{\partial t} - \frac{\partial^2 U}{\partial x^2} \right)_{i,j} . \tag{3.6.26}
$$

Since U is the solution of Eq. (3.6.15), the limiting value of M and therefore E_j is zero. As $|U_{i,j} - u_{i,j}| < E_j$, this proves that u converges to U as h and k both tend to zero.

3.6.2. *Treatment of Stability*

There are two standard methods of computing the growth of errors in the operations needed to investigate the stability of an approximating scheme. In one we express the equations in matrix form and examine the eigenvalues of the associated matrix; in the other we use a finite Fourier series. The Fourier method is the easier one of the two in that it avoids all knowledge of matrix algebra but is the less rigorous because it neglects the boundary conditions.

3.6.2.1. Matrix Method. Consider the equation

$$
\frac{\partial u}{\partial t} = \frac{\partial^2 u}{\partial x^2} \quad 0 < x < 1 \tag{3.6.27}
$$

$$
u = u_o(x) \quad \text{at} \quad t = 0 \tag{3.6.28}
$$

$$
u = 0 \quad \text{at} \quad x = 0 \quad \text{and} \quad x = 1, \quad t > 0 . \tag{3.6.29}
$$

Example 1. The explicit finite-difference approximation of Eq. (3.6.27)

$$
u_{i,j+1} = ru_{i-1,j} + (1-2r)u_{i,j} + ru_{i+1,j} , \tag{3.6.30}
$$

which, with Eq. (3.6.29), leads to

$$u_{1,j+1} = 0 + (1-2r)u_{1,j} + ru_{2,j}$$

$$u_{2,j+1} = ru_{1,j} + (1-2r)u_{2,j} + ru_{3,j}\ , \tag{3.6.31}$$

$$u_{N,j+1} = ru_{N-1,j} + (1-2r)u_{N,j} + 0$$

where $(N+1)h = 1$, and $u_{o,j} = u_{N+1,j} = 0$. These can be written in matrix form as

$$\begin{Bmatrix} u_{1,j+1} \\ u_{2,j+1} \\ u_{3,j+1} \\ \cdot \\ \cdot \\ \cdot \\ u_{N,j+1} \end{Bmatrix} = \begin{bmatrix} (1-2r) & r & \cdot & \cdot & \cdot & \cdot \\ r & (1-2r) & r & \cdot & \cdot & \cdot \\ \cdot & r & (1-2r) & r & \cdot & \cdot \\ & & & & & \\ & & & & & \\ & & & & & \\ \cdot & \cdot & \cdot & \cdot & r & (1-2r) \end{bmatrix} \begin{Bmatrix} u_{1,j} \\ u_{2,j} \\ u_{3,j} \\ \cdot \\ \cdot \\ \cdot \\ u_{N,j} \end{Bmatrix} \tag{3.6.32}$$

or as

$$\mathbf{u}_{j+1} = \mathbf{A}\mathbf{u}_j\ . \tag{3.6.33}$$

Hence,

$$\mathbf{u}_j = \mathbf{A}\mathbf{u}_{j-1} = \mathbf{A}(\mathbf{A}\mathbf{u}_{j-2}) = \ldots = \mathbf{A}^j\mathbf{u}_o\ , \tag{3.6.34}$$

where $\mathbf{u}_o$ is the vector of initial values. Now suppose we introduce errors at every pivotal point along $t = 0$ and start the computation with the vector of values $\mathbf{u}_o^*$. We shall then calculate

$$\mathbf{u}_1^* = \mathbf{A}\mathbf{u}_o^*, \quad \mathbf{u}_1^* = \mathbf{A}\mathbf{u}_2^* = \mathbf{A}^2\mathbf{u}_o^*, \quad \mathbf{u}_j^* = \mathbf{A}^j\mathbf{u}_o^*. \tag{3.6.35}$$

Define the error vector **e** by

$$\mathbf{e} = \mathbf{u} - \mathbf{u}^* \tag{3.6.36}$$

then

$$\mathbf{e}_j = \mathbf{u}_j - \mathbf{u}_j^* = \mathbf{A}^j(\mathbf{u}_o - \mathbf{u}_o^*) = \mathbf{A}^j\mathbf{e}_o \tag{3.6.37}$$

showing that the formula for the propagation of errors is the same as that for the calculation of **u**. This also shows, immediately, that when the finite-difference equations are linear we need consider the propagation of only one line of errors because the overall effect of several lines will be given by the addition of the effect produced by each line considered separately.

The finite-difference scheme will be stable when $\mathbf{e}_j$ remains bounded as j increases infinitely. This can be investigated by expressing the error vector in terms of the eigenvectors of **A**.

Assume that the N eigenvalues λ_s (s = 1, 2, ..., N) of **A** are all different. Then the corresponding N eigenvectors $\mathbf{v}_s$ form a linearly independent set of vectors. Hence the error vector $\mathbf{e}_o$ can be expressed uniquely in terms of the N eigenvectors. This is easily seen because if we write out the equation

$$\mathbf{e}_o = \sum_{s=1}^{N} c_s \mathbf{v}_s \tag{3.6.38}$$

in full, as below

$$\begin{Bmatrix} e_{1,0} \\ e_{2,0} \\ \cdot \\ \cdot \\ e_{N,0} \end{Bmatrix} = c_1 \begin{Bmatrix} v_{1,1} \\ v_{2,1} \\ \cdot \\ \cdot \\ v_{N,1} \end{Bmatrix} + c_2 \begin{Bmatrix} v_{1,2} \\ v_{2,2} \\ \cdot \\ \cdot \\ v_{N,2} \end{Bmatrix} + \ldots + c_N \begin{Bmatrix} v_{1,N} \\ v_{2,N} \\ \cdot \\ \cdot \\ v_{N,N} \end{Bmatrix} \tag{3.6.39}$$

we have N equations for the N unknowns $c_1, c_2, \ldots, c_N$, since the e's and v's are known and are independent.

The errors along the time-row, t = k, resulting from the initial error $\mathbf{e}_o$ will then be given by

$$\mathbf{e}_1 = \mathbf{A}\mathbf{e}_o = \mathbf{A}\sum c_s \mathbf{v}_s = \sum c_s \lambda_s \mathbf{v}_s \tag{3.6.40}$$

by the definition of an eigenvalue. Similarly,

$$\mathbf{e}_j = \sum c_s \lambda_s^j \mathbf{v}_s \,. \tag{3.6.41}$$

Equation (3.6.41) shows that the errors will not increase exponentially with j provided the eigenvalue with the largest modulus has a modulus less than or equal to unity.

The matrix **A** can be written as

$$\mathbf{A} = \mathbf{I} + r\mathbf{T}\,, \tag{3.6.42}$$

in which **I** and **T** are given by

$$\mathbf{I} = \begin{bmatrix} 1 & 0 & 0 & & & \\ 0 & 1 & 0 & & & \\ & & \cdot & \cdot & \cdot & \\ & & & \cdot & \cdot & \cdot \\ & & & 0 & 1 & 0 \\ & & & & 0 & 1 \end{bmatrix} \quad \textit{and} \quad \mathbf{T} = \begin{bmatrix} -2 & 1 & & & & \\ 1 & -2 & 1 & & & \\ & \cdot & \cdot & \cdot & & \\ & & \cdot & \cdot & \cdot & \\ & & & 1 & -2 & 1 \\ & & & & 1 & -2 \end{bmatrix}, \tag{3.6.43}$$

where **T** is an N × N matrix whose eigenvalues λ_s and eigenvectors $\mathbf{v}_s$ are given by

$$\lambda_s = -4\sin^2\left(\frac{s\pi}{2(N+1)}\right), \quad s = 1,2,...N \tag{3.6.44}$$

$$\mathbf{v}_s = \left(\sin\frac{s\pi}{N+1}, \ \sin\frac{2s\pi}{N+1}, ..., \ \sin\frac{Ns\pi}{N+1}\right) . \tag{3.6.45}$$

These values can be verified by substitution into $\mathbf{T}\mathbf{v}_s = \lambda_s \mathbf{v}_s$.

Recall that when the eigenvalue of **B** is λ, then the eigenvalue of f(**B**) is f(λ), and it follows that the eigenvalues of **A** are

$$1 + r\left\{-4\sin^2\left(\frac{s\pi}{2(N+1)}\right)\right\} . \tag{3.6.46}$$

Therefore, the condition for the stability of the explicit scheme is

$$\left|1-4r\sin^2\left(\frac{s\pi}{2(N+1)}\right)\right| < 1 . \tag{3.6.47}$$

The only useful inequality is

$$-1 \le 1-4r\sin^2\left(\frac{s\pi}{2(N+1)}\right) , \tag{3.6.48}$$

which gives

$$r \le \frac{1}{2\sin^2\dfrac{s\pi}{2(N+1)}} > \frac{1}{2} \tag{3.6.49}$$

Thus, the explicit scheme is stable when $r < 1/2$.

Example 2. The Crank-Nicolson scheme is given by

$$\frac{u_{i,j+1}-u_{i,j}}{k} = \frac{1}{2}\left\{\frac{u_{i+1,j+1}-2u_{i,j+1}+u_{i-1,j+1}}{h^2}+\frac{u_{i+1,j}-2u_{i,j}+u_{i-1,j}}{h^2}\right\} , \tag{3.6.50}$$

which gives

$$\begin{aligned} -ru_{i-1,j+1} + (2+2r)u_{i,j+1} - ru_{i+1,j+1} \\ = ru_{i-1,j} - (2-2r)u_{i,j} + ru_{i+1,j} . \end{aligned} \tag{3.6.51}$$

Equation (3.6.51) written in matrix form is

$$(2\mathbf{I}-r\mathbf{T})\mathbf{u}_{j+1} = (2\mathbf{I}+r\mathbf{T})\mathbf{u}_j \tag{3.6.52}$$

that is,

$$\begin{bmatrix} (2+2r) & -r & & & \\ -r & (2+2r) & -r & & \\ & \cdot & \cdot & & \\ & & -r & (2+2r) & -r \\ & & & -r & (2+2r) \end{bmatrix} \begin{Bmatrix} u_{1,j+1} \\ u_{2,j+1} \\ \cdot \\ \cdot \\ u_{N,j+1} \end{Bmatrix}$$

$$= \begin{bmatrix} (2-2r) & r & & & \\ r & (2-2r) & r & & \\ & \cdot & \cdot & \cdot & \\ & & r & (2-2r) & r \\ & & & r & (2-2r) \end{bmatrix} \begin{Bmatrix} u_{1,j} \\ u_{2,j} \\ \cdot \\ \cdot \\ u_{N,j} \end{Bmatrix}. \qquad (3.6.53)$$

Hence

$$\mathbf{u}_{j+1} = (2\mathbf{I} - r\mathbf{T})^{-1}(2\mathbf{I} + r\mathbf{T})\mathbf{u}_j \,. \qquad (3.6.54)$$

By the previous argument these finite-difference equations will be stable when the moduli of the eigenvalue of

$$\bar{\bar{A}} = (2\bar{\bar{I}}-r\bar{\bar{T}})^{-1}(2\bar{\bar{I}}+r\bar{\bar{T}}) \qquad (3.6.55)$$

are each less than 1. The eigenvalues of **T** are

$$-4\sin^2\left(\frac{s\pi}{2(N+1)}\right), \qquad (3.6.56)$$

the eigenvalues of **A** are

$$\frac{2-4r\sin^2\left(\frac{s\pi}{2(N+1)}\right)}{2+4r\sin^2\left(\frac{s\pi}{2(N+1)}\right)}, \qquad (3.6.57)$$

and these are clearly less than 1 for all positive values of r. Thus, the Crank-Nicolson scheme is unconditionally stable.

The key to the stability analysis with the matrix method lies in one's ability to obtain the eigenvalues of a matrix. The following two theorems should prove useful in finding the bounds of eigenvalues.

Gerschgorin's Theorem. The modulus of the largest eigenvalue of the square matrix A cannot exceed the largest sum of the moduli of the terms along any row or any column, that is,

$$|\lambda| \le \sum_{j=1}^{N} |a_{i,j}| \quad \text{for all} \quad i \tag{3.6.58}$$

or

$$|\lambda| \le \sum_{i=1}^{N} |a_{i,j}| \quad \text{for all} \quad j\,. \tag{3.6.59}$$

Brauer's Theorem. Let P_s be the sum of the moduli of the terms along the s-th row excluding the diagonal elements $a_{s,s}$. Then every eigenvalue of a square matrix A lies inside or on the boundary of at least one of the circles $|\lambda - a_{s,s}| = P_s$, that is,

$$|\lambda_i - a_{s,s}| \le \sum_{j=1, j=s}^{N} |a_{s,j}|\,. \tag{3.6.60}$$

3.6.2.2. Fourier Series Method. Suppose that the error E as function of x is introduced; this error can be decomposed into Fourier series as

$$E = \sum_n a_n e^{\iota\sigma_n x}\,, \tag{3.6.61}$$

where $\iota = \sqrt{-1}$. Since a finite number of points, say N, exists in the x-direction, the number of terms of this decomposition equals N. The system under consideration is a linear system, and thus the behavior of only one term of the Fourier series can be considered. The coefficient a_n is time dependent, and in order to satisfy the expression of the particular error at t = 0, the coefficient must take the form

$$a_n(t) = a_n^* e^{\iota\beta_n t}\,, \tag{3.6.62}$$

where a_n^* and β_n are constants. Thus the express of error at (x,t) should satisfy the form

$$E = a^* e^{\iota\beta t} e^{\iota\sigma x}\,. \tag{3.6.63}$$

It is assumed that the errors are perturbations imposed on the solution of the linear system. If we subtract the exact solutions from the difference equations with the perturbation, we obtain among the errors' components a set of relations that is identical to the relationship for the components of the u, as we are dealing with a linear system.

Example 1. Investigate the stability of the explicit scheme

$$u_{i,j+1} = r u_{i-1,j} + (1-2r)u_{i,j} + r u_{i+1,j}\,. \tag{3.6.64}$$

Substitution of Eq. (3.6.3) into (3.6.64) yields

$$e^{\iota\beta k} = r e^{-\iota\sigma h} + (1-2r) + r e^{\iota\sigma h}\,. \tag{3.6.65}$$

If r < 1/2, Eq. (3.6.65) gives

$$\lambda = |e^{\iota\beta k}| = 1 + 2r(\cos\sigma h - 1) . \tag{3.6.66}$$

Obviously, if $r < 1/2$, λ is less than 1. Thus, the stability condition for the explicit scheme is $r < 1/2$ as given before by the matrix method.

Example 2. Investigate the stability of the Crank-Nicolson Scheme:

$$-ru_{i-1,j+1} + (2+2r)u_{i,j+1} - ru_{i+1,j+1} = ru_{i-1,j} + (2-2r)u_{i,j} + ru_{i+1,j} . \tag{3.6.67}$$

Substitution of Eq. (3.6.63) into (3.6.67) yields

$$\lambda = |e^{\iota\beta k}| = \frac{1-2r\sin^2(\sigma h/2)}{1+2r\sin^2(\sigma h/2)} , \tag{3.6.68}$$

which is always less than or equal to 1. Thus, the Crank-Nicolson scheme is unconditionally stable.

Example 3. Investigate the stability of the implicit scheme:

$$-ru_{i-1,j+1} + (1+2r)u_{i,j+1} - ru_{i+1,j+1} = u_{i,j} . \tag{3.6.69}$$

Substitution of Eq. (3.6.63) into Eq. (3.6.69) yields

$$\lambda = |e^{\iota\beta k}| = \frac{1}{1+4r\sin^2(\sigma h/2)} , \tag{3.6.70}$$

which is also always less than or equal to 1. Thus, the implicit scheme is also unconditionally stable.

3.7. Saturated-Unsaturated Flows of Essentially Incompressible Fluids

Let us consider a simple flow system in which the temperature field is constant and the solute concentration will not affect the flow. Under these conditions, Eqs. (3.1.3) and (3.1.4) are reduced to

$$\frac{\partial \rho_f n_e S}{\partial t} + \nabla\cdot(\rho_f n_e S \mathbf{V}_s) + \nabla\cdot(\rho_f n_e S \mathbf{V}_{fs}) = \rho_f^* q \tag{3.7.1}$$

$$\frac{\partial(1-n_e)}{\partial t} + \nabla\cdot[(1-n_e)\mathbf{V}_s] = 0 . \tag{3.7.2}$$

Equations (3.7.1) and (3.7.2) are derived based on the continuity law of fluid and solid grains, with the assumptions that the grains are incompressible and interphase mass exchange is zero. These two equations involve five state variables, ρ_f, n_e, $\mathbf{V}_{fs}$, $\mathbf{V}_s$, and S. Up to this point, no empiricism has been introduced in the derivation of the governing equations Eqs. (3.7.1) and (3.7.2). However, the number of state variables exceeds the number of equations. Thus, constitutive relationships must be empirically or theoretically established among n_e, ρ_f, $\mathbf{V}_{fs}$, $\mathbf{V}_s$, and S.

The first constitutive relationship is the empirical Darcy's law, stating that the relative fluid velocity $\mathbf{V}_{fs}$ is proportional to the pressure gradient. This empirical law, in fact, can be derived theoretically from the continuity of fluid momentum with the assumptions of neglecting the inertial forces and linearizing the frictional force (Polubarinova-Kochina, 1962). The derivation of Darcy's law was carried out in Section 3.1 for a generalized system in which the fluid flow depends on both temperature and solute concentration. For the present system when the fluid flow does not depend on temperature and solute concentration, Eq. (3.1.43) is reduced to

$$\mathbf{V}_{fs} = -\frac{1}{\mu_f S n_e}\mathbf{P}\cdot(\nabla p - \rho_f \mathbf{g}) \;, \qquad (3.7.3)$$

where $\mathbf{P} = (-\mathbf{R})^{-1}(Sn_e)^2\mu_f$ is the intrinsic permeability tensor of the media (L^2) and μ_f is the dynamic viscosity of the fluid ($ML/T^2/L_2/T$).

A new state variable p in Eq. (3.7.3) has been introduced. Thus, we will need a constitutive relationship to take care of the new state variable p. Thus, the second constitutive relationship will be the empirical, thermodynamic equation of state. In general, the thermal equation of state describes that the density is a function of pressure, temperature, and mole fractions of all chemical species as given in Eq. (3.1.42). For our simplified system, the fluid density ρ_f is a function of the pressure p only; that is,

$$\rho_f = \rho_f(p) \;. \qquad (3.7.4)$$

The third constitutive relationship can be derived based on the principle of solid momentum (Hassanizadeh and Gray, 1980). However, if we assume that the initial force of solids can be neglected, we can obtain the consolidation law of the media to relate the solid velocity $\mathbf{V}_s$ to pressure p. The three-dimensional consolidation equation developed by Biot (1941) can be modified to include gravity force as

$$(\lambda_s + 2\mu_s)\nabla^2 e = \nabla^2 p + \nabla\cdot[(\rho_f n_e S + \rho_b)\mathbf{g}] \;, \qquad (3.7.5)$$

where λ_s is the Lame first constant $[(ML/T^2)/L^2]$, μ_s is the Lame second constant $[(ML/T^2)/L^2]$, e is the dilatation of the media (dimensionless), and $\rho_b = \rho_s(1 - n_e)$ is the bulk density. The dilatation e and the solid velocity are defined by

$$e = \nabla\cdot\mathbf{U} \quad \text{and} \quad \mathbf{V}_s = \frac{\partial \mathbf{U}}{\partial t} \;, \qquad (3.7.6)$$

where $\mathbf{U}$ is the displacement of the media (L). Taking the divergence of $\mathbf{V}_s$, and from Eq. (3.7.5), we have

$$\nabla\cdot\mathbf{V}_s = \frac{\partial}{\partial t}(\nabla\cdot\mathbf{U}) = \frac{\partial e}{\partial t} \;. \qquad (3.7.7)$$

Equations (3.7.4) and (3.7.6) have implicitly established the constitutive relationship between $\mathbf{V}_s$ and the pressure p.

To complete the final, fourth constitutive relationship, experimental evidence has shown that the degree of saturation is a function of pressure, as

$$S = S(p) \;. \qquad (3.7.8)$$

Equations (3.7.1), (3.7.2), (3.7.3), (3.7.4), (3.7.5), (3.7.7), (3.7.8), and (3.7.21) (representing continuity of fluid, continuity of solids, Darcy's law, the equation of state, consolidation of the medium, the definition of $\mathbf{V}_s$, and experimental evidence between S and p, respectively) contain seven variables ρ_f, n_e, $\mathbf{V}_{fs}$, p, e, $\mathbf{V}_s$, and S. Hence the number of equations is equal to the number of unknowns. The system is complete, and a mathematical statement is posed. However, we can combine these seven equations into a single one to simplify the problem. The simplification is demonstrated below.

Expanding Eqs. (3.7.1) and (3.7.2), we have

$$S\rho_f\left[\frac{\partial n_e}{\partial t}+\nabla\cdot(n_e\mathbf{V}_s)\right]+n_eS\frac{\partial\rho_f}{\partial t}+n_e\rho_f\frac{\partial S}{\partial t} = -(n_e\mathbf{V}_s)\cdot\nabla(S\rho_f)-\nabla\cdot(\rho_f S n_e\mathbf{V}_{fs})+\rho_f^* q \tag{3.7.9}$$

and

$$\frac{\partial n_e}{\partial t} = -\nabla\cdot(n_e\mathbf{V}_s) + \nabla\cdot\mathbf{V}_s\,, \tag{3.7.10}$$

respectively. Neglecting the second-order term, $(n_e\mathbf{V}_s)\cdot\nabla(S\rho_f)$, and substituting Eq. (3.7.10) into (3.7.9), we obtain

$$n_e\rho_f\frac{\partial S}{\partial t}+n_eS\frac{\partial\rho_f}{\partial t}+S\rho_f\nabla\cdot\mathbf{V}_s = -\nabla\cdot(\rho_f S n_e\mathbf{V}_{fs})+\rho_f^* q\,. \tag{3.7.11}$$

Define the pressure head by

$$h = \int_{p_o}^{p}\frac{dp}{\rho_f g}\,, \tag{3.7.12}$$

where h is the pressure head and p_o is the datum pressure $[(ML/T^2)/L^2]$. Taking the gradient of Eq. (3.7.12), we obtain

$$\nabla h = \rho_f g\nabla p\,. \tag{3.7.13}$$

The gravitational field vector can be expressed in terms of a potential

$$\mathbf{g} = -g\nabla z\,, \tag{3.7.14}$$

where $g = |\mathbf{g}|$ is the acceleration of the gravity (L/T^2) and z is the potential head (L). Substituting Eqs (3.7.13) and (3.7.14) into Eq. (3.7.3), we rewrite Darcy's law as

$$Sn_e\mathbf{V}_{fs} = -\mathbf{K}\cdot(\nabla h+\nabla z) = -\mathbf{K}\cdot\nabla H\,, \tag{3.7.15}$$

where $\mathbf{K} = (\rho_f g/\mu_f)\mathbf{P}$ is the hydraulic conductivity tensor, (L/T) and H = (h + z) is the total head (L).

From Eq. (3.7.4), a compressibility of the fluid, β, is defined by

$$\beta = \frac{1}{\rho_f}\frac{d\rho_f}{dp} . \tag{3.7.16}$$

Rewriting Eq. (3.7.28) in the following form,

$$d\rho_f = \beta\rho_f dp , \tag{3.7.17}$$

and using Eq. (3.7.21), we have

$$\frac{\partial \rho_f}{\partial t} = \rho_f\beta\frac{\partial p}{\partial t} = \rho_f\beta'\frac{\partial h}{\partial t} , \tag{3.7.18}$$

where

$$\beta' = \rho_f g\beta \tag{3.7.19}$$

is the modified compressibility of water.

Assume that the gravitational force is small in comparison with the pressure force in the consolidation of the media; then we can integrate Eq. (3.7.5) to yield the following:

$$(\lambda_s+2\mu_s) = p+f , \tag{3.7.20}$$

where f is the integration function. The integration function f must satisfy the Laplace's equation for all time. To simplify the matter further, we will consider only vertical consolidation. Under this condition, it has been shown (Verruijt, 1969) that the integration function f is equal to 0. It then follows from Eq. (3.7.20) that

$$\frac{\partial e}{\partial t} = \alpha\frac{\partial p}{\partial t} , \tag{3.7.21}$$

where

$$\alpha = \frac{1}{(\lambda_s+2\mu_s)} , \tag{3.7.22}$$

in which α is the coefficient of consolidation of the media. Substituting Eqs. (3.7.13) and (3.7.21) into Eq. (3.7.7), we obtain

$$\nabla\cdot V_s = \frac{\partial e}{\partial t} = \alpha\frac{\partial \rho_f}{\partial t} = \alpha\rho_f g\frac{\partial h}{\partial t} = \alpha'\frac{\partial h}{\partial t} , \tag{3.7.23}$$

where α' is the modified compressibility of the media.

From Eq. (3.7.8), we get

$$\frac{\partial S}{\partial t} = \frac{dS}{dh}\frac{\partial h}{\partial t} . \tag{3.7.24}$$

Substituting Eqs. (3.7.15), (3.7.18), (3.7.23), and (3.7.24) into (3.7.11), we obtain

$$\rho_f \frac{d\theta}{dh}\frac{\partial h}{\partial t} + \rho_f \theta \beta' \frac{\partial h}{\partial t} + S\rho_f \alpha' \frac{\partial h}{\partial t} = \nabla \cdot (\rho_f \mathbf{K} \cdot \nabla H) + \rho_f^* q \ . \qquad (3.7.25)$$

Expanding the fourth term of Eq. (3.7.25), neglecting the second-order term $[(K.\nabla H).(\rho_f)]$, and assuming $\rho_f^* = \rho_f$, we finally have the following governing equation for saturated-unsaturated media:

$$F\frac{\partial h}{\partial t} = \nabla \cdot [\mathbf{K} \cdot \nabla h + \nabla z)] + q \ , \qquad (3.7.26)$$

in which

$$F = \alpha' \frac{\theta}{n_e} + \beta' \theta + n_e \frac{dS}{dh} \ . \qquad (3.7.27)$$

The first two terms in Eq. (3.7.27) make Eq. (3.7.26) a modified form of the Richards equation.

To complete the mathematical formulation of the saturated-unsaturated problems, Eq. (3.7.26) must be supplemented with initial and boundary conditions. The initial condition is stated mathematically as

$$h = h_i(\mathbf{x}) \quad \text{in} \quad R \ , \qquad (3.7.28)$$

where R is the region of interest and h_i is the prescribed initial condition, which can be obtained by either field measurements or by solving the steady state version of Eq. (3.7.26) with time-invariant boundary conditions

As in saturated flows, three basic types of boundary conditions and a river boundary condition can be specified for variably saturated flows. In addition, a variable boundary condition normally at the air-media interface can be specified. These boundary conditions are stated mathematically as follows.

Dirichlet Conditions. On a Dirichlet boundary, the pressure head is prescribed as

$$h = h_d(\mathbf{x}_b, t) \quad \text{on} \quad B_d \ , \qquad (3.7.29)$$

where $\mathbf{x}_b$ is the spatial coordinate on the boundary, h_d is the prescribed head, and B_d is the Dirichlet boundary.

Neumann Conditions. On a Neumann boundary, which normally is the drainage boundary, the gradient flux is specified as

$$-\mathbf{n} \cdot \mathbf{K} \cdot \nabla h = q_n(\mathbf{x}_b, t) \quad \text{on} \quad B_n \ , \qquad (3.7.30)$$

where $\mathbf{n}$ is an outward unit vector normal to the boundary, q_n is the prescribed Neumann flux, and B_n is the Neumann boundary.

Cauchy Conditions. On a Cauchy boundary, which normally is an infiltration boundary, the volume flux is prescribed as

$$-\mathbf{n} \cdot (\mathbf{K} \cdot \nabla h + \mathbf{K} \cdot \nabla z) = q_c(\mathbf{x}_b, t) \quad \text{on} \quad B_c \ , \qquad (3.7.31)$$

where q_c is the prescribed Cauchy flux and B_c is the Neumann boundary.

Variable Conditions - During Precipitation Period. On a variable boundary, which is normally the air-media interface, either infiltration, ponding, or seepage can occur during precipitation periods. If infiltration occurs, the maximum amount of infiltration rate is the excess precipitation rate. As to which condition is prevalent, it cannot be determined as *a priori.* Rather it must be determined in a cyclic iterative procedure. These physical conditions lead to the mathematical representations

$$h = h_p(\mathbf{x}_b,t) \quad \textit{iff} \quad -\mathbf{n}\cdot\mathbf{K}\cdot\nabla H \geq q_p \quad \text{on } B_v \tag{3.7.32}$$

or

$$-\mathbf{n}\cdot\mathbf{K}\cdot\nabla H = q_p(\mathbf{x}_b,t) \quad \text{iff} \quad h \leq h_p \quad \text{on} \quad B_v\,, \tag{3.7.33}$$

where B_v is the variable boundary, h_p is the allowed ponding depth, and q_p (numerically negative) is the excess precipitation. Either Eq. (3.7.32) or (3.7.33), but not both, is used at any point on the variable boundary at any time during precipitation periods.

Variable Conditions - During Nonprecipitation Period. During nonprecipitation periods, either seepage or evaporation can occur on a variable boundary. If evaporation occurs, the maximum amount of evaporation is the potential evaporation. These physical considerations lead to the following mathematical statements:

$$h = h_p(\mathbf{x}_b,t) \quad \textit{iff} \quad -\mathbf{n}\cdot\mathbf{K}\cdot\nabla H \geq q_p \quad \text{on } B_v \tag{3.7.34}$$

$$h = h_m(\mathbf{x}_b,t) \quad \text{iff} \quad -\mathbf{n}\cdot\mathbf{K}\cdot\nabla H \leq q_e(\mathbf{x}_b,t) \text{ on } B_v \tag{3.7.35}$$

or

$$-\mathbf{n}\cdot\mathbf{K}\cdot\nabla H = q_e(\mathbf{x}_b,t) \quad \text{iff} \quad h \geq h_m \quad \text{on} \quad B_v\,, \tag{3.7.36}$$

where h_m is the allowed minimum pressure on the variable boundary and q_e is the allowed maximum evaporation rate on the variable boundary, which is the potential evaporation. Only one of Eqs. (3.7.34) through (3.7.36) is used at any point on the variable boundary at any time during non-precipitation periods.

River Boundary Conditions. At river-media interfaces, two types of boundary conditions can be specified depending on physical conditions. If there are sediment layers around the wet perimeter of the river-media interfaces, then a radiation type of river boundary conditions can be imposed:

$$-\mathbf{n}\cdot\mathbf{K}\cdot(\nabla h + \nabla z) = -\frac{K_R}{b_R}(h_R - h) \quad \text{on} \quad B_r(x_b,t)\,, \tag{3.7.37}$$

where K_R is the hydraulic conductivity of the river bottom sediment layer, b_R is the thickness of the river bottom sediment layer, and h_R is the depth of the river bottom measured from the river surface to the top of the sediment layer.

3.8. Finite-Element Modeling of Saturated-Unsaturated Flows

We approximate the hydraulic head h in Eq. (3.7.26) by

$$h \approx \hat{h} = \sum_{j=1}^{N} h_j N_j \,, \tag{3.8.1}$$

where N is the total number of nodes in the region and N_j and h_j are the base function and the amplitude of h, respectively, at nodal point j. It is interesting to note that the fundamental distinction between finite-element methods (FEMs) and finite-difference methods (FDMs) is that the former is based directly on approximation of the function as in Eq. (3.8.1), whereas the latter is based on approximation of derivative (Hilderbrand, 1968).

Since h is only an approximate solution of Eq. (3.7.26), it will not satisfy the equation. Thus, we define a residual R_r by

$$R = F\frac{\partial h}{\partial t} - \nabla\cdot[\mathbf{K}\cdot(\nabla h + \nabla z)] - q \tag{3.8.2}$$

According to the principle of the FEM, we choose the coefficient h_j's in Eq. (3.8.1) such that the residual in Eq. (3.8.2) weighted by the set of weighting functions is zero. For the Galerkin finite-element method, we select the set of weighting functions as the same set of base functions. Applying the principle of weighted residual, we obtain

$$\int_R N_i \left\{ F\frac{\partial h}{\partial t} - \nabla\cdot[\mathbf{K}\cdot(\nabla h + \nabla z)] - q \right\} dR \,, \quad i = 1, 2, \ldots, N \,. \tag{3.8.3}$$

Substituting Eq. (3.8.1) into Eq. (3.8.3) and integrating by part, we obtain

$$\left[\int_R N_i F N_j dR\right]\left\{\frac{dh_j}{dt}\right\} + \left[\int_R (\nabla N_i)\cdot\mathbf{K}\cdot(\nabla N_j) dR\right]\{h_j\} = \int_R N_i q dR - \int_R (\nabla N_i)\cdot\mathbf{K}\cdot(\nabla z) dR + \int_B \mathbf{n}\cdot\mathbf{K}\cdot(\nabla H) N_i dR \,, \quad i = 1, 2, \ldots, N \,. \tag{3.8.4}$$

Equation (3.8.4) written in matrix form is

$$[M]\left\{\frac{dh}{dt}\right\} + [S]\{h\} = \{G\} + \{Q\} + \{B\} \,, \tag{3.8.5}$$

where {dh/dt} and {h} are the column vectors containing the values of dh/dt and h, respectively, at all nodes; [M] is the mass matrix resulting from the storage term; [S] is stiff matrix resulting from the action of conductivity; {G}, {Q}, and {B} are the load vectors from the gravity force, internal source/sink, and boundary conditions, respectively. The matrices, [M] and [S], are given by

$$M_{ij} = \sum_{e \in M_e} \int_{R_e} N_\alpha^e F N_\beta^e dR \tag{3.8.6}$$

$$S_{ij} = \sum_{e \in M_e} \int_{R_e} \left[(\nabla N_\alpha^e) \cdot \mathbf{K} \cdot (\nabla N_\beta^e) \right] dR \,, \tag{3.8.7}$$

where R_e is the region of element e, M_e is the set of elements that have a local side α-β coinciding with the global side i-j and N_α^2 is the α-th local basis function of element e. Similarly, the load vectors {G}, {Q} and {B} are given by

$$G_i = -\sum_{e \in M_e} \int_{R_e} (\nabla N_\alpha^e) \cdot \mathbf{K} \cdot \nabla z dR \tag{3.8.8}$$

$$Q_i = \sum_{e \in M_e} \int_{R_e} N_\alpha^e q dR \tag{3.8.9}$$

and

$$B_i = -\sum_{e \in N_{se}} \int_{B_e} N_\alpha^e \cdot [-\mathbf{K} \cdot (\nabla H)] dB \,, \tag{3.8.10}$$

where B_e is the length of boundary segment e and N_{se} is the set of boundary segments that have a local node α coinciding with the global node i.

The reduction of the partial differential equation (PDE), Eq. (3.8.2), to the set of ordinary differential equations (ODE), Eq. (3.8.5), simplifies to the evaluation of integrals on the right-hand side of Eqs. (3.8.6) through (3.8.10) for every element or boundary segment e to yield the element mass matrix $[M^e]$ and stiff matrix $[S^e]$ as well as the element gravity column vector, $\{G^e\}$, source/sink column vector $\{Q^e\}$, and boundary column vector $\{B^e\}$ as

$$M_{\alpha\beta}^e = \int_{R_e} N_\alpha^e F N_\beta^e dR \tag{3.8.11}$$

$$S_{\alpha\beta}^e = \int_{R_e} \left[(\nabla N_\alpha^e) \cdot \mathbf{K} \cdot (\nabla N_\beta^e) \right] dR \tag{3.8.12}$$

$$G_\alpha^e = -\int_{R_e} (\nabla N_\alpha^e) \cdot \mathbf{K} \cdot \nabla z dR \tag{3.8.13}$$

$$Q_\alpha^e = \int_{R_e} N_\alpha^e q dR \qquad (3.8.14)$$

and

$$B_\alpha^e = -\int_{B_e} N_\alpha^e \cdot [-\mathbf{K}\cdot(\nabla H)] dB, \qquad (3.8.15)$$

where the superscript or subscript e denote the element. The evaluation of Eqs. (3.8.11) through (3.8.15) depends on the dimensionality and the types of elements used. In the following, we discussed the one-, two-, and three-dimensional problems. For one-dimensional problems, we use linear line elements. For two-dimensional problems, we use hybrid isoparametric quadrilateral and triangular elements. For three-dimensional problems, we use isoparametric hexahedral elements.

3.8.1. One-Dimensional Problems

For a linear element, Eqs. (3.8.11) through (3.8.14) can be computed analytically as follows:

$$[M^e] = \Delta x \begin{bmatrix} \frac{F_1}{4} + \frac{F_2}{12} & \frac{F_1}{12} + \frac{F_2}{12} \\ \frac{F_1}{12} + \frac{F_2}{12} & \frac{F_1}{12} + \frac{F_2}{4} \end{bmatrix} \qquad (3.8.16)$$

$$[S^e] = \frac{1}{\Delta x} \begin{bmatrix} \bar{K} & -\bar{K} \\ -\bar{K} & \bar{K} \end{bmatrix} \qquad (3.8.17)$$

$$\{G^e\} = \begin{Bmatrix} \bar{K} \\ -\bar{K} \end{Bmatrix} \qquad (3.8.18)$$

$$\{Q^e\} = \frac{\Delta x}{2} \begin{Bmatrix} q \\ q \end{Bmatrix}, \qquad (3.8.19)$$

where $\bar{K}$ is the average hydraulic conductivity of the element e.

3.8.2. Two-Dimensional Problems

For a quadrilateral element, Eqs. (3.8.11) through (3.8.14) are computed by Gaussian quadrature (Conte, 1965) because it is not easy to evaluate these equations analytically. In general, the base functions are given in terms of local coordinates. In the local coordinate, the element is square regardless of the shape of the quadrilateral in the global coordinate. The global coordinates at any point within an element e are given in terms of the local coordinate

$$x = \sum_{j=1} x_j N_j(\xi,\eta), \quad z = \sum_{j=1} z_j N_j(\xi,\eta) , \tag{3.8.20}$$

where x_j and z_j are the global coordinates of the nodes and N_j, which depends on the local coordinate ξ and η, is the shape function for node j. When the coordinate transformation uses the base functions, the element is termed the *isoparametric* element; we shall use the isoparametric elements throughout the discussion.

Now using the chain rule, we obtain

$$\frac{\partial N_i}{\partial \xi} = \frac{\partial N_i}{\partial x}\frac{\partial x}{\partial \xi} + \frac{\partial N_i}{\partial z}\frac{\partial z}{\partial \xi}, \qquad \frac{\partial N_i}{\partial \eta} = \frac{\partial N_i}{\partial x}\frac{\partial x}{\partial \eta} + \frac{\partial N_i}{\partial z}\frac{\partial z}{\partial \eta} . \tag{3.8.21}$$

Written in matrix notation, Eq. (3.8.21) is

$$\begin{Bmatrix} \dfrac{\partial N_i}{\partial \xi} \\ \dfrac{\partial N_i}{\partial \eta} \end{Bmatrix} = \begin{bmatrix} \dfrac{\partial x}{\partial \xi} & \dfrac{\partial z}{\partial \xi} \\ \dfrac{\partial x}{\partial \eta} & \dfrac{\partial z}{\partial \eta} \end{bmatrix} \begin{Bmatrix} \dfrac{\partial N_i}{\partial x} \\ \dfrac{\partial N_i}{\partial z} \end{Bmatrix} = [J] \begin{Bmatrix} \dfrac{\partial N_i}{\partial x} \\ \dfrac{\partial N_i}{\partial z} \end{Bmatrix} . \tag{3.8.22}$$

Inversion of Eq. (3.8.22) yields

$$\begin{Bmatrix} \dfrac{\partial N_i}{\partial x} \\ \dfrac{\partial N_i}{\partial z} \end{Bmatrix} = [J]^{-1} \begin{Bmatrix} \dfrac{\partial N_i}{\partial \xi} \\ \dfrac{\partial N_i}{\partial \eta} \end{Bmatrix} , \tag{3.8.23}$$

where

$$[J] = \begin{bmatrix} \dfrac{\partial x}{\partial \xi} & \dfrac{\partial z}{\partial \xi} \\ \dfrac{\partial x}{\partial \eta} & \dfrac{\partial z}{\partial \eta} \end{bmatrix} \qquad [J]^{-1} = \begin{bmatrix} \dfrac{\partial z}{\partial \eta} & -\dfrac{\partial z}{\partial \xi} \\ -\dfrac{\partial x}{\partial \eta} & \dfrac{\partial x}{\partial \xi} \end{bmatrix} . \tag{3.8.24}$$

The determinant of the Jacobian is given by

$$|J| = \det[J] = \left(x_j \frac{\partial N_j}{\partial \xi}\right)\left(z_k \frac{\partial N_k}{\partial \eta}\right) - \left(z_j \frac{\partial N_j}{\partial \xi}\right)\left(x_k \frac{\partial N_k}{\partial \eta}\right) . \tag{3.8.25}$$

Finally, the integration of a differential area can be written as

$$\int_e dxdz = \int_{-1}^{1}\int_{-1}^{1} |J| d\xi d\eta . \tag{3.8.26}$$

For triangular elements, the transformation between global and local coordinates is written as

$$L_i = \frac{1}{2A}\left(a_i + b_i x + c_i z\right) , \tag{3.8.27}$$

where A is the area of the triangular element and a_i, b_i, and c_i are given by

$$\begin{aligned} a_1 &= x_2 z_3 - x_3 z_2 , \quad b_1 = z_2 - z_3 , \quad c_1 = x_3 - x_2 \\ a_2 &= x_3 z_1 - x_1 z_3 , \quad b_2 = z_3 - z_1 , \quad c_2 = x_1 - x_3 \\ a_3 &= x_1 z_2 - x_1 z_2 , \quad b_3 = z_1 - z_2 , \quad c_3 = x_2 - x_1 . \end{aligned} \tag{3.8.28}$$

The derivatives in terms of global coordinates can also be transformed to the derivatives in terms of natural coordinates (area coordinates) as follows:

$$\begin{aligned} \frac{\partial N_i}{\partial x} &= \frac{\partial N_i}{\partial L_1}\frac{\partial L_1}{\partial x} + \frac{\partial N_i}{\partial L_2}\frac{\partial L_2}{\partial x} + \frac{\partial N_i}{\partial L_3}\frac{\partial L_3}{\partial x} = \sum_{k=1}^{3} b_k \frac{\partial N_i}{\partial L_k} \\ \frac{\partial N_i}{\partial z} &= \frac{\partial N_i}{\partial L_1}\frac{\partial L_1}{\partial z} + \frac{\partial N_i}{\partial L_2}\frac{\partial L_2}{\partial z} + \frac{\partial N_i}{\partial L_3}\frac{\partial L_3}{\partial z} = \sum_{k=1}^{3} c_k \frac{\partial N_i}{\partial L_k} . \end{aligned} \tag{3.8.29}$$

Integration of area coordinates over the element is simple with the aid of the following formula

$$\int_e L_1^{\alpha} L_2^{\beta} L_3^{\gamma} dA = 2A \frac{\alpha!\beta!\gamma!}{(\alpha+\beta+\gamma+2)!} \tag{3.8.30}$$

3.8.3. Three-Dimensional Problems

For a hexahedron element, Eqs. (3.8.11) through (3.8.14) are computed by Gaussian quadrature (Conte, 1965) similar to the case of two-dimensional problems. The base functions are given in terms of local coordinates. In the local coordinate, the element is a cubic regardless of the shape of the hexahedron in the global coordinate. The global coordinates at any point within an element e are given in terms of the local coordinate

$$x = \sum_{j=1} x_j N_j(\xi,\eta,\zeta), \quad y = \sum_{j=1} y_j N_j(\xi,\eta,\zeta), \quad z = \sum_{j=1} z_j N_j(\xi,\eta,\zeta) , \tag{3.8.31}$$

where (x_j, y_j, z_j) are the global coordinates of the j-th node and N_j, which depends on the local coordinate (ξ, η, ζ), is the shape function for node j. When the coordinate transformation uses the base functions, the element is termed the "isoparametric" element; we shall use the *isoparametric* elements throughout the discussion.

Now using the chain rule, we obtain

$$\frac{\partial N_i}{\partial \xi} = \frac{\partial N_i}{\partial x}\frac{\partial x}{\partial \xi} + \frac{\partial N_i}{\partial y}\frac{\partial y}{\partial \xi} + \frac{\partial N_i}{\partial z}\frac{\partial z}{\partial \xi}$$

$$\frac{\partial N_i}{\partial \eta} = \frac{\partial N_i}{\partial x}\frac{\partial x}{\partial \eta} + \frac{\partial N_i}{\partial y}\frac{\partial y}{\partial \eta} + \frac{\partial N_i}{\partial z}\frac{\partial z}{\partial \eta} \tag{3.8.32}$$

$$\frac{\partial N_i}{\partial \zeta} = \frac{\partial N_i}{\partial x}\frac{\partial x}{\partial \zeta} + \frac{\partial N_i}{\partial y}\frac{\partial y}{\partial \zeta} + \frac{\partial N_i}{\partial z}\frac{\partial z}{\partial \zeta}.$$

Written in matrix notation, Eq. (3.8.32) is

$$\begin{Bmatrix} \dfrac{\partial N_i}{\partial \xi} \\ \dfrac{\partial N_i}{\partial \eta} \\ \dfrac{\partial N_i}{\partial \zeta} \end{Bmatrix} = \begin{bmatrix} \dfrac{\partial x}{\partial \xi} & \dfrac{\partial y}{\partial \xi} & \dfrac{\partial z}{\partial \xi} \\ \dfrac{\partial x}{\partial \eta} & \dfrac{\partial y}{\partial \eta} & \dfrac{\partial z}{\partial \eta} \\ \dfrac{\partial x}{\partial \zeta} & \dfrac{\partial y}{\partial \zeta} & \dfrac{\partial z}{\partial \zeta} \end{bmatrix} \begin{Bmatrix} \dfrac{\partial N_i}{\partial x} \\ \dfrac{\partial N_i}{\partial y} \\ \dfrac{\partial N_i}{\partial z} \end{Bmatrix} = [J] \begin{Bmatrix} \dfrac{\partial N_i}{\partial x} \\ \dfrac{\partial N_i}{\partial y} \\ \dfrac{\partial N_i}{\partial z} \end{Bmatrix}, \tag{3.8.33}$$

where [J] denotes

$$[J] = \begin{bmatrix} \dfrac{\partial x}{\partial \xi} & \dfrac{\partial y}{\partial \xi} & \dfrac{\partial z}{\partial \xi} \\ \dfrac{\partial x}{\partial \eta} & \dfrac{\partial y}{\partial \eta} & \dfrac{\partial z}{\partial \eta} \\ \dfrac{\partial x}{\partial \zeta} & \dfrac{\partial y}{\partial \zeta} & \dfrac{\partial z}{\partial \zeta} \end{bmatrix}. \tag{3.8.34}$$

Inversion of Eq. (3.8.33) yields

$$\begin{Bmatrix} \dfrac{\partial N_i}{\partial x} \\ \dfrac{\partial N_i}{\partial y} \\ \dfrac{\partial N_i}{\partial z} \end{Bmatrix} = [J]^{-1} \begin{Bmatrix} \dfrac{\partial N_i}{\partial \xi} \\ \dfrac{\partial N_i}{\partial \eta} \\ \dfrac{\partial N_i}{\partial \zeta} \end{Bmatrix}, \tag{3.8.35}$$

where $[J]^{-1}$ is the inverse of [J]. Denoting the determinant of the Jacobian by $|J|$, we have the integration of a differential area given by

$$\int_e dxdydz = \int_{-1}^{1}\int_{-1}^{1}\int_{-1}^{1} |J|\, d\xi d\eta d\zeta \ . \tag{3.8.36}$$

For a tetrahedral element, Eqs. (3.8.11) through (3.8.14) can be computed analytically. Basis functions are written in terms of volume coordinates, L_1, L_2, L_3, and L_4. The transformation between global and natural coordinates (volume coordinates) is written as

$$L_i = \frac{1}{6V}\left(a_i + b_i x + c_i y + d_i z\right) \ , \tag{3.8.37}$$

where V is the volume of the tetrahedral element and a, b, c, and d are given by

$$a_1 = \begin{vmatrix} x_2 & y_2 & z_2 \\ x_3 & y_3 & z_3 \\ x_4 & y_4 & z_4 \end{vmatrix}, \quad b_1 = -\begin{vmatrix} 1 & y_2 & z_2 \\ 1 & y_3 & z_3 \\ 1 & y_4 & z_4 \end{vmatrix} ,$$

$$c_1 = \begin{vmatrix} x_2 & 1 & z_2 \\ x_3 & 1 & z_3 \\ x_4 & 1 & z_4 \end{vmatrix}, \quad d_1 = \begin{vmatrix} x_2 & y_2 & 1 \\ x_3 & y_3 & 1 \\ x_4 & y_4 & 1 \end{vmatrix} . \tag{3.8.38}$$

The derivatives in terms of global coordinates can also be transformed to the derivatives in terms of natural coordinates (area coordinates) as follows:

$$\frac{\partial N_i}{\partial x} = \frac{\partial N_i}{\partial L_1}\frac{\partial L_1}{\partial x} + \frac{\partial N_i}{\partial L_2}\frac{\partial L_2}{\partial x} + \frac{\partial N_i}{\partial L_3}\frac{\partial L_3}{\partial x} + \frac{\partial N_i}{\partial L_4}\frac{\partial L_4}{\partial x} = \sum_{k=1}^{4} b_k \frac{\partial N_i}{\partial L_k}$$

$$\frac{\partial N_i}{\partial y} = \frac{\partial N_i}{\partial L_1}\frac{\partial L_1}{\partial y} + \frac{\partial N_i}{\partial L_2}\frac{\partial L_2}{\partial y} + \frac{\partial N_i}{\partial L_3}\frac{\partial L_3}{\partial y} + \frac{\partial N_i}{\partial L_4}\frac{\partial L_4}{\partial y} = \sum_{k=1}^{4} c_k \frac{\partial N_i}{\partial L_k} \tag{3.8.39}$$

$$\frac{\partial N_i}{\partial z} = \frac{\partial N_i}{\partial L_1}\frac{\partial L_1}{\partial z} + \frac{\partial N_i}{\partial L_2}\frac{\partial L_2}{\partial z} + \frac{\partial N_i}{\partial L_3}\frac{\partial L_3}{\partial z} + \frac{\partial N_i}{\partial L_4}\frac{\partial L_4}{\partial z} = \sum_{k=1}^{4} d_k \frac{\partial N_i}{\partial L_k} \ .$$

Integration of area coordinates over the element is simple with the aid of the following formula:

$$\int_e L_1^{\alpha} L_2^{\beta} L_3^{\gamma} L_4^{\delta} dV = 6V \frac{\alpha!\beta!\gamma!\delta!}{(\alpha+\beta+\gamma+\delta+3)!} \ . \tag{3.8.40}$$

For a triangular element, Eqs. (3.8.11) through (3.8.14) are computed by Gaussian quadrature (Conte, 1965) similar to the case of a hexahedral element. The base functions can be written in terms of a two-dimensional area coordinate (L_1, L_2, L_3) and a one-dimensional local coordinate ζ. Only two of the three area coordinates L_1, L_2, and L_3 are independent. Thus, we can chose any two, say L_1 and L_2, along with the local coordinate ζ to form a "mixed" area-local coordinate (L_1, L_2, ζ). In the area-local coordinate, the element is a regular shape regardless of the shape of the triangular prism in the global coordinate. The global coordinates at any point within an element e are given in terms of the local coordinate

$$x = \sum_{j=1} x_j N_j(L_1,L_2,\zeta), \quad y = \sum_{j=1} y_j N_j(L_1,L_2,\zeta), \quad z = \sum_{j=1} z_j N_j(L_1,L_2,\zeta) \, , \tag{3.8.41}$$

where (x_j, y_j, z_j) are the global coordinates of the j-th node and N_j, which depends on the area-local coordinate (L_1, L_2, ζ), is the shape function for node j.

Now using the chain rule, we obtain

$$\begin{aligned} \frac{\partial N_i}{\partial L_1} &= \frac{\partial N_i}{\partial x}\frac{\partial x}{\partial L_1} + \frac{\partial N_i}{\partial y}\frac{\partial y}{\partial L_1} + \frac{\partial N_i}{\partial z}\frac{\partial z}{\partial \xi} \\ \frac{\partial N_i}{\partial L_2} &= \frac{\partial N_i}{\partial x}\frac{\partial x}{\partial L_2} + \frac{\partial N_i}{\partial y}\frac{\partial y}{\partial L_2} + \frac{\partial N_i}{\partial z}\frac{\partial z}{\partial \eta} \\ \frac{\partial N_i}{\partial \zeta} &= \frac{\partial N_i}{\partial x}\frac{\partial x}{\partial \zeta} + \frac{\partial N_i}{\partial y}\frac{\partial y}{\partial \zeta} + \frac{\partial N_i}{\partial z}\frac{\partial z}{\partial \zeta} . \end{aligned} \tag{3.8.42}$$

Written in matrix notation, Eq. (3.8.42) is

$$\begin{Bmatrix} \dfrac{\partial N_i}{\partial L_1} \\ \dfrac{\partial N_i}{\partial L_2} \\ \dfrac{\partial N_i}{\partial \zeta} \end{Bmatrix} = \begin{bmatrix} \dfrac{\partial x}{\partial L_1} & \dfrac{\partial y}{\partial L_1} & \dfrac{\partial z}{\partial L_1} \\ \dfrac{\partial x}{\partial L_2} & \dfrac{\partial y}{\partial L_2} & \dfrac{\partial z}{\partial L_2} \\ \dfrac{\partial x}{\partial \zeta} & \dfrac{\partial y}{\partial \zeta} & \dfrac{\partial z}{\partial \zeta} \end{bmatrix} \begin{Bmatrix} \dfrac{\partial N_i}{\partial x} \\ \dfrac{\partial N_i}{\partial y} \\ \dfrac{\partial N_i}{\partial z} \end{Bmatrix} = [J] \begin{Bmatrix} \dfrac{\partial N_i}{\partial x} \\ \dfrac{\partial N_i}{\partial y} \\ \dfrac{\partial N_i}{\partial z} \end{Bmatrix} , \tag{3.8.43}$$

where [J] denotes

$$[J] = \begin{bmatrix} \dfrac{\partial x}{\partial L_1} & \dfrac{\partial y}{\partial L_1} & \dfrac{\partial z}{\partial L_1} \\ \dfrac{\partial x}{\partial L_2} & \dfrac{\partial y}{\partial L_2} & \dfrac{\partial z}{\partial L_2} \\ \dfrac{\partial x}{\partial \zeta} & \dfrac{\partial y}{\partial \zeta} & \dfrac{\partial z}{\partial \zeta} \end{bmatrix} . \tag{3.8.44}$$

Inversion of Eq. (3.8.43) yields

$$\begin{Bmatrix} \dfrac{\partial N_i}{\partial x} \\ \dfrac{\partial N_i}{\partial y} \\ \dfrac{\partial N_i}{\partial z} \end{Bmatrix} = [J]^{-1} \begin{Bmatrix} \dfrac{\partial N_i}{\partial L_1} \\ \dfrac{\partial N_i}{\partial L_2} \\ \dfrac{\partial N_i}{\partial \zeta} \end{Bmatrix} , \tag{3.8.45}$$

where $[J]^{-1}$ is the inverse of [J]. Denoting the determinant of the Jacobian by $|J|$, we have the integration of a differential area given by

$$\int_e dxdydz = \int_{-1}^{1}\int_{0}^{1}\int_{0}^{1} |J| dL_1 dL_2 d\zeta \ . \tag{3.8.46}$$

The surface integration of Eq. (3.8.15) in three-dimensional space is not as straightforward as the line integration in two-dimensional space. This issue has been addressed earlier in Section 3.1.

The evolution of pressure head can be obtained by integration of Eq. (3.8.5) with respect to time. As with the groundwater flow modeling, three time-marching schemes have been used: the implicit finite difference, the Crank-Nicolson central difference, and the middifference. After time discretization of Eq. (3.8.5) and incorporation of boundary conditions, we obtain the following matrix equation

$$[C]\{h\} = \{R\} , \tag{3.8.47}$$

where [C] is the coefficient matrix and {R} is the known vector of the right-hand side. For the saturated-unsaturated flow simulation, [C] is a highly nonlinear function of the pressure head {h}. The Picard method is used to linearize the matrix equation. For one-dimensional problems, the matrix equation is solved with a direct band-matrix solver. For two-dimensional problems, two options are provided to solve the matrix equation: one is the direct-band matrix method and the other is the basic point iteration method. For three-dimensional problems, a subregional block iteration method or a basic point iteration method has been used. The use of the subregional block iteration method requires two pointer arrays: one is the mapping between the subregional and global node numbering systems and the other is the stencil of subregional nodes. The implementation of these two pointer arrays is rather involved (Yeh, 1987a, 1987b).

3.8.4. Numerical Implementation of Variable Boundary Conditions

Numerical implementation of Cauchy, Neuman, Dirichlet, and river boundary conditions is straightforward. They can be done similar to those for saturated flows as presented in Section 3.3.4 or 3.5.4. The implementation of the variable type boundary conditions for variably saturated flow is, however, more involved. This type of boundary condition normally occurs at the air-soil interface. During precipitation periods, we will assume that only seepage and infiltration can occur for any point on the air-soil interface. No evapotranspiration is allowed. If seepage occurs, the Dirichlet boundary condition, Eq. (3.7.32), must be imposed. On the other hand, if infiltration occurs, either the Dirichlet boundary condition, Eq. (3.7.32), or the Cauchy boundary condition, Eq. (3.7.33), may be specified, depending on the soil property and the excess precipitation rate q_p in Eq. (3.7.33). The problem is which equation, Eq. (3.7.32) or Eq. (3.7.33), should be used for a point on the boundary. This problem is settled by iteration. The procedure adopted is as follows. At each iteration, we examine the solution at each node along the variable boundary and test whether the existing boundary condition is still consistent. Specifically, if the existing condition is Eq. (3.7.33) (Cauchy boundary condition), we compute the pressure head at the boundary node. If the head is greater than the allowed ponding depth h_p in Eq. (3.7.32), too much water has been forced into the region through the node. In other words, the excess precipitation rate is greater than that which the media can absorb. To account for this, the boundary condition is changed to Eq. (3.7.32), which in practice should result in infiltration at a rate less that q_p in Eq. (3.7.33) or result in seepage. If the computed head is less than the ponding depth, the media is capable of adsorbing all the excess precipitation and no change of boundary condition is required. On the other, if the existing boundary condition is Eq. (3.7.32) (Dirichlet boundary condition), we compute Darcy's flux at the node. If the computed Darcy's flux is going out of the region (seepage) or into the region (infiltration) but its magnitude is less than q_p in Eq. (3.7.33), no change of boundary condition is needed. However, if the computed Darcy's flux is directed into the region (infiltration) with a rate greater than the excess precipitation rate q_p, a change of boundary condition to Eq. (3.7.33) is required since Eq. (3.7.32) would force more water than available into the region. By changing the boundary condition to Eq. (3.7.33), in practice, a pressure head less than h_p should result. The iteration outlined above is discontinued when no change-over of boundary condition is encountered along the entire boundary.

Similarly, during nonprecipitation periods, we will assume that only evapotranspiration or seepage can occur and no infiltration is allowed. If the seepage actually occurs at a node, Eq. (3.7.34) (Dirichlet boundary condition with ponding depth) must be specified at a node. On the other hand, if evapotranspiration happens, either Eq. (3.7.35) (Dirichlet boundary condition with minimum pressure) or Eq. (3.7.36) (Cauchy boundary condition) may be imposed at the node. The question is again which of the three equations should be used as boundary conditions. The iteration procedure is used to resolve the question. First, if the existing boundary condition is Eq. (3.7.34), we calculate the Darcy's flux. When the computed Darcy's flux is going out of the region, the existing boundary condition is consistent and no change of boundary condition is necessary. When the Darcy flux is directed into the region (remember no infiltration is allowed), the application of Eq. (3.7.34) implies infiltration and prohibits evapotranspiration. Hence, the boundary condition is changed to Eq. (3.7.36), which in practice would generate evapotranspiration and would result in a pressure head lower than the ponding depth in Eq. (3.7.34). Second, if the existing boundary condition is Eq. (3.7.35), we compute Darcy's flux. Since the minimum pressure is prescribed on the boundary, it is unlikely that this computed Darcy's flux will be directed into the region. Thus, when the computed outgoing Darcy's flux is less than q_e in Eq. (3.7.36), the existing boundary condition is consistent and no change of boundary condition is needed. When the computed Darcy's flux is greater than q_e in Eq. (3.7.36), the application of Eq. (3.7.35) implies the imposition of too much suction at the node. Hence the boundary condition is changed to Eq. (3.7.36), which in practice should result in a pressure head greater than h_m in Eq. (3.7.35). Third, if the existing boundary condition is Eq. (3.7.36), we calculate the pressure head at the node. If the computed pressure head is not lower than h_m in Eq. (3.7.36), the boundary condition is consistent and no change is required. However, if the computed pressure head is lower than h_m in Eq. (3.7.35), the application of Eq. (3.7.36) implies too much water is removed through the node yielding too low a pressure. Hence the boundary condition is changed to Eq. (3.7.35), which should yield an evapotranspiration rate less than q_e in Eq. (3.7.36). This iteration process is completed only when consistent boundary conditions have been applied to all nodes on the variable boundary.

3.8.5. Computation of Darcy's Velocity

Finally, after we have obtained the pressure field with Eq. (3.8.47) for each time step, we apply the Galerkin finite-element method to Darcy's law to yield the following matrix equations governing the velocity:

$$\begin{bmatrix} [U] & [0] & [0] \\ [0] & [U] & [0] \\ [0] & [0] & [U] \end{bmatrix} \begin{Bmatrix} \{V_x\} \\ \{V_y\} \\ \{V_z\} \end{Bmatrix} = \begin{Bmatrix} \{R_x\} \\ \{R_y\} \\ \{R_z\} \end{Bmatrix}, \tag{3.8.48}$$

where [0] is the zero block matrix; $\{V_x\}$, $\{V_y\}$, and $\{V_z\}$ are block vectors whose components are the x-, y-, and z-velocity at all nodes; [U] is a block matrix whose ij-th component is given by

$$U_{i,j} = \sum_{e \in M_e} \int_{R_e} N_\alpha^e N_\beta^e dR , \tag{3.8.49}$$

and $\{R_x\}$, $\{R_y\}$, and $\{R_z\}$ are the right-hand block vector, whose i-th entries are given by

$$R_{x_i} = \sum_{e \in M_e} \int_{M_e} N_\alpha^e \left[-\left(K_{xx}\frac{\partial H}{\partial x} + K_{xy}\frac{\partial H}{\partial y} + K_{xz}\frac{\partial H}{\partial z} \right) \right] dR \tag{3.8.50}$$

$$R_{y_i} = \sum_{e \in M_e} \int_{M_e} N_\alpha^e \left[-\left(K_{yx}\frac{\partial H}{\partial x} + K_{yy}\frac{\partial H}{\partial y} + K_{yz}\frac{\partial H}{\partial z} \right) \right] dR \tag{3.8.51}$$

$$R_{z_i} = \sum_{e \in M_e} \int_{M_e} N_\alpha^e \left[-\left(K_{zx}\frac{\partial H}{\partial x} + K_{zy}\frac{\partial H}{\partial y} + K_{zz}\frac{\partial H}{\partial z} \right) \right] dR \tag{3.8.52}$$

Clearly, Eq. (3.8.48) can be used to solve the spatial distribution of three-velocity components independent of each other. Sometimes, to save the computational effort, the matrix [U] is lumped similar to the lumping of the mass matrix [M] presented earlier (see Eq. (3.3.34)). The application of finite-element methods to Darcy's Law ensures that the computed velocity will be continuous over the element boundary and a unique velocity at every node is obtained (Yeh, 1981). Traditionally, the velocity was computed by taking the derivative of the simulated pressure field with respect to the spatial coordinate. Obviously, using the traditional approach would result in a discontinuity in velocity at every node and across element interfaces (Yeh, 1981). The application of the finite-element method to solve Darcy's Law rather than simply taking the derivatives of the computed pressure head is very desirable. It was shown that this approach yielded accurate, continuous flow fields, which had enhanced numerical stabilities in the computation of natural convection flow and phase change (O'Neil and Albert, 1984).

3.9. Computer Implementation of Finite-Element Modeling of Saturated-Unsaturated Flows

Two computer codes have been designed to solve the saturated-unsaturated flow equations, Eq. (3.7.26) subject to the initial and boundary conditions, Eqs. (3.7.28) through (3.7.33). These two computer models were named FEMWATER and 3DFEMWATER for dealing with two- and three-dimensional problems, respectively. Verification of these models is described below.

3.9.1. FEMWATER

To test the consistency of the revised FEMWATER, three illustrative examples reported elsewhere (Yeh and Ward, 1980; Huyakorn, 1986) are presented. The first two examples represent simple one- and two-dimensional problems, respectively (Huyakorn, 1986). The third example is the seepage pond problem reported in the original FEMWATER (Yeh and Ward, 1980).

3.9.1.1. One-Dimensional Column Problem. This example is selected to represent the simulation of a one-dimensional problem with the revised FEMWATER. The column is 200 cm long and 50 cm wide (Fig. 3.9.1). The column is assumed to contain soil with a saturated hydraulic conductivity of 10 cm/day, a porosity of 0.45 and a field capacity of 0.1. The unsaturated characteristic hydraulic properties of the soil in the column are given as

$$\theta = \theta_s - (\theta_0 - \theta_r)\left(\frac{h - h_s}{h_b - h_a}\right) \tag{3.9.1}$$

and

$$K_r = \frac{\theta - \theta_r}{\theta_s - \theta_r}, \tag{3.9.2}$$

where θ is the water content, θ_s is the saturated water content, θ_r is the residual water content or the field capacity, h is the pressure head, h_b and h_a are the parameters used to compute the water content and the relative hydraulic conductivity, and K_r is the relative hydraulic conductivity.

The initial conditions are assumed: a pressure head of -90 cm is imposed on the top surface of the column, 0 cm on the bottom surface of the column, and -97 cm elsewhere. The boundary conditions are given as: no flux is imposed on the left, front, right, and back surfaces of the column; pressure head is held at 0 cm on the bottom surface; and variable condition is used on the top surface of the column with a ponding depth of zero, minimum pressure of -90 cm, and a rainfall of 5 cm/day for the first ten days and a potential evaporation of 5 cm/day for the second ten days.

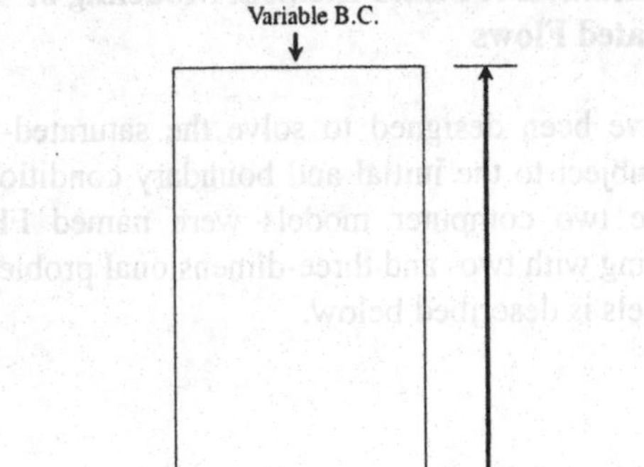

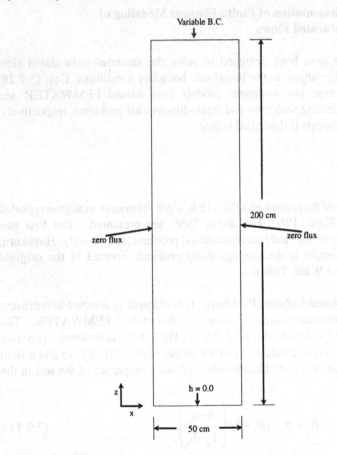

Fig. 3.9.1 Problem Definition for the One-Dimensional Transient Flow in a Soil Column.

The region of interest, that is, the whole column, will be discretized with 1 x 40 = 40 elements with element size = 50 x 5 cm, resulting in 2 x 41 = 82 node points (Fig. 3.9.2). A variable time step size is used. The initial time step size is 0.05 days and each subsequent time step size is increased by 0.2 times with a maximum time step size not greater than 1.0 day. Because there is an abrupt change in the flux value from 5 cm/day (infiltration) to -5 cm/day (evaporation) imposed on the top surface at day 10, the time step size is automatically reset to 0.05 day on the tenth day. A 20-day simulation was made with FEMWATER. With the time step size described above, 44 time steps are needed.

Fig. 3.9.2 Finite-Element Discretization for the One-Dimensional Transient Flow Problem in a Soil Column.

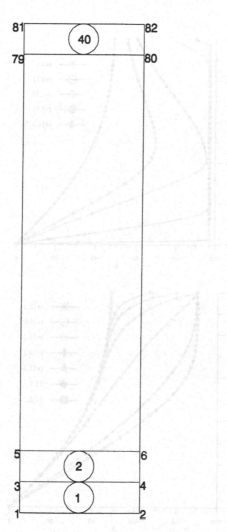

The pressure head tolerance for nonlinear iteration is $2 \cdot 10^{-2}$ cm, and the relaxation factors is set equal to 0.5. The pressure head simulated with FEMWATER is plotted in Fig. 3.9.3. As expected, during the rainfall period, the pressure head is gradually built up, starting from the top surface of the column. In the subsequent evaporation period, the pressure head is gradually reduced. The minimum allowable pressure head of -90 cm was first reached at the top surface of the column and eventually propagated into the soil column. These numerical results are almost identical to that given by Huyakorn (1986). Thus, the consistency of the computer code is partially verified.

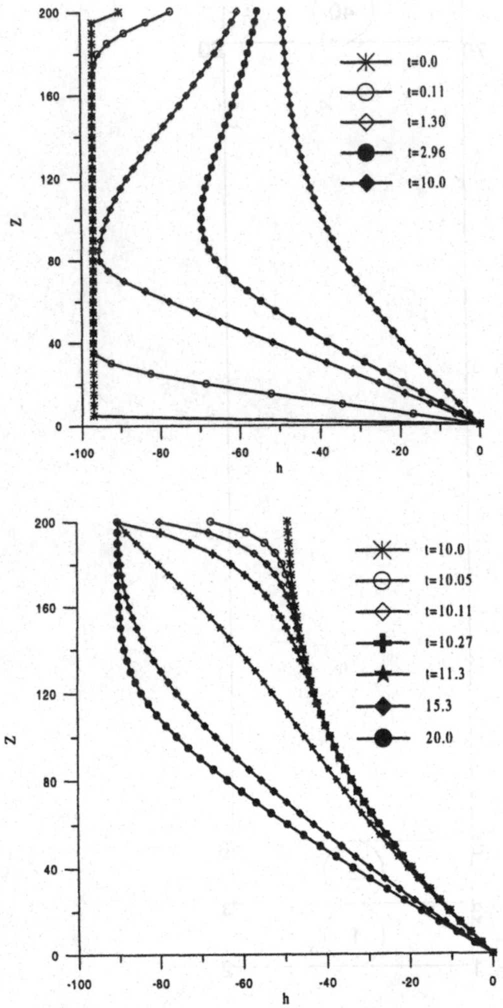

Fig. 3.9.3 Simulated Pressure Heads at Various Times: Infiltration (top) and Evaporation (bottom)

3.9.1.2. Two-Dimensional Drainage Problem. This example is selected to represent the simulation of a two-dimensional problem with FEMWATER. The region of interest is bounded on the left and right by parallel drains fully penetrating the medium, on the bottom by an impervious aquifuge, and on the top by an air-soil interface (Fig. 3.9.4). The distance between two the drains is 20 m (Fig. 3.9.4). The medium is assumed to have a saturated hydraulic conductivity of 0.01 m/day, a porosity of 0.25 and a field capacity of 0.05. The unsaturated characteristic hydraulic properties of the medium are given as

$$\theta = \theta_s + (\theta_s - \theta_r)\frac{A}{A + |h - h_a|^B} \tag{3.9.3}$$

and

$$K_r = \left(\frac{\theta - \theta_r}{\theta_s - \theta_r}\right)^n , \tag{3.9.4}$$

where h_a, A, and B are the parameters used to compute the water content and n is the parameter to compute the relative hydraulic conductivity.

Because of the symmetry, the region for numerical simulation will be taken as $0 < x < 10$ m and $0 < z < 10$ m. The boundary conditions are given as follows: no flux is imposed on the left ($x = 0$) and bottom ($z = 0$) sides of the region; pressure head is assumed to vary from zero at the water surface ($z = 2$) to 2 m at the bottom ($z = 0$) on the right side ($x = 10$); and variable conditions are used elsewhere. Ponding depth is assumed to be zero m on the whole variable boundary. Fluxes on the top side of the variable boundary are assumed equal to 0.006 m/day and on the right side above the water surface are equal to zero. Three case studies are conducted here: (1) a one-step steady state solution, (2) a transient simulation with prescribed initial conditions, and (3) a transient state simulation using the steady-state solution resulting from zero flux on the top as the initial condition. For all three case studies, a preinitial (for Cases (1) and (3)) or initial condition (for Case (2)) is set as $h = 10 - z$.

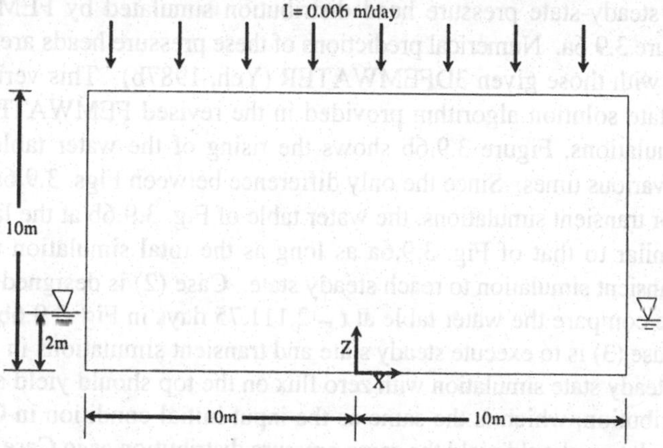

Fig. 3.9.4 Problem Sketch for Two-Dimensional Flow.

The region of interest is discretized with 10 x 10 = 100 elements with element size = 1 x 1 cm, resulting in 11 x 11 = 121 node points (Fig. 3.9.5). The pressure head tolerance for nonlinear iteration is $2 \cdot 10^{-3}$ m. The relaxation factor for the nonlinear iteration is set equal to 0.5. For transient (Case (2)) or transient state (Case (3)) simulations, the initial time step size is 0.25 day and each subsequent time step size is increased with a multiplier of 2.0 with the maximum time step size of less than or equal to 64 days. A total of 40 time steps and 2,112 days was simulated.

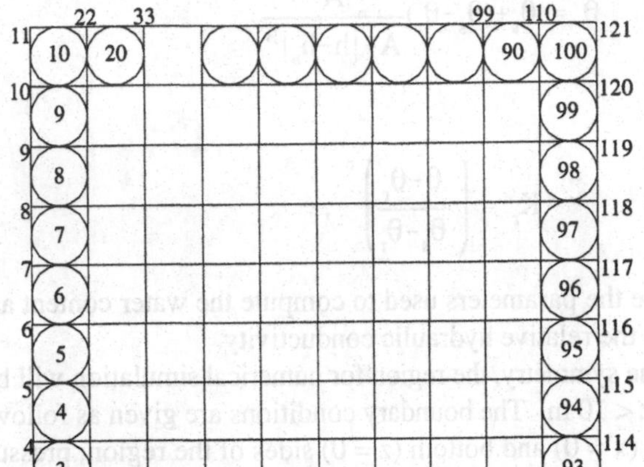

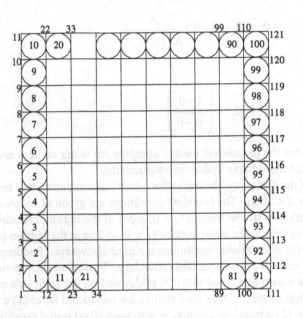

Fig. 3.9.5 Finite-Element Discretization for Two-Dimensional Drainage Problem.

The steady-state pressure head distribution simulated by FEMWATER is plotted in Figure 3.9.6a. Numerical predictions of these pressure heads are remarkably in agreement with those given 3DFEMWATER (Yeh, 1987b). This verifies the one step steady-state solution algorithm provided in the revised FEMWATER. For the transient simulations, Figure 3.9.6b shows the rising of the water table caused by infiltration at various times. Since the only difference between Figs. 3.9.6a and 3.9.6b is the steady or transient simulations, the water table of Fig. 3.9.6b at the last time step should be similar to that of Fig. 3.9.6a as long as the total simulation time is long enough for transient simulation to reach steady state. Case (2) is designed to meet this purpose as we compare the water table at t = 2,111.75 days in Fig. 3.9.6b with that in Fig. 3.9.6a. Case (3) is to execute steady state and transient simulations in a single job. Because the steady state simulation with zero flux on the top should yield a hydrostatic pressure distribution, which is the same as the input initial condition in Case (2), the combined simulation should yield the same pressure distribution as in Case (2). Indeed, the simulation of Case (3) yields identical results as that of Case (2), as expected.

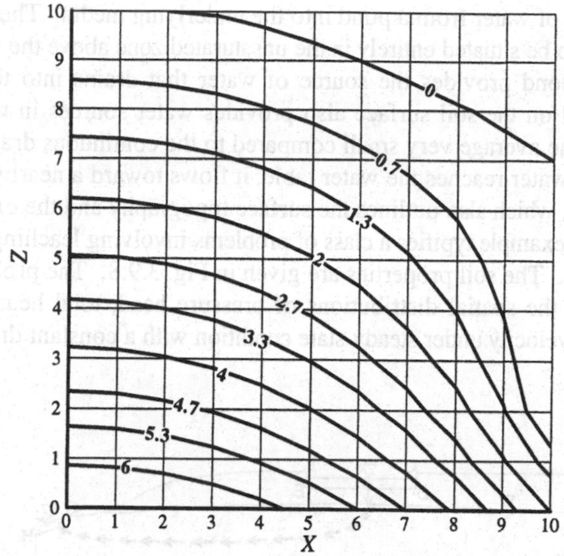

Fig. 3.9.6a. Simulated Steady-State Water Table and Pressure Head with FEMWATER.

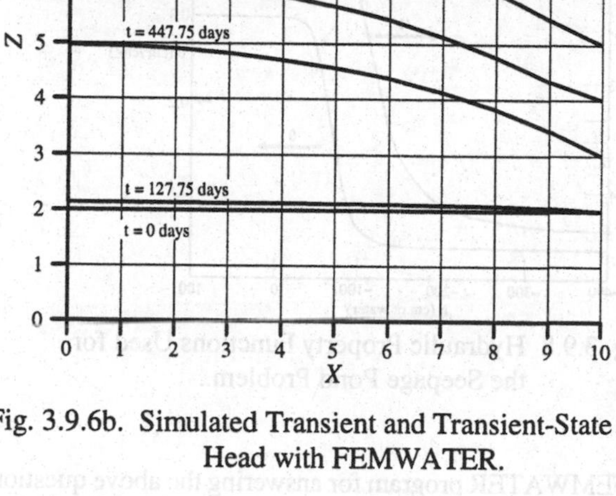

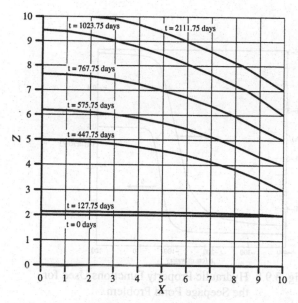

Fig. 3.9.6b. Simulated Transient and Transient-State Pressure Head with FEMWATER.

3.9.1.3. Seepage Pond Problem. This example is selected to compare the simulation results by FEMWATER with those obtained by the original version. The problem involved the seepage of water from a pond into the underlying media. The pond near a stream is assumed to be situated entirely in the unsaturated zone above the water table (Fig. 3.9.7). This pond provides the source of water that drains into the aquifer. Although the rainfall on the soil surface also provides water sources in the form of infiltration, it is on the average very small compared to the continuous drainage from the pond. After the water reaches the water table, it flows toward a nearby stream as depicted in Fig. 3.9.7, which also outlines the surface topography and the extent of the aquifer system. This example typifies a class of problems involving leaching of wastes from storage lagoons. The soil properties are given in Fig. 3.9.8. The problem being addressed is to find the spatial distributions of pressure head, total head, moisture content, and Darcy's velocity under steady-state condition with a constant drainage rate of 1.44 cm/h.

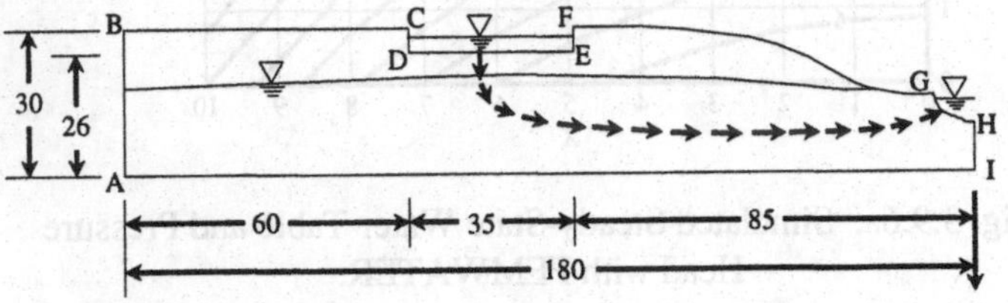

Fig. 3.9.7 Problem Definition for the Seepage Pond Problem.

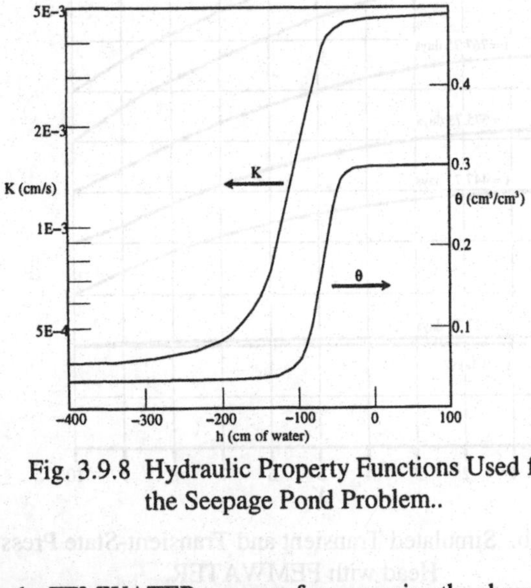

Fig. 3.9.8 Hydraulic Property Functions Used for the Seepage Pond Problem..

To use the FEMWATER program for answering the above question, the region of interest in Fig. 3.9.7 is discretized by 595 nodes and 528 elements as shown in Fig.

3.9.9. For the finite-element computation, the seven nodal points on the stream-soil interface are designated as Dirichlet nodes (Fig. 3.9.9). Seven nodal points on the bottom of the seepage pond namely, nodal points numbers 152, 164, 172, 180, 188, 196, and 204 are considered as constant Cauchy flux points and are assigned a constant infiltration rate of 4.0 x 10^{-4} cm/s. The top side of all elements on the sloping surface, except for the two elements immediately to the right of the seepage pond, are considered the seepage-evaporation-rainfall boundary surface. In other words, the nodal points on this surface are either Dirichlet or Cauchy points with the infiltration rate equal to the throughout rainfall rate, which is assumed to be zero.

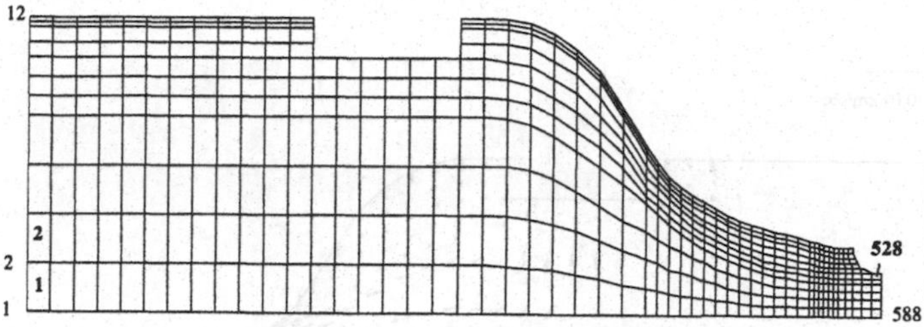

Fig. 3.9.9 Finite-Element Discretization for the Seepage Pond Problem.

Figures 3.9.10 show the distribution of pressure head, the Darcy velocity plot, and streamlines, as simulated by FEMWATER. The simulation is almost identical to that obtained by the original version. This verifies the revised algorithm of FEMWATER.

3.9.2. 3DFEMWATER

To verify 3DFEMWATER, three illustrative examples reported elsewhere (Huyakorn, 1986) are presented. Problem 1 is a transient response of a column due to infiltration and evaporation. Problem 2 is a drainage problem in a parallel plate. Problem 3 is designed to simulate the response of the subsurface media under pumping conditions. These represent the one-, two-, and three-dimensional problems, respectively. Problems 1 and 2 have the same setup as the first two example problems simulated with FEMWATER. The purpose is to demonstrate that a three-dimensional code can be used to investigate one- and two-dimensional problems. The third problem is to demonstrate that 3DFEMWATER can deal with three-dimensional saturated-unsaturated flow problems and to compare the simulation results with those reported by others (e.g., Huyakorn, 1986).

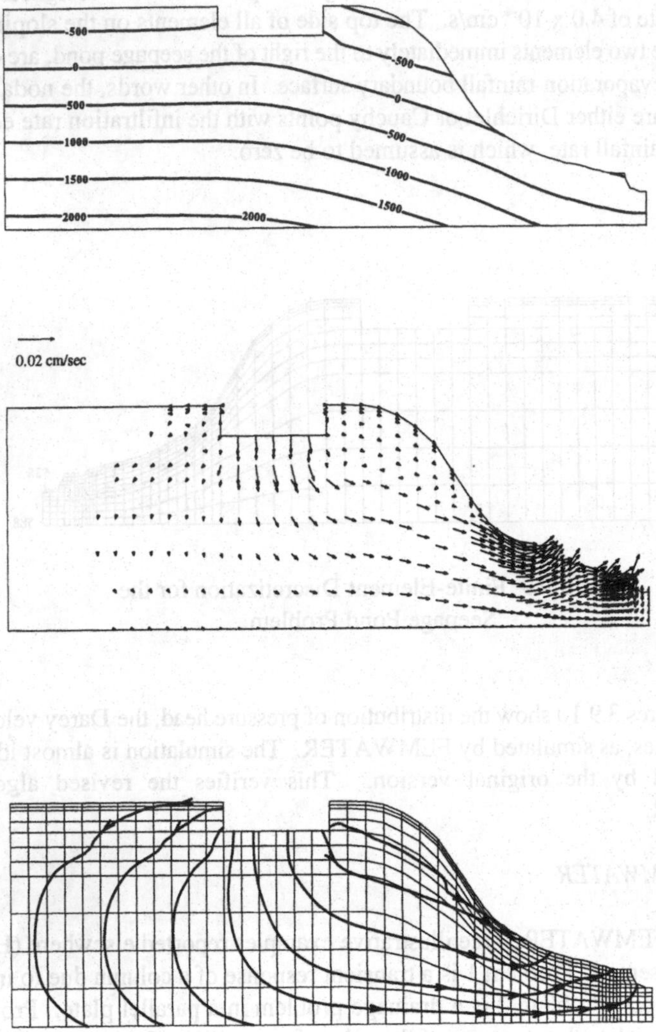

Fig. 3.9.10 Flow Variables of Seepage Pond as Simulated by the Revised FEMWATER. Top: Distribution of Pressure Head, middle: Darcy's Velocity, and bottom: Streamlines.

3.9.2.1. One-Dimensional Column Problem. This example is selected to represent the simulation of a one-dimensional problem with 3DFEMWATER. The column is 200 cm long and 50 by 50 cm in cross-section (Fig. 3.9.11). The column is assumed to contain the soil with a saturated hydraulic conductivity of 10 cm/d, a porosity of 0.45

and a field capacity of 0.1. The unsaturated characteristic hydraulic properties of the soil are the same as those in Problem 1 of FEMWATER. The initial conditions assumed are a pressure head of -90 cm imposed on the top surface of the column, 0 cm on the bottom surface of the column, and -97 cm elsewhere. The boundary conditions are given as: no flux is imposed on the left, front, right, and back surfaces of the column; pressure head is held at 0 cm on the bottom surface; and variable condition is used on the top surface of the column with a ponding depth of zero, minimum pressure of -90 cm, and a rainfall of 5 cm/d for the first 10 days and a potential evaporation of 5 cm/d for the second 10 days.

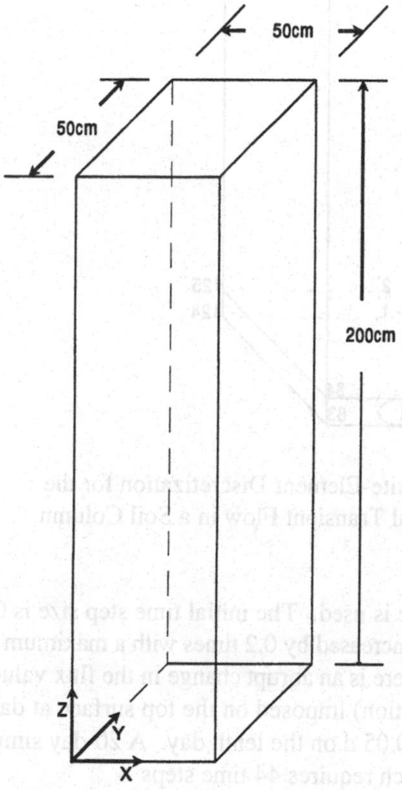

Fig. 3.9.11 Problem Definition and Sketch for the One-Dimensional Transient Flow in a Soil Column.

The region of interest, that is, the whole column, will be discretized with 1 x 1 x 40 = 40 elements with element size = 50 x 50 x 5 cm, resulting in 2 x 2 x 41 = 164 node points (Fig. 3.9.12). For 3DFEMWATER simulation, each of the four vertical lines will be considered a subregion. Thus, a total of four subregions, each with 41 node points, is used for the subregional block iteration simulation.

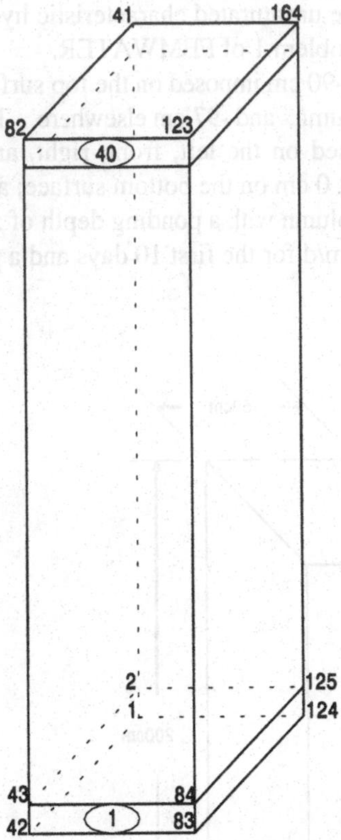

Fig. 3.9.12 Finite-Element Discretization for the One-Dimensional Transient Flow in a Soil Column.

A variable time step size is used. The initial time step size is 0.05 days, and each subsequent time step size is increased by 0.2 times with a maximum time step size not greater than 1.0 d. Because there is an abrupt change in the flux value from 5 cm/d (infiltration) to -5 cm/d (evaporation) imposed on the top surface at day 10, the time step size is automatically reset to 0.05 d on the tenth day. A 20-day simulation will be made with 3DFEMWATER, which requires 44 time steps

The pressure head tolerance is $2 \cdot 10^{-2}$ cm for nonlinear iteration and is $1 \cdot 10^{-2}$ cm for block iteration. The relaxation factors for both the nonlinear iteration and block iteration are set equal to 0.5. The pressure head simulated with 3DFEMWATER is plotted in Figs. 3.9.13 and 3.9.14. As expected, during the rainfall period the pressure head is gradually built up, starting from the top surface of the column. In the subsequent evaporation period, the pressure head is gradually reduced. The minimum allowable pressure head of -90 cm was first reached at the top surface of the column and eventually propagated into the soil column. These numerical results are almost identical to those given by Huyakorn (1986). Thus, the consistency of the computer code is partially verified.

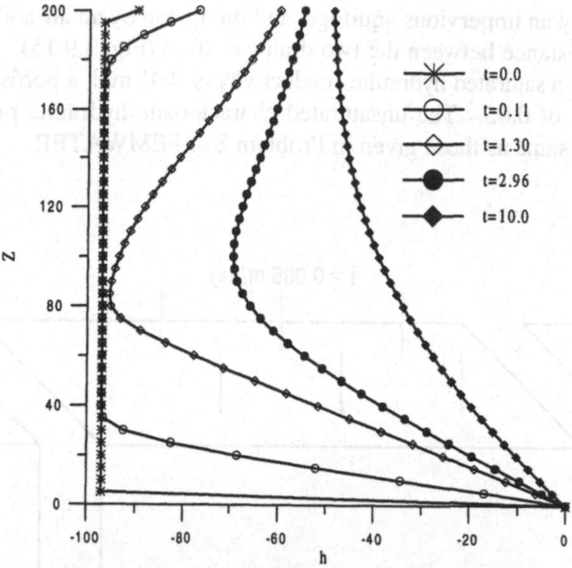

Fig. 3.9.13 Simulated Pressure Head Profiles at Various Times During Precipitation Period.

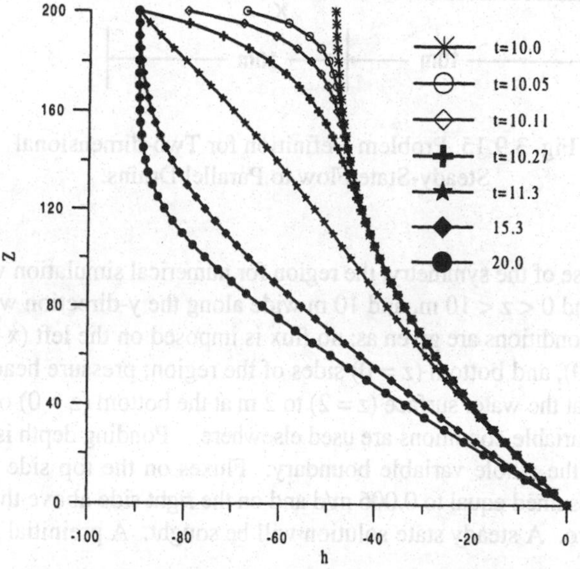

Fig. 3.9.14 Simulated Pressure Head Profiles at Various Times During Evaporation Period.

3.9.2.2. Two-Dimensional Drainage Problem. This example is selected to represent the simulation of a two-dimensional problem with 3DFEMWATER. The region of interest is bounded on the left and right by parallel drains fully penetrating the medium, on the bottom by an impervious aquifuge, and on the top by an air-soil interface (Fig. 3.9.15). The distance between the two drains is 20 m (Fig. 3.9.15). The medium is assumed to have a saturated hydraulic conductivity of 0.01 m/d, a porosity of 0.25, and a field capacity of 0.05. The unsaturated characteristic hydraulic properties of the medium are the same as those given in Problem 2 of FEMWATER.

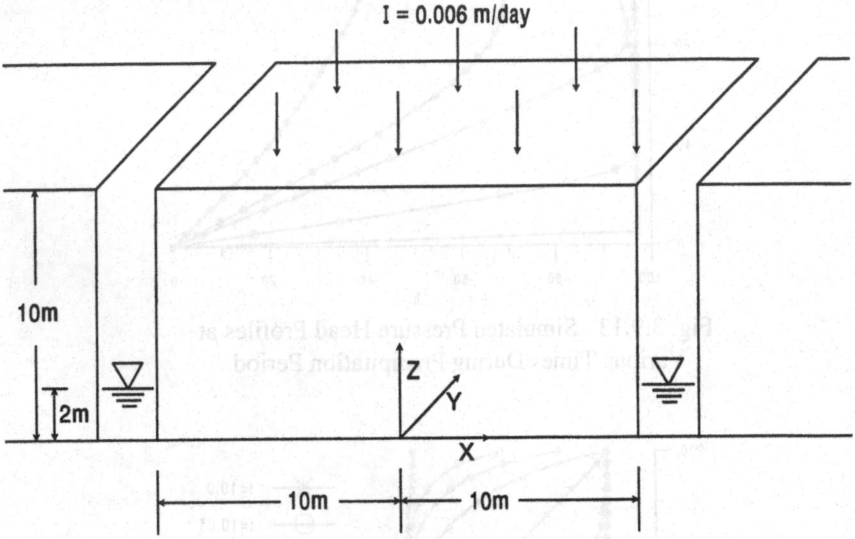

Fig. 3.9.15 Problem Definition for Two-dimensional Steady-State Flow to Parallel Drains.

Because of the symmetry, the region for numerical simulation will be taken as 0 < x < 10 m and 0 < z < 10 m, and 10 m wide along the y-direction will be assumed. The boundary conditions are given as: no flux is imposed on the left (x = 0), front (y = 0), back (y = 10), and bottom (z = 0) sides of the region; pressure head is assumed to vary from zero at the water surface (z = 2) to 2 m at the bottom (z = 0) on the right side (x = 10); and variable conditions are used elsewhere. Ponding depth is assumed to be zero meter on the whole variable boundary. Fluxes on the top side of the variable boundary are assumed equal to 0.006 m/d and on the right side above the water surface are equal to zero. A steady state solution will be sought. A preinitial condition is set as h = 10 - z.

The region of interest is discretized with 10 x 1 x 10 = 100 elements with element size = 1 x 10 x 1 cm, resulting in 11 x 2 x 11 = 242 node points (Fig. 3.9.16). For 3DFEMWATER simulation, each of the two vertical planes will be considered a subregion. Thus, the total of two subregions, each with 121 node points, is used for the subregional block iteration simulation.

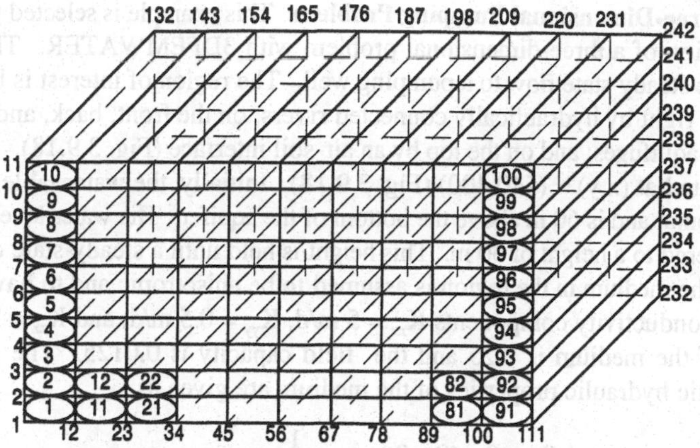

Fig. 3.9.16 Finite-Element Discretization for Two-Dimensional Steady-State Flow to Parallel Drains.

The pressure head tolerance is $2 \cdot 10^{-3}$ m for nonlinear iteration and is 10^{-3} m for block iteration. The relaxation factors for both the nonlinear iteration and block iteration are set equal to 0.5. The pressure heads simulated with 3DFEMWATER are plotted in Fig. 3.9.17. Numerical predictions of these pressure heads are in agreement with those given by Huyakorn (1986). This verifies the one-step steady-state solution algorithm provided in 3DFEMWATER. The slight difference between the predictions of 3DFEMWATER and those of Huyakorn results from the implementation of the variable boundary conditions. While 3DFEMWATER allows the whole amount of flux to infiltrate if the media is capable of doing so; the model given by Huyakorn only allows a fraction of flux to infiltrate as long as the ponding depth condition is not violated.

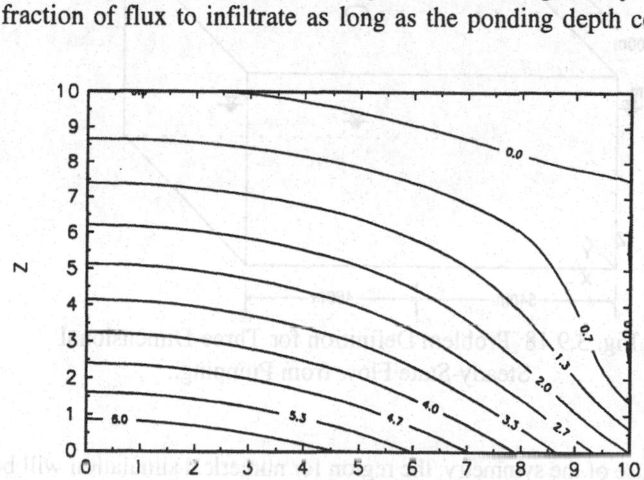

Fig. 3.9.17 Simulated Steady-State Water Table and Base Pressure Head for Two-Dimensional Drainage Problem with 3DFEMWATER.

3.9.2.3. Three-Dimensional Pumping Problem. This example is selected to represent the simulation of a three-dimensional problem with 3DFEMWATER. The problem involves the steady-state flow to a pumping well. The region of interest is bounded on the left and right by hydraulically connected rivers; on the front, back, and bottom by impervious aquifuges; and on the top by an air-soil interface (Fig. 3.9.18). A pumping well is located at (x,y) = (540,400) (Fig. 3.9.18). Initially, the water table is assumed to be horizontal and is 60 m above the bottom of the aquifer. The water level at the well is then lowered to a height of 30 m. This height is held until a steady-state condition is reached. The medium in the region is assumed to be anisotropic and to have saturated hydraulic conductivity components K_{xx} = 5 m/d, K_{yy} = 0.5 m/d, and K_{zz} = 2 m/d. The porosity of the medium is 0.25 and the field capacity is 0.0125. The unsaturated characteristic hydraulic properties of the medium are given as

$$\theta = \theta_s + (\theta_s - \theta_r)\frac{1}{1+\left(\alpha\left|h_a - h\right|\right)^{\beta}} \tag{3.9.5}$$

and

$$K_r = \left(\frac{\theta - \theta_r}{\theta_s - \theta_r}\right)^2 , \tag{3.9.6}$$

where h_a, α, and β are the parameters used to compute the water content and the relative hydraulic conductivity.

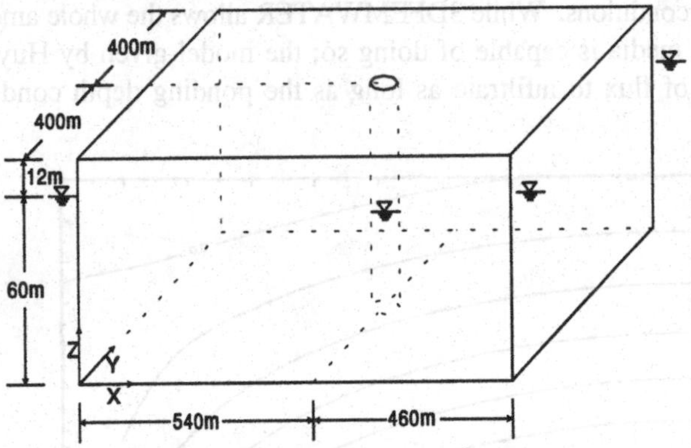

Fig. 3.9.18 Problem Definition for Three-Dimensional Steady-State Flow from Pumping..

Because of the symmetry, the region for numerical simulation will be taken as $0 < x < 1000$ m, $0 < y < 400$ m, and $0 < z < 72$ m. The boundary conditions are given as: pressure head is assumed hydrostatic on two vertical planes located at x = 0 and $0 < z < 60$, and x = 1,000 and $0 < z < 60$, respectively; no flux is imposed on all other boundaries of the flow regime. A steady-state solution will be sought. A preinitial condition is set as h = 60 - z.

The region of interest is discretized with 20 x 8 x 10 = 1600 elements resulting in 21 x 9 x 11 = 2079 node points (Fig. 3.9.19). The nodes are located at x = 0, 70, 120, 160, 200, 275, 350, 400, 450, 500, 540, 570, 600, 650, 750, 800, 850, 900, 950, and 1,000 in the x-direction, and at z = 0, 15, 30, 35, 40, 45, 50, 55, 60, 66, and 72 m in the z-direction, as reported by Huyakorn (1986). In the y-direction, nodes are spaced evenly at $\Delta z = 50$ m. For 3DFEMWATER simulation, each of the nine vertical planes perpendicular to the y-axis will be considered a subregion. Thus, a total of nine subregions, each with 231 node points, is used for the subregional block iteration simulation.

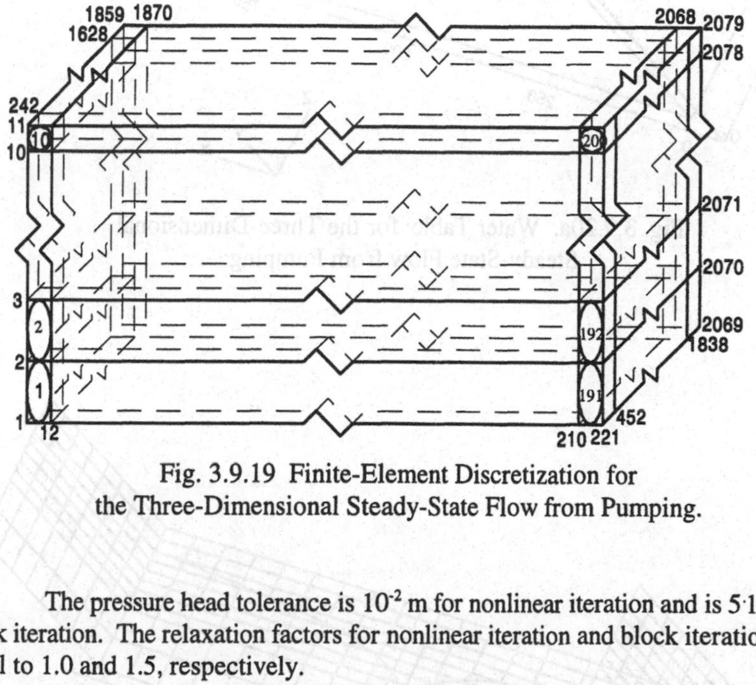

Fig. 3.9.19 Finite-Element Discretization for the Three-Dimensional Steady-State Flow from Pumping.

The pressure head tolerance is 10^{-2} m for nonlinear iteration and is $5 \cdot 10^{-3}$ m for block iteration. The relaxation factors for nonlinear iteration and block iteration are set equal to 1.0 and 1.5, respectively.

The water tables simulated with 3DFEMWATER are plotted in Figs. 3.9.20a and 3.9.20b, respectively, for a three-dimensional perspective view and a transverse section through the pumping well. Numerical predictions of these pressure heads are in good agreement with those given by Huyakorn (1986). This verifies the three-dimensional algorithm of 3DFEMWATER. There are some differences between the two model predictions around the pumping well, which are probably due to different boundary conditions imposed on the pumping well. Because it is not clear what boundary conditions were imposed on the pumping well by Huyakorn, we have imposed a constant total head on the three nodes in the pumping well. The velocity distributions are plotted in Figs. 3.9.21a and 3.9.21b, respectively, for the domain of interest and a transverse section through the pumping well.

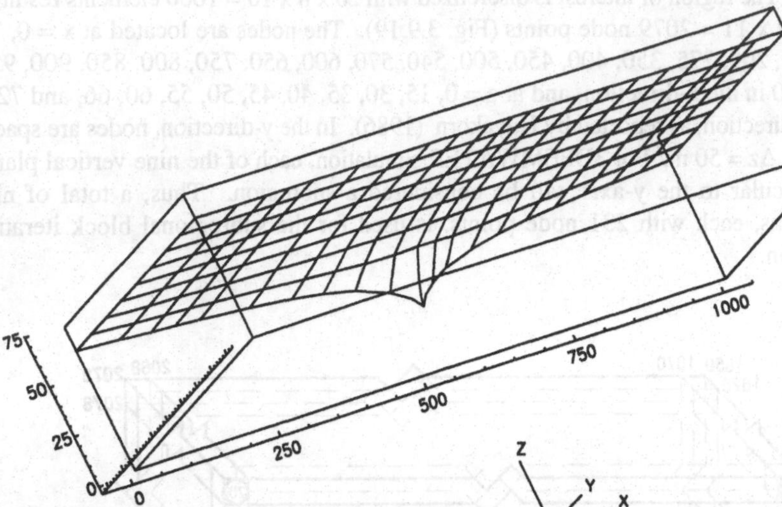

Fig. 3.9.20a. Water Table for the Three-Dimensional Steady-State Flow from Pumping.

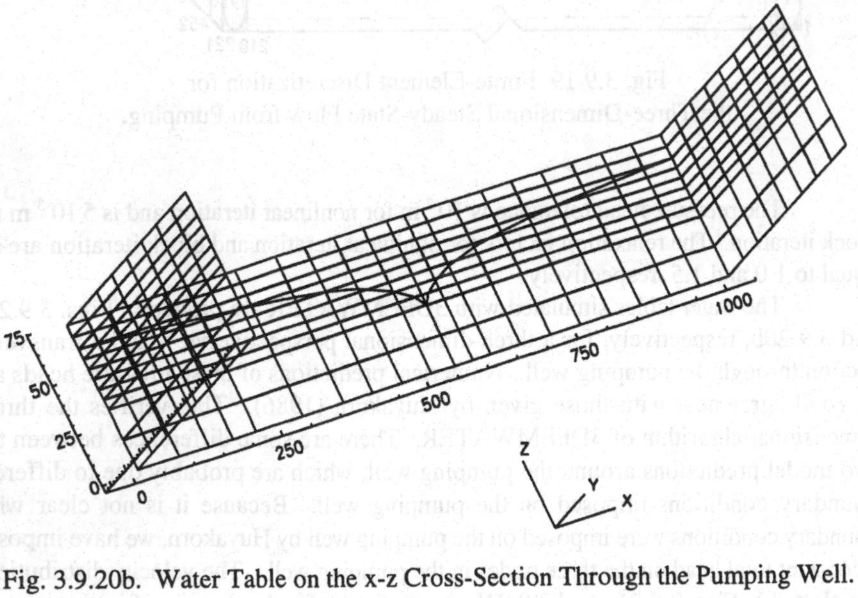

Fig. 3.9.20b. Water Table on the x-z Cross-Section Through the Pumping Well.

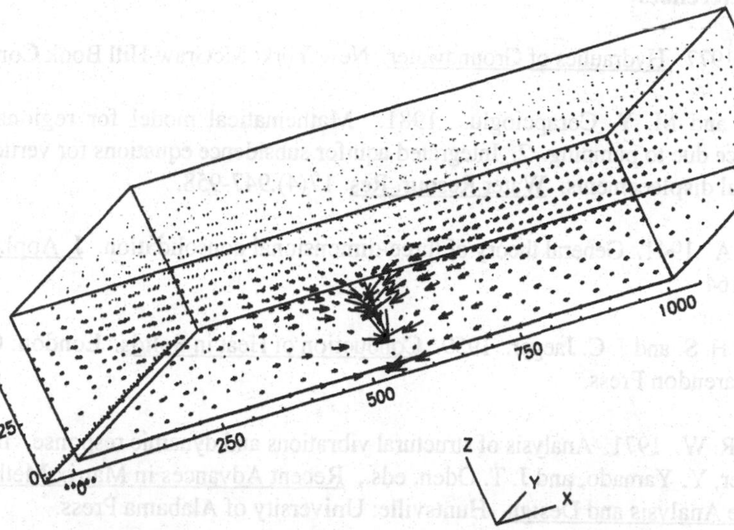

Fig. 3.9.21a. Velocity Distribution Throughout the Domain for the Three-Dimensional Pumping Problem.

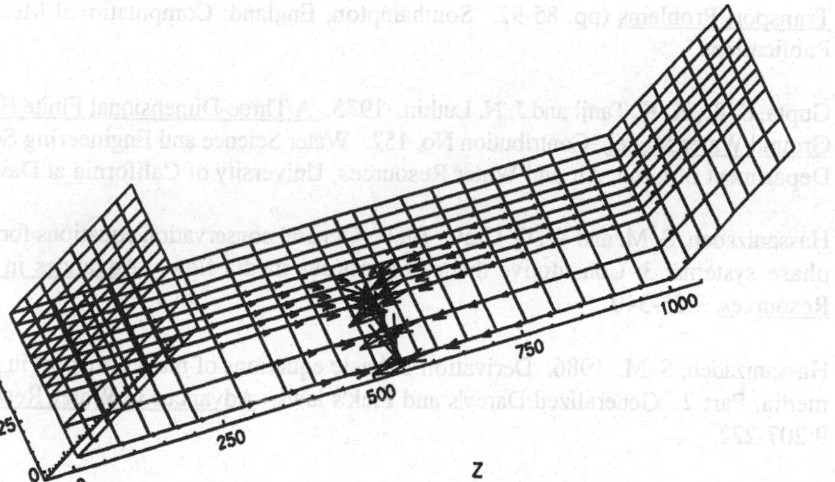

Fig. 3.9.21b. Velocity Distribution on the x-z Cross-Section Through the Well for the Three-Dimensional Pumping Problem.

3.10. References

Bear, J. 1977. Hydraulics of Groundwater. New York: McGraw-Hill Book Company.

Bear, J. and M. Y. Corapcioglu. 1981. Mathematical model for regional land subsidence due to pumping. 2. Integrated aquifer subsidence equations for vertical and horizontal displacements. Water Resour. Res. 17(4):947-958.

Biot, M. A. 1941. General theory of three-dimensional consolidation. J. Appl. Phys., 12:155-164.

Carslaw, H. S. and J. C. Jaeger. 1959. Conduction of Heat in Solids. London: Oxford at the Clarendon Press.

Clough, R. W. 1971. Analysis of structural vibrations and dynamic response. In R. H. Gallagher, Y. Yamado, and J. T. Oden, eds., Recent Advances in Matrix Methods of Structure Analysis and Design. Huntsville: University of Alabama Press.

Conte, S. D. 1965. Elementary Numerical Analysis. New York: McGraw-Hill Book Company.

Diersch, H.-J. G. and I. Michels. 1996. Moving finite element meshes for simulating three-dimensional transient free surface groundwater flow and transport processes. In A. A. Aldama et al., eds., Volume 1: Computational Methods in Subsurface Flow and Transport Problems (pp. 85-92. Southampton, England: Computational Mechanics Publications.

Gupta, S. K., K. K. Tanji and J. N. Luthin. 1975. A Three-Dimensional Finite Element Ground Water Model. Contribution No. 152. Water Science and Engineering Section, Department of Land, Air and Water Resources, University of California at Davis.

Hassanizadeh, S. M. and W. G. Gray. 1980. General conservation equations for multi-phase systems: 3. Constitutive theory for porous media flow. Advances in Water Resources, 3:25-340.

Hassanizadeh, S. M. 1986. Derivation of basic equations of mass transport in porous media, Part 2. Generalized Darcy's and Fick's laws. Advances in Water Resources, 9:207-222.

Hilderbrand, F. B. 1968. Finite-Difference Equations and Simulations. Englewood Cliffs, NJ: Prentice-Hall, Inc.

Huyakorn, P. S. 1986. A three-dimensional finite element model for simulating water flow in variably saturated porous media. Water Resources Res., 22(13):1790-1809.

Kipp, K. L. 1987. HST3D: A Computer Code for Simulation of Heat and Solute Transport in Three-Dimensional Ground-Water Flow Systems. U.S. Geological Survey.

Water Resources Investigations Report 86-4095, Denver.

Knupp, P. 1996. A moving mesh algorithm for 3-D regional groundwater flow with water table and seepage face. Advances in Water Resources, 19(2):83-95.

McDonald, M. G. and A. W. Harbough. 1985. A Modular Three-dimensional Finite-Difference Ground-Water Flow Model. U.S. Geological Survey. Open File Report 83-875.

O'Neil K. and M. R. Albert. 1984. Computation of porous media natural convection flow and phase change. In J. P. Laible, C. A. Brebbia, W. Gray, and G. Pinder, eds. Finite Elements in Water Resources (pp. 213-229). New York: Springer-Verlag.

Owczarek, J. A. 1964. Fundamental of Gas Dynamics. Scranton, PA: International Textbook Company.

Polubarinova-Kochina, P. Ya. 1962. Theory of Groundwater Movement. Princeton: Princeton University Press.

Reeves, M., D. S. Ward, N. D. Johns, and R. M. Cranwell. 1986. Theory and Implementation for SWIFT II: The Sandia Waste-Isolation Flow and Transport Model for Fractured Media. SAND83-1159. Albuquerque, N.M: Sandia National Laboratory.

Richtmyer, R. D. 1957. Difference Method for Initial-Value Problems, New York: Interscience.

Sears, F. W. 1952. An Introduction to Thermodynamics, The Kinetic Theory of Gases, and Statistical Mechanics. Reading, MA: Addison-Wesley Publishing Company, Inc.

Smith, G. D. 1965. Numerical Solution of Partial Differential Equations. London: Oxford University Press.

Trescott, P. C., G. F. Pinder, and S. P. Larson. 1976. Finite-difference model for aquifer simulation in two dimensions with results of numerical experiments. In Technique of Water-Resources Investigations of the United States Geological Survey (Bk. 7, Ch. C1). Reston, VA: U.S. Geological Survey.

Verruijt, A., 1969. Elastic storage of aquifers. In R. J. M. DeWiest, ed., Flow Through Porous Media (pp. 331-376). New York: Academic Press.

Wang, J. D. and J. J. Connor. 1975. Mathematical modeling of near-coastal circulation. Report No. MITSG 75-13. Cambridge, MA: MIT.

Yeh, G. T. and D. S. Ward. 1980. FEMWTER: A Finite Element Model of WATER Flow through Saturated-Unsaturated Porous Media. ORNL-5567. Oak Ridge, TN: Oak Ridge National Laboratory.

Yeh, G. T. 1981. On the computation of Darcian velocity and mass balance in finite element modeling of groundwater flow. Water Resources Res. 17:1529-34.

Yeh, G. T. 1987a. 3DFEMWTER: A Three-dimensional Finite Element Model of WATER Flow through Saturated-Unsaturated Porous Media. ORNL-6386. Oak Ridge, TN: Oak Ridge National Laboratory.

Yeh, G. T. 1987b. FEMWTER: A Finite Element Model of WATER Flow through Saturated-Unsaturated Porous Media - First Revision. ORNL-5567/R1. Oak Ridge, TN: Oak Ridge National Laboratory.

Yeh, G. T. and D. D. Huff. 1983. FEWA: A Finite Element Model of Water Flow through Aquifers. ORNL-5976. Oak Ridge, TN: Oak Ridge National Laboratory.

Yeh, G. T., J. R. Cheng, and J. A. Bensabat. 1994. A three-dimensional finite-element model of transient free surface flow in aquifers. In X. A. Peters el al., eds., Computational Methods in Water Resources (pp 131-138). Amsterdan: Kluwer Academic Pub.

Index